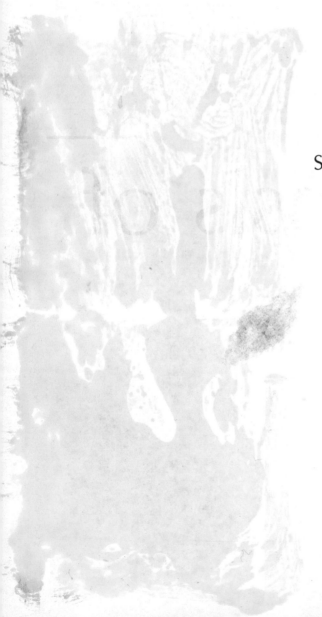

Synopsis of Gross Anatomy

Synopsis of

John B. Christensen, Ph.D.

Chairman and Professor of Anatomy, St. George's University School of Medicine, Grenada, West Indies

Ira R. Telford, Ph.D.

Visiting Professor of Anatomy, Uniformed Services University of the Health Sciences, Bethesda, Maryland; Professor Emeritus of Anatomy, George Washington University School of Medicine and Health Sciences, Washington, D.C.

Gross Anatomy

FOURTH EDITION
revised and enlarged

HARPER & ROW, PUBLISHERS
PHILADELPHIA

Cambridge
New York
Hagerstown
San Francisco

1817

London
Mexico City
São Paulo
Sydney

Acquisitions Editor: William Burgower
Sponsoring Editor: Richard Winters
Manuscript Editor: Shirley Kuhn
Indexer: Ann Cassar
Art Director: Maria S. Karkucinski
Designer: Lawrence Didona
Production Supervisor: N. Carol Kerr
Production Assistant: Charlene Catlett Squibb
Compositor: American–Stratford Graphic Services, Inc.
Printer/Binder: R. R. Donnelly & Sons

The authors and publisher have exerted every effort to ensure that drug selection and dosage set forth in this text are in accord with current recommendations and practice at the time of publication. However, in view of ongoing research, changes in government regulations, and the constant flow of information relating to drug therapy and drug reactions, the reader is urged to check the package insert for each drug for any change in indications and dosage and for added warnings and precautions. This is particularly important when the recommended agent is a new or infrequently employed drug.

3 5 6 4 2

Library of Congress Cataloging in Publication Data

Christensen, John B., DATE
  Synopsis of gross anatomy.
  Includes index.

  1. Anatomy, Human. I. Telford, Ira Rockwood, DATE. II. Title. [DNLM: 1. Anatomy. QS4 C549s]
QM23.2.C52 1982   611   81-7172
ISBN 0-06-140632-5   AACR2

Printed in the United States of America

Contents

Preface

Encouraged by the enthusiastic acceptance of our clinically oriented third edition of *Synopsis of Gross Anatomy,* we have retained the same format for this new edition. Some of the correlation statements have been repositioned to coincide more appropriately with the anatomy text. Through the generosity of our publishers, the illustrations of the fourth edition have been greatly enhanced by shading separation to more clearly differentiate vasculature, nerves, and muscles. Some of the artwork of the previous edition has been deleted, while forty new illustrations have been added.

A new feature of the present edition is the inclusion of nerve tables. These have been organized on a regional basis, which follows the format of the muscle tables. Hopefully, this will facilitate the study of each region by providing a concise grouping of the nerves for each area of the body.

We have incorporated corrective criticisms, freely offered by students and colleagues. Selectively, sections have been rewritten to provide a clearer, more complete, yet succinct, overview of the essentials of gross anatomy. We solicit your suggestions on this new edition.

We wish to acknowledge the contribution of Dr. Shane R. Christensen for his careful review of the clinical correlation and suggestions for additional material, the excellent work of Virginia L. Schoonover for her rendering of most of the new artwork, and finally, the many specific suggestions and editorial comments offered by the students of St. George's University School of Medicine.

John B. Christensen, Ph.D.
Ira R. Telford, Ph.D.

Preface to the first edition

This synoptic volume of regional anatomy presents the basic facts and concepts in the study of gross anatomy considered essential for students of medicine and associated sciences. It meets the need of the student for a concise, straightforward textbook, uncluttered by minutiae. This synopsis is intended not to replace selective reading in large conventional textbooks, but rather to give the student an initial appreciation of important body structures and relations.

Many original illustrations have been especially prepared to enhance this epitomized approach to the study of anatomy. The line drawings are keyed to the text and can be readily correlated with regional dissections.

For the student who finds gross anatomy difficult, this compact text may provide all that one can or need comprehend of the subject. However, for those who wish to pursue the subject more deeply, it will serve as a framework for the building of a broader and firmer foundation in anatomy.

Because of its regional approach, we suggest that this book could be used as 1) a study guide in conjunction with anatomic atlases and larger textbooks, 2) a companion text in gross dissection, or 3) a review of the fundamentals of gross anatomy.

In brief, we have endeavored to present, in the most succinct form, the essentials of human gross anatomy that we believe every medical student should know.

We gratefully acknowledge the kindness of our colleagues, Dr. Frank D. Allen, Associate Professor of Anatomy, George Washington University, and Dr. W. Montague Cobb, Chairman and Professor of Anatomy, Howard University, in reading and offering constructive suggestions for improvement of our original text.

We are deeply indebted to Mr. David S. Kern and his daughter Bonnie for the excellent rendering of most of the illustrations, to Mr. Michael S. Murtaugh for his splendid diagrammatic sketches and drawings, and to Dr. William A. Rush, Jr., Mrs. Margaret Dupree, and Miss Joan Ruback for their contributions to the artwork.

By the kind permission of various authors and their publishers, we have borrowed a few illustrations from the sources acknowledged in the individual legends.

J.B.C.
I.R.T.

Washington, D.C.

Synopsis of Gross Anatomy

1

introduction

Man has a natural curiosity about his body. It is first expressed when the infant becomes fascinated by his own hand movements. The study of gross anatomy is the continuation of this innate interest but it goes far beyond bodily movements; it is the formal identification and study of the dissectable structures of the body and their interrelationships. To the degree that this inborn interest is cultivated and developed, anatomy will be a stimulating and rewarding study.

The purpose of gross anatomic dissection and study is to obtain direct exposure to the three-dimensional relationships of the body and to be able to visualize just how we are put together. This does not negate the utilization of oral descriptions or explanations in the form of lectures, written text material, illustrations, or various visual aids such as models and films. These, however, should be considered only as adjunct tools to the firsthand information obtained through dissection as the student moves toward the goal of gross anatomy study—to secure a working knowledge of, and appreciation for, the structure and form of the human body.

In gross anatomy a voluminous amount of factual data must be acquired during a short time. This introductory chapter will present certain basic concepts as a foundation to aid the student to assimilate the deluge of information that will follow. As regional dissection and study of the body proceed, these general concepts of the different systems should become recurrent themes. Understanding them will make the subsequent acquisition of information more meaningful and rewarding.

The brief accounts given here of concepts of the skeletal system and associated joints, muscular system, fasciae, body cavities, lymphatics, cardiovascular system, and the nervous system are in no sense complete, nor are they intended to be. Rather, they constitute an introductory, conceptual approach to your study of anatomy.

Anatomic Terminology

Anatomy introduces the student to the language of medicine and dentistry. It has been estimated that this language, which the physician must master, comprises about 10,000 terms, three-fourths of which are encountered in anatomy. The word roots or stems, prefixes and suffixes are largely derived from Latin and Greek. Those who have studied these languages are well-equipped to understand and use anatomic terminology. If, however, the student's background in classical languages is deficient, he can overcome this handicap by learning certain fundamentals of vocabulary and linguistic principles. For example, the stem *myo* (Greek μῦς) in the terms myocardium (*cardium* heart), myometrium (*metrium* uterus), myoglobin (*globin* protein), myoblast (*blast* immature cell), myocele (*cele* hernia), myoma (*oma* tumor), all refer to muscle. Thus, the astute student will pay special attention as he encounters new terms to make certain he defines new words and stems as he encounters them.

TERMS OF REFERENCE

Terms of reference to the human body are standardized to refer to a rather arbitrary concept called the **anatomic position,** in which the body is erect, the face forward, the arms at the sides with the palms of the hands turned forward (Fig. 1-1). The terms listed in Table 1-1 are used to indicate the location of structures in the body

Fig. 1-1. Anatomic position (Langley LL, Telford IR, Christensen JB: Dynamic Anatomy and Physiology, 5th ed. New York, McGraw-Hill, 1980)

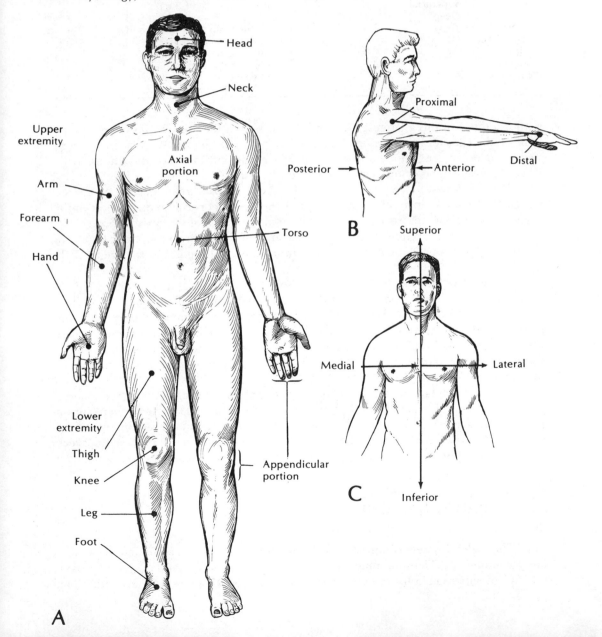

Table 1-1. Terms of Reference

Term	Synonym	Definition
Superior	Cranial, cephalic	Toward the head
Inferior	Caudal	Toward the feet
Anterior	Ventral, volar, palmar (latter two refer to hand)	Toward the front of the body
Posterior	Dorsal	Toward the back of the body
Medial		Toward the midline of the body
Lateral		Toward the side of the body
External	Superficial	Toward the surface of the body
Internal	Deep	Away from the surface of the body
Proximal		Toward the main mass of the body
Distal		Away from the main mass of the body
Central		Toward the center of the body
Peripheral		Away from the center of the body
Plantar		Sole of the foot

with reference to this anatomic position irrespective of the position the body of a patient or the cadaver might assume. This does away with the necessity of using words such as over, under, below, above, all of which can indicate two directions in the three-dimensional body and may thereby be confusing.

It is important to keep in mind that the above terms of reference locate a structure in its relationship to other structures of the body. For example, the descending aorta in the chest is located anterior to (in front of) the necks of the ribs and posterior to (behind) the heart; it is inferior to (below) the arch of the aorta and superior to (above) the diaphragm. It lies lateral to the vertebral column but medial to the angle of the ribs.

Terms denoting planes of the body also refer to the anatomic position. Of the three basic planes of the body, the **sagittal** and **coronal** (frontal) **planes** are both vertical along the long axis of the body, while the **transverse** (horizontal or cross-sectional) **plane** is at right angles to the longitudinal axis (Fig. 1-2). These planes are used either in reference to the whole body, to a specific region of the body, or to a separate organ. In the latter case, if the structure has been removed from the body, the terms longitudinal and transverse may be substituted.

The sagittal plane separates the body into right and left segments. If sectioned in the median sagittal plane, the body would be divided into equal halves, except for unpaired organs; if in a parasagittal plane, unequal portions. A coronal section would separate the anterior (front) part of the body from the posterior (back) part. A transverse section would bisect a superior (upper) segment from an inferior (lower) segment.

Skeletal System

The skeletal system comprises approximately 206 bones and a number of cartilaginous components. The total number of bones is an approximation because a variable number of supernumerary or accessory bones may also be present. These additional

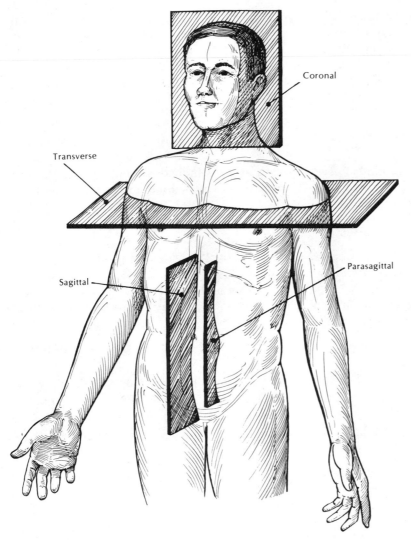

Fig. 1-2. Anatomic planes of reference. (Langley LL, Telford IR, Christensen JB: Dynamic Anatomy and Physiology, 5th ed. New York, McGraw-Hill 1980)

elements most frequently occur as small (wormian) bones between the flat bones of the skull, as additional carpal or tarsal bones in the hands and feet, or as sesamoid bones within tendons.

CARTILAGE

Three types of cartilage occur in the body: hyaline, elastic, and fibrous. **Hyaline cartilage** covers the articular surface of bones at the distal ends of ribs and in the fetal skeleton. **Elastic cartilage,** which has a greater resiliency due to embedded elastic fibers in its ground substance, is present in structures that undergo functional distortion, as for example, the external ear and epiglottic cartilage. **Fibrocartilage** has an increased strength due to a preponderance of collagen fibers in its ground substance. It is found in structures subjected to excessive stress, such as the weight-bearing intervertebral discs.

BONES

Although the number of bones may vary from individual to individual, the average complement, as tabulated below, may be subdivided into appendicular and axial portions of the skeleton.

Functions of Bones

Bones provide the supportive framework of the body and protect vital organs. Their marrow cavities are the primary site of blood formation in the adult (especially the flat bones of the skull, scapulae, vertebrae, and ilia). They also afford storage of minerals, principally calcium and phosphorus. The primary interests of bones to the student in gross anatomy are the following:

1. They are sites of muscular attachments and thereby act as levers to provide movement.
2. Their morphologic characteristics and markings including sexual differences.
3. The relationships of other structures to bones, principally vessels and nerves, which become clinically important in fractures.
4. The sequential appearance and fusion of epiphyses during the growth period, which are indicative of normal or pathologic development.

Axial skeleton		
Skull	22	
Ear ossicles	6	
Hyoid bone	1	
Vertebral column	26	
Ribs and sternum	25	
		80
Appendicular skeleton		
Upper extremity	64	
Lower extremity	62	
		126
Total		206

Morphologic Characteristics

Bones are classified as to shape, as for example, long, short, flat, or irregular. In **long bones** the length exceeds the width, as in most bones of the extremities. Long bones consist of 1) a **shaft** (diaphysis) having an elongated marrow cavity with little internal trabeculation, 2) two ends or **extremities** (epiphyses), which may or may not be separated from the diaphysis during the growth period by a plate of cartilage, the epiphyseal disc, and, 3) the **metaphysis,** the zone between the diaphysis and the epiphysis which flares out from the shaft of the bone toward the epiphysis. During the growth period, the cartilaginous **epiphyseal disc** provides new cells for the increase in length of long bones. When growth ceases, this area ossifies and the epiphysis becomes continuous with the diaphysis. The metaphysis and the epiphysis have extensive internal trabeculation. The trabeculae are usually aligned in the direction of the stress and strain placed on the bone.

Short bones are approximately equal in all three dimensions. Examples of the short bones are the carpal and tarsal bones of the wrist and ankle. Internal trabeculation, similar to that seen in the epiphyses of long bones, is also present in short bones.

Flat bones include the scapulae, ribs, sternum, and bones of the cranium. They are formed by two thin plates of compact bone with a minimal interval of trabecular bone between them that forms the interval of the marrow cavity. In the flat bones of the skull this area is referred to as the diploë.

Irregular bones, as their name implies, have a complicated configuration with numerous processes. Examples of irregular bones are the individual vertebrae of the vertebral column or the sphenoid and ethmoid bones of the skull.

Sesamoid bones are very small bones (with the exception of the relatively large patella) embedded within certain tendons. They usually occur in a tendon as it passes over an articulation. They can act as fulcrums to increase the mechanical advantage of the muscle action over the joint as well as provide for extra strength to limit trauma to the tendon from pressure or friction. The most constant sesamoid bones are the patella and those associated with the metacarpophalangeal joint of the thumb and metatarsophalangeal joint of the great toe.

Bone processes are discrete projections from the main body of the bone. The naming of processes indicates their morphologic characteristics. For example, crests or ridges are lineal elevations; sulci and grooves are lineal depressions; tubercles, tuberosities, or trochanters are circumscribed, roughened elevations of increasing size, and styloid processes or spines are spike-like projections. Foramina are holes that may pass either entirely through the bone or only through the cortex. They provide for the passage of nerves and vessels. Fissures are clefts between adjacent bones.

Experimental evidence indicates that the definitive form of a bone is dependent upon both genetic or intrinsic factors, and physical or external factors. Embryologic transplant studies have shown that a given bone, for example, the femur, will develop its characteristic processes irrespective of chemical, hormonal, or extrinsic factors. Bone markings, however, are modified by mechanical factors. As a response to increased stress of the muscular attachments to bone, well-developed muscular individuals have more prominent processes than do less muscular individuals. Moreover, the orthopedic surgeon is able, by varying stress or tension on a given bone, to compensate to a degree for developmental abnormalities.

Sexual Characteristics

Sexual differences are reflected by 1) the degree of massiveness of the skeleton and 2) by functional modifications. In the first instance although variations occur in races and between individuals, the male skeleton, in general, is more massive, with more pronounced and larger processes, than is the female skeleton.

The most pronounced example of functional adaptation occurs in the female pelvis. The innominate bones are modified to provide minimal osseous impedence in the birth channel to facilitate passage of the fetus through the pelvic cavity in parturition. Features of this adaptation include a greater relative as well as absolute width of the pelvic openings. This results in a circular inlet of the female pelvis, in contrast to an oval configuration in the male, as well as a displacement of the ischial spines and coccyx to increase the dimensions of the pelvic outlet.

Bones as Levers

The physician is also interested in the interplay of bones at articulations. The articular ends of a bone reflect the type of activity at the joint. Thus, the structure of bones associated with freely movable articulations are covered with hyaline (articular) cartilage which has a smooth, glassy appearance. In contrast, bones forming nonmovable joints have rough surfaces at their junctional sites.

The disparity in area of smoothness of two bones forming a freely movable joint is also meaningful. For example, the extent of the articular surfaces at the distal end of the femur and the proximal end of the tibia suggests that the knee is not a simple hinge joint, but rather that the surfaces must also slide across one another in movement.

Relationships of Bones to Other Structures

Bone markings frequently provide clues to the relationships of bones to other anatomic structures. If a muscle attaches to a bone over a relatively extensive area, the surface of the bone will appear smooth. Attachment of a muscle to a limited area, as for example, in the attachment of the tendon of the deltoid muscle, results in a definitive elevation, the deltoid tuberosity. Moreover, the size of the elevation will usually indicate the magnitude of stress the muscle produces.

Study of the foramina of the skull should always parallel study of the nerves, arteries, and veins that traverse these foramina.

Sulci or grooves almost invariably reflect the relationship of soft structures to bone. For example, the intertubercular sulcus of the humerous lodges the tendon of the long head of the biceps. The depth of a sulcus that lodges a tendon is, moreover, related to the strength of the pull of the muscle upon that tendon. Grooves may also lodge nerves or vessels. The depression along the posterior aspect in the midshaft of the humerus delineates the path of the radial nerve as it spirals around the bone. The relatively deep channels on the internal surface of the skull reflect the course of the meningeal blood vessels.

Organs also form depressions on bones. The bilateral, shallow concavities on the internal aspect of the occipital bone, are due to the convexity of the cerebellar hemispheres situated in this area.

Centers of Ossification

Embryologically, bones may arise either by direct transformation of mesenchyme into bone, as occurs in the flat bones of the skull, or through **endochondral bone forma-**

tion. In the latter, a precursor cartilage model of the developing bone is laid down with a subsequent bone replacement of this cartilage model. In long bones such bone transformation from cartilage occurs initially at the midpoint of the diaphysis and is referred to as the **primary center of ossification. Secondary ossification centers** appear later in the epiphyses at the ends of the long bones. The cartilage between the primary and secondary centers decreases progressively in relative amount, but persists as the epiphyseal disc, a plate-like zone, as long as growth in length is taking place.

SLIPPED EPIPHYSIS

Injury to the extremities of long bones, in a young person, may cause an epiphyseal displacement. Such an injury may be serious since the epiphyseal plate (disc) is the growth center for the bone. If such injuries go untreated, the longitudinal growth of the affected bone may be retarded, or even arrested, resulting in permanent shortening of the limb.

Some basic concepts of bone growth are summarized below:

1. The age of a growing individual may be reliably estimated from an assessment of his ossification centers.
2. Centers normally present at birth include the distal femur, proximal tibia, calcaneus, talus, cuboid, and the proximal end of the humerus.
3. Most epiphyses have fused (*i.e.*, the growth zone cartilage has become ossified) in the male by the age of 20. Both the appearance and fusion of epiphyses of the female precedes those of the male by about 2 years.
4. A given long bone may have epiphyses at both ends of the bone or only at one end. The latter is seen at the metacarpals, metatarsals, and phalanges.
5. If a long bone has a single epiphysis, it usually occurs at the end of the bone that undergoes the greatest excursion in movement.
6. In long bones that have two epiphyses, the epiphysis that appears first is usually the last to fuse with the shaft and it contributes most to the growth in length of the bone.
7. The more rapidly growing ends of bones of the extremities are at the knee (for the femur, fibula, and tibia), at the shoulder (for the humerus), and at the wrist (for the radius and ulna).
8. Nutrient arteries entering the diaphyses reflect this disparity in growth rate at the two epiphyses by angling away from the more rapidly growing end (*mnemonic:* From the knee I flee, to the elbow I go).

Joints or Articulations

In gross anatomy our primary interest in articulations has to do with the degree of motion occurring at joints. In studying joint movements we are concerned with: 1) the

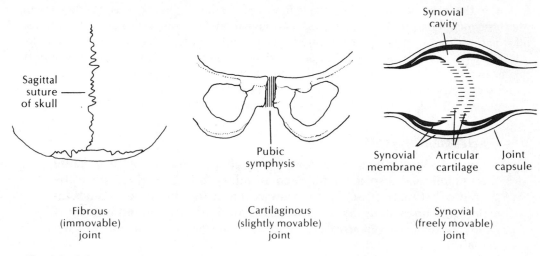

Fig. 1-3. Joints.

type of joint (hinge, ball and socket, gliding); 2) the accessory structures associated with joints, which may act primarily to stabilize, or conversely, allow maximal motion; 3) the muscles or their tendons that cross joints, and are the movers of the bones forming the joint; and 4) the blood and nerve supply of joints.

JOINT CLASSIFICATION

An articulation, or joint, is the contact or union between two or more bones or cartilages. They are classified by the degree of motion that occurs at the joint, namely, immovable, slightly movable, and freely movable joints (Fig. 1-3). This classification also reflects, in the same order, the nature of the tissue between the bony surfaces, for example, fibrous, cartilaginous or synovial.

Fibrous Joints

These joints are characterized by bones united by a minimal amount of fibrous tissue (sutural and gomphoses) or by a sheet of fibrous tissue (syndesmosis). Three types of **sutural** joints are recognized, serrated, squamosal, and plane. In **serrated** sutures the margins of the bones interlock like the cogs of a wheel. The sagittal suture of the skull is an example of a serrated suture. In the articulation of the temporal bone with the parietal bone of the calvarium the margins of the bones overlap each other, this is an example of a **squamosal** suture. In the **plane** sutural joints smooth edges of the bones abut against each other as seen in the articulation of the vomer with the perpendicular plate of the ethmoid in the nasal septum.

The teeth encased in the sockets of the alveolar processes of the maxilla and mandible are examples of **gomphoses.** The interosseus membrane uniting the radius and ulnar bones of the forearm typifies a **syndesmosis.**

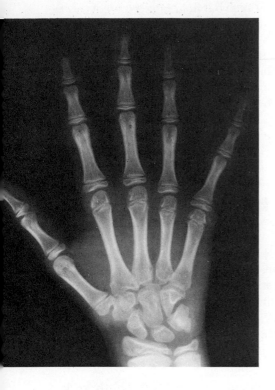
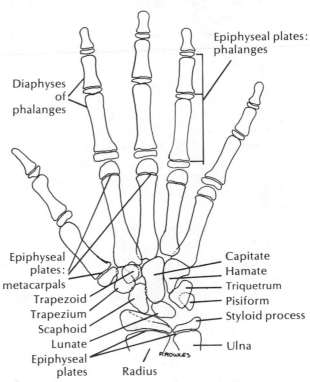

Fig. 1-4. Epiphyseal centers.

Cartilaginous Joints

Cartilaginous unions are differentiated as primary (synchondroses) or secondary (symphyses) cartilaginous joints. **Primary cartilaginous** joints are transitory and are frequently not considered to be true joints. They include the epiphyseal growth plates located between the epiphyses and diaphysis in growing bones (Fig. 1-4). Primary interest in these centers, apart from their being indicators of growth, is that they may be disrupted in traumatic injury with a consequent separation of the epiphysis from the diaphysis, often resulting in arrested growth and deformity of the bone.

In **secondary cartilaginous joints** the bones are united by fibrocartilage. This type of union allows only a limited degree of movement. Symphyses are present between the adjacent vertebral bodies where fibrocartilage forms the intervertebral disc, and at the pubic symphysis where the intervening fibrocartilage plate permits a slight widening of the infrapubic angle during parturition.

Synovial or Freely Movable Joints

The articular surfaces of bones forming a **synovial joint** are surrounded by an outer sleeve of connective tissue, the **joint capsule.** The joint capsule is lined by a **syno-**

vial membrane that secretes a viscous, lubricating synovial fluid into the synovial cavity. The synovial fluid is interposed between articular surfaces which, in all instances, are covered by hyaline (articular) cartilage. Such cartilage is not uniform in thickness, but is relatively thin in the central area of the articular surface and thicker toward the periphery. Examples of diarthrodial joints are the joints of the extremities.

A variety of subclassifications of synovial joints is described. Subtypes primarily reflect the shape of the articular surfaces, which is the prime determinant of the motion that can occur at a joint. The surfaces may be either ovoid or saddle-shaped. In ovoid joints one surface is concave, the other convex. The convex surface has invariably a greater surface area than does the concave surface. The opposed surfaces of saddle-shaped joints are reciprocally shaped to be congruous with a "male" or a "female" component. While few synovial joints provide movement in only one direction, most of them can be classified according to their major movement as follows:

1. **Hinge** or **ginglymus joints** move essentially in a single axis at right angles to the long axes of the bones. The elbow is an example of this type of joint.
2. **Pivot** or **trochoid joints** allow movement in the vertical axis only. An example of this type of motion occurs in supination and pronation (turning over the hand) in which the head of the radius rotates (pivots) against the capitulum of the humerus.
3. **Condylar joints** permit movement in two planes at right angles to each other. This is seen in the temporomandibular joint where the act of biting is combined with the grinding movement of the molars.
4. **Plane joints** provide a sliding movement in any direction over slightly curved surfaces. While the movement at these joints is multiaxial, their excursion is usually limited. Plane joints occur between the carpal and tarsal bones.
5. **Ball and socket** or **enarthrodial joints** allow maximum freedom of movement in all directions. The hip and shoulder joints are examples.

The range of motion in joints is also determined by the degree of laxity of the joint capsule and the amount of slack present in the ligaments that bind adjacent bones together. Strain on a joint is taken up predominately by the ligaments. These reinforcements may be merely thickenings of the fibrous connective tissue of the joint capsule, or they may be bands of connective tissue separate from the capsule. The strength of a ligament is proportional to the strain it bears. One of the strongest ligaments of the body, the iliofemoral ligament of the hip joint, is under constant strain in the erect position as it aids in the counteraction of the force of gravity.

The direction of the fibers as well as the composition of a ligament also indicate its function. The direction of the fibers of the interosseous membrane in the forearm is oriented to counteract stress placed upon the radius and ulna. The ligamenta flava of the vertebral column, composed of predominately elastic tissue, aids the erector spinae muscles in counteracting the force of gravity.

The degree of laxity of the joint capsule and associated ligaments determines to some extent the freedom of motion at a joint. As a limb is moved at a synovial joint in one direction, the capsule on the opposite side becomes taut. At the outset of a given action sufficient slack must exist to permit a normal range of movement. For example, in abduction at the shoulder, slack must be present on the medial aspect of the joint capsule to permit the movement.

The degree of tonicity or laxity of muscles which have tendons extending over a joint, as well as the amount of soft tissue adjacent to a joint, may physically impede its motion.

Intracapsular structures are present in many joints. These may be in the form of strengthening internal ligaments, as in the cruciate ligaments of the knee joints, or they may be associated with the functional muscles of the joint, as occurs in the long head of the biceps in the shoulder joint. They may exist as intra-articular discs, as in temporo-mandibular joint, or they may modify the articular surfaces as in the menisci cartilages of the knee. In some joints elaborations of the synovial lining increases its surface area and thereby increases its secretory capability.

Muscular System

Three types of muscle tissue are present in the body: **smooth muscle,** associated with blood vessels and organs; **cardiac muscle,** which forms the walls of the heart, and **skeletal muscle,** which forms the voluntary muscles of the body. Types of muscle cells have specific histologic and functional characteristics, and differ in their innervation. In gross anatomy our interest is centered on skeletal muscle. Interest in smooth and cardiac muscle tissue is limited to the distribution of sympathetic and parasympathetic nerves that supply the organs containing these types of muscle.

The muscular system comprises approximately 650 definitive skeletal or voluntary muscles, which afford us the only conscious control we have over our external environment. They are the motors of the body and produce their effect by virtue of their ability to contract. Muscles can decrease their length by as much as 50%. Inasmuch as both ends of a muscle are usually fixed to bones, movement of a bone results from contraction. Movement primarily occurs at the distal attachment of the muscle called the **insertion.** However, the attachment of insertion can sometimes be stabilized so as to produce movement at the proximal end of the muscle, the attachment of **origin.**

All voluntary muscles, however, are not under conscious control, for example, striated muscles of the middle ear and the musculature of the upper portion of the esophagus.

Basic Unit of Function and Structure

The functional and anatomic unit of the voluntary muscles, the **muscle fiber,** is an elongated, cylindrical, multinucleated cell covered with a tenuous membrane, the sarcolemma. With the light microscope, this cell (muscle fiber) can be seen to possess alternate light and dark cross-striated bands, and elongated, oval, peripherally located nuclei. These cells are the units that constitute the fleshy belly of a definitive voluntary muscle. Muscle fibers are separated from each other by a delicate connective tissue

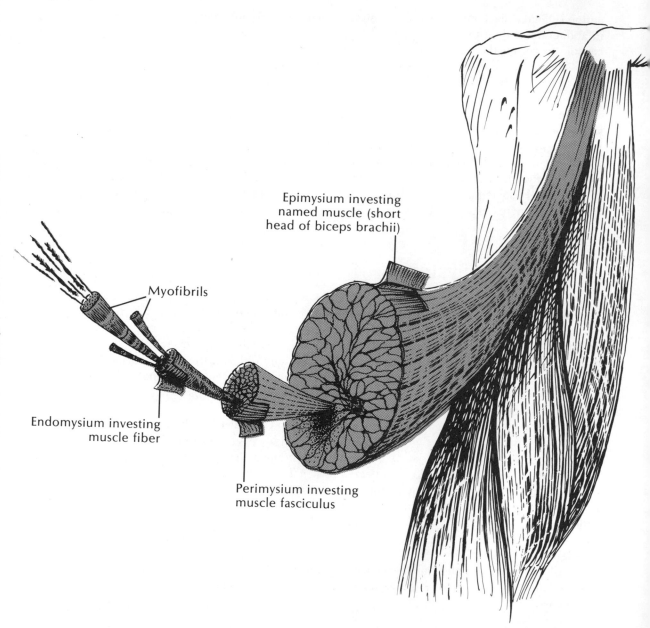

Epimysium investing
named muscle (short
head of biceps brachii)

Myofibrils

Endomysium investing
muscle fiber

Perimysium investing
muscle fasciculus

Fig. 1-5. Investments of structural subunits of a muscle.

covering, the **endomysium.** Groups of muscle fibers are in turn bound together
as a fasciculus or muscle bundle by a connective tissue investment, the **perimysium.**
In addition to these connective tissue elements, the entire muscle is surrounded by
the **epimysial fascial envelope,** a thickened connective tissue portion of deep fascia
(Fig. 1-5).

Attachment of Muscles

Towards the end of a muscle, the muscle fibers are replaced by collagen fibers which form **tendons.** They attach the muscle belly to structures to be moved, usually a bone. The manner of attachment of muscles varies considerably. Some muscles appear to rise directly from the surface of a bone, such as the intercostal muscles in which the tendinous attachment is very short. Muscle bellies may also share an attachment in common with adjacent muscles, for example, the flexor and extensor muscle masses at the medial and lateral aspects of the elbow joint. Some of the forearm muscles have long cylindrical tendons that extend from the bellies near the elbow to the digits. These tendons may be longer than their muscle bellies. Such thinning-out over the wrist and digits reduces tissue bulk opposite these joints, which otherwise would impede joint movement. Broad flat tendons are called **aponeuroses.** Examples of these are the aponeuroses of the anterolateral abdominal wall muscles which extend from the inferior extent of the sternum to the level of the pubic symphysis, as they pass to their insertion into the linea alba in the midline of the abdomen.

Intrinsic Architecture of Muscles

Muscles may be classified according to the arrangement of their fasciculi to their tendons of attachment (Fig. 1-6). A **fusiform muscle** is a cigarshaped muscle with the fleshy belly tapering at both ends as tendons. The fascicles of this type of muscle are arranged essentially in parallel so that its functional length approximates its actual length. Contraction of a muscle with this pattern results in decreasing its overall length by one-third to one-half. The movement of bones is maximal when such a muscle contracts. The power or strength that a contracting muscle is capable of generating is also a function of the arrangement of its fascicles. In some muscles the tendon extends into the fleshy belly and the fibers are oriented obliquely as they attach to their tendon. If the tendon is located along one side of the belly and all of the fasciculi are oriented in the same oblique direction into the tendon, the muscle is described as being **unipennate.** In such a muscle the range of movement is decreased by the pattern of fasciculi,

Fig. 1-6. Gross muscular patterns.

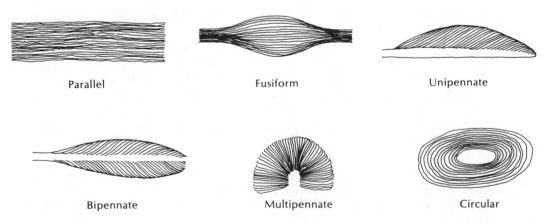

Parallel   Fusiform   Unipennate

Bipennate   Multipennate   Circular

but its power is increased; moreover the direction of pull will be deviated toward the attachment of its fibers. In **bipennate muscles** the fasciculi attach obliquely to both sides of a centrally placed tendon in a feather-like pattern. Such fasciculi do not distort the directional pull of the tendon inasmuch as they exert equal pull from both sides of the tendon.

In the **multipennate muscles,** such as the deltoideus, the fasciculi radiate out from a tendon in the central axis of a muscle. The arrangement of the fasciculi of this muscle permits it to perform many different actions at the shoulder joint. The more anteriorly placed fascicles working independently can flex or internally rotate the humerus; the central fascicles abduct the humerus; while the anteriormost and posteriormost fascicles working together act to adduct the humerus. Fascicles may also be arranged in parallel to form a flattened muscle, or circularly to form a **sphincteral** muscle.

Basic Muscle Actions

The action of a given muscle is designated according to the direction it moves a part of the body away from the anatomic position (Fig. 1-7). These movements are designated as flexion, extension, abduction, adduction, medial or lateral rotation, circumduction, supination, pronation, inversion, and eversion.

In the upper extremity, trunk, and at the hip joint, **flexion** is the action that opposes (brings together) anterior surfaces of segments of the body, thus reducing the angle at the joint. For example, flexion at the hand brings the anterior surface of the digits to the anterior surface of the palm, as in making a fist. At the elbow, flexion approximates the anterior surface of the forearm to the anterior surface of the arm.

Extension in all instances is the opposite action from flexion. In the upper extremity, extensor muscles open the clenched fist or straighten out the elbow to return the forearm to the anatomical position.

Due to the developmental rotation of the lower extremity, flexion at the knee opposes the posterior aspect of the leg to the posterior aspect of the thigh. At the ankle, the terms plantar flexon and dorsiflexion are substituted for flexion and extension to avoid confusion. **Plantar flexion** brings the foot in line with the leg, whereas **dorsiflexion** approximates the dorsum of the foot on the tibia (shin bone).

Abduction is the action of moving the extremity laterally or away from the body, while **adduction** is the movement of a part toward the midline of the body. In movement of the digits in abduction and adduction a different line of reference is used. In the hand, the line of reference is the middle finger, in the foot the second toe. Thus, abduction spreads the fingers and toes; adduction brings them together.

Rotation is a pivoting movement. If this action is toward the body it is internal or medial rotation; if it is away from the body it is lateral or external rotation.

Circumduction is a combination movement in which the motion of an extremity describes a cone, the apex being at the shoulder or hip, the base being at the hand or foot.

Pronation and supination are actions limited to the hand and forearm. **Pronation** turns the palm downward; **supination** turns the palm upward.

Inversion and eversion are similarly limited to the foot. **Inversion** is turning the

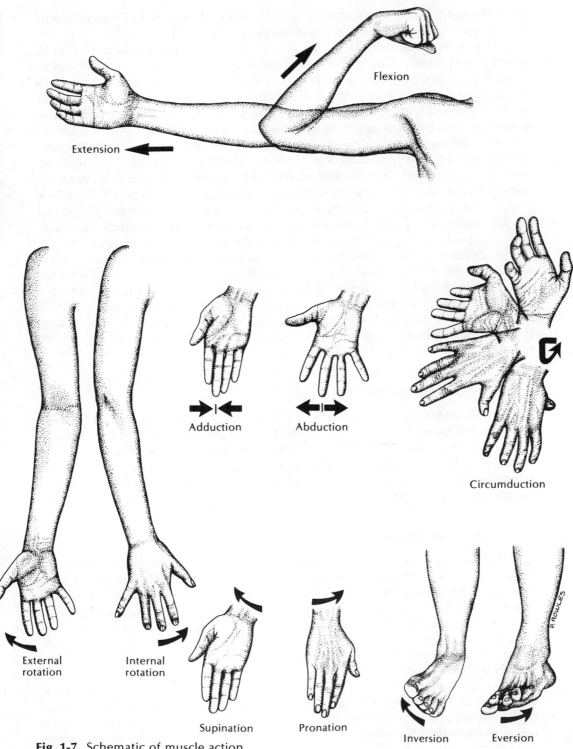

Fig. 1-7. Schematic of muscle action.

sole of the foot inward, as in standing on the lateral side of the foot. **Eversion** is turning the sole outward as in standing on the medial side of the foot.

In a given action several muscles must act in concert if the movement is to be smoothly performed. Thus, in describing motion at a given joint, additional assignments beyond flexion, extension, abduction, and adduction are given to muscles. Although several muscles may contract to perform a given action, one muscle usually predominates and is called the prime mover. **Prime movers,** moreover, are subdivided as primary if they initiate an action, or secondary if they reinforce the action against resistance. Of the several actions a muscle may perform, the action in which it is most efficient is given as its prime function. All other muscles contributing to the performance of a given action are referred to as **synergistic muscles.** Muscles that essentially oppose a given motion are called **antagonistic muscles.** For example, at the elbow joint, the biceps brachii, and brachialis muscles act synergistically in flexing the elbow; the antagonistic triceps brachii muscle must relax concomitantly to balance the force of the contracting biceps and brachialis muscles for smooth movement to occur.

Stabilizing or **fixating muscles** are also necessary to perform many body movements. This is especially important in muscles that cross two joints. For example, in the action of the biceps brachii, flexion of the elbow can only occur after the shoulder has been fixed, which gives the biceps a rigid base from which it can exert its pull.

In different movements a given muscle may function in all four roles given above, that is, it may perform as a prime mover, an antagonist, a synergist, or as a stabilizing muscle.

Muscular Compartments

In the extremities, muscles with similar actions are usually contained within a fascial compartment and are usually innervated by a common nerve. For example, in the thigh there are three well-defined compartments: an anterior compartment for the extensors of the leg, a posterior compartment for the flexors of the leg, and a medial compartment for the adductors of the thigh. The primary innervation of these muscle compartments are, respectively, the femoral, sciatic, and obturator nerves.

Muscle Nomenclature

The names of muscles are orientation aids to the beginning student in anatomy if one considers the implication of the terms in the name. Most muscles are named from anatomic characteristics that indicate shape, size, location, action, attachment, or fiber direction. The following are examples: the rhomboid muscles attached to the scapula are so named from their shape; they are differentiated from each other on the basis of their size, as the rhomboideus major and rhomboideus minor. The muscles on the dorsum of the scapula are named according to their location, with respect to the spine of the scapula, as the supraspinatus and the infraspinatus. The name of the levator scapulae muscle indicates its action, the elevation of the scapula. The sternocleidomastoideus is named from its origin on the sternum and the clavicle (cleido) and its insertion on the mastoid process. The external abdominal oblique muscle of the abdominal wall is named because of its superficial location and the oblique direction of its muscle fibers.

Fasciae

Fasciae occur mostly as membranous sheets that are present between the tela subcutanea (superficial fascia) beneath the skin, and the subserous fascia adjacent to the lining membranes of the serous body cavities. Fascia is also present as loose, connective-tissue packing material containing variable amounts of fat that fills in the spaces between adjacent organs. Elsewhere fascia forms discrete membranous sheets or tubular enclosures, which are grossly dissectable. These sheets may separate organs or may enclose groups of organs. In the latter instance, fasciae function to hold structures in their proper locations.

Fasciae have three major subdivisions: the superficial (subcutaneous) fascia, the deep fascia, and the subserous fascia. Typically, the superficial fascia consists of loose areolar connective tissue, whereas deep fasciae form membranous sheets of varying density, which often enclose muscles or surround organs. The continuity of all fasciae can be demonstrated; however, by fusion and splitting they form cleavage planes or compartments throughout the body.

Superficial Fascia

The **superficial fascia** (subcutaneous tissue or tela subcutanea) is immediately deep to the skin. It invests the entire body and varies in thickness in different areas. Deep to the skin on the back of the hand, this layer is quite sparse, while over the lower abdominal wall it is markedly increased in thickness as a heavy layer of fat, the **panniculus adiposus.** Here two distinct layers may be identified, the superficial **fatty** layer (**Camper's** fascia), and a deeper, **membranous** layer (**Scarpa's** fascia). In obesity, fat in the superficial layer increases throughout the body.

OBESITY

In obesity, while fat accumulates in the body cavity, for example, around and between organs, in mesenteries, where it is most clinically significant, it is most apparent in the superficial fascia. In the "fat man" at the circus, fat in this location is responsible for his sixty-inch waist.

Superficial arteries, veins, lymphatics, and nerves course in the superficial fascia. Hair follicles, and sebaceous, sweat, and mammary glands, and the muscles of facial expression are all embedded in this layer.

Deep Fasciae

The **deep fasciae** of the body are present as a series of laminae. The most superficial layer is the **external investing layer** of deep fascia; the deepest layer is the **internal investing layer.** These two layers form, respectively, a continuous external covering of the body and a continuous lining of the body cavities. Intervening or intermediate laminae between the external and internal investing layers vary in number depending upon the number of structures that constitute the body wall at a given area.

While deep fascia is continuous, individual laminae are encountered throughout the study of gross anatomy. This apparent paradox of the continuity of deep fascia with designation of individual fascial laminae may create difficulty in gaining a correct concept of deep fascia. An example of this arrangement is the make-up of the abdominal wall in which there is a lineal fusion of the deep fascia along the lateral border of the quadratus lumborum muscle (Fig. 1-8). Anterior to this line of fusion (which extends between the twelfth rib and the iliac crest), six laminae split off to envelop each of the three muscles that form the lateral abdominal wall. Anteriorly, a similar vertical line of fusion occurs at the lateral extent of the rectus abdominis muscle as the semilunar line. In the interval between these lineal fusions, **fascial envelopes** surround each of the lateral abdominal wall muscles. The **fascial clefts** created between contiguous layers of the superimposed muscles permit each muscle to contract independently.

Essentially bloodless planes exist between the fascial clefts. The arteries, nerves, and veins supplying a given muscle usually enter at a definitive site. Thus, considerable surgical manipulation can be undertaken without damage to the primary blood and nerve supply of a muscle.

LINES OF FUSION OF FASCIAL SHEETS

Lines of fusion of fascial sheets are essentially avascular and are often sites for surgical incisions because the operating field is relatively free of blood, for example, linea alba. Surgeons favor fascial junctional areas for anchoring of sutures because of the inherent tensile strength of fascia and the strong, fibrous union that results from wound healing.

In other regions of the body, splitting and fusion of contiguous fascial layers form **fascial compartments.** In the upper and lower extremities deep extensions from the external investing fascial layer extend to bones to form intermuscular septa. These septa separate muscles of the extremities into functional groups.

Over joints, the external investing layer of fascia is modified. At the knee this modification forms the retinaculi, which help to support this joint. At the wrist and ankle joints, the deep fascia forms thickened transverse bands that bind the long flexor and extensor tendons in place.

FASCIAL CLEFTS

Clinically, fascial clefts provide routes for spread of infection. Fusion of laminae of deep fascia also form pockets or compartments where the accumulation of pus, tissue fluid, or blood may occur. An understanding of fascial compartmentalization is important in the diagnosis and treatment of the accumulation of these fluids. Shortening of fascial components in the extremities can cause contracture deformities of the hand or foot. Fascial membranes may be utilized in corrective surgery, for example, the iliotibial tract in the thigh may be used to rebuild the anterior abdominal wall in hernial repair.

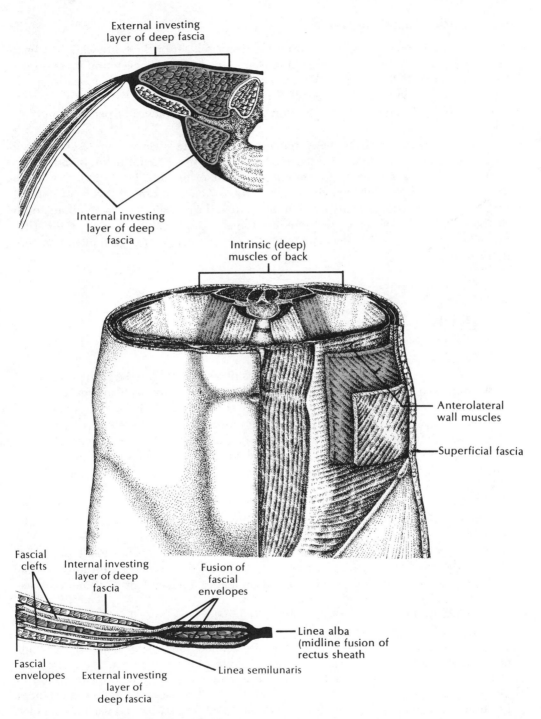

External investing
layer of deep fascia

Internal investing
layer of deep
fascia

Intrinsic (deep)
muscles of back

Anterolateral
wall muscles

Superficial fascia

Fascial
clefts

Internal investing
layer of deep
fascia

Fusion of
fascial
envelopes

Linea alba
(midline fusion of
rectus sheath

Fascial
envelopes

External investing
layer of
deep fascia

Linea semilunaris

Fig. 1-8. Typical body wall section.

Internal Investing Fascia

This continuous fascial layer forms the deepest layer of the body wall as it lines all body cavities. In the cervical region it is described as the **prevertebral layer;** in the thorax as the **endothoracic;** in the abdomen as the **endoabdominal** or transversalis layer; and in the pelvis as the **endopelvic** or supra-anal layer. It is the layer in the thoracic, abdominal, and pelvic cavities to which the parietal pleura or peritoneum is attached.

The **subserous fascia,** between the internal investing layer of deep fascia and the serous membrane, is composed of loose areolar tissue. It varies in thickness and is continuous via visceral ligaments (mesenteries) onto the external surfaces of viscera in the thorax, abdomen, and pelvic cavities. Clinically, the composition of this layer permits the surgeon to separate, usually by blunt dissection, the parietal serous membrane from the body wall. Thus, this layer is present in the abdominal cavity between the peritoneum and endoabdominal fascia and in the thoracic cavity between the pleura and endothoracic fascia.

Body Cavities

Except for certain hollow organs and air sinuses in the skull, the entire area of the body internal to the skin is occupied by tissues. Thus, terms such as tissue spaces, fascial spaces, serous cavities, and body cavities refer not to voids, but rather to areas filled with fluids, tissues, or organs. Extracellular or tissue spaces exist between cells, between contiguous layers of tissues, and between adjacent structures, such as organs, muscles, tendons, and fascial membranes. In abnormal conditions, such an area may fill with excess tissue fluid, which results in edema or swelling, often accompanied by severe pain.

Facilitating mechanisms, which prevent friction between adjacent structures that move against one another, are modifications of the above so-called spaces. The simplest are the fascial spaces, or more correctly, the **fascial clefts.** These clefts, present between fascial membranes, are filled by loose connective tissue and tissue fluid. Where adjacent fascial membranes surround muscles, the resultant cleft between the contiguous muscle fasciae permits a limited amount of independent movement of muscles.

In addition to this functional role, fascial clefts are clinically significant in that they permit the surgeon to identify, isolate, manipulate, or transect a specific muscle. They also provide a bloodless plane for surgical incisions since major arteries and nerves to a muscle cross the fascial cleft at definite sites. Most fascial clefts also form channels or pathways that may facilitate the spread of infection, blood, tissue fluid, or air; conversely, they may fuse to form pockets to limit the spread of these substances. For example, urine extravasated from a ruptured urethra may extend via a fascial cleft into the abdominal wall, but is limited from extending laterally into the thigh by the fusion of superficial fascia to deep fascia along the ischiopubic rami.

Bursae

Bursae are more elaborate structures that facilitate motion. They are modifications of tissue spaces that coalesce to form a closed sac, lined with a secretory serous (synovial) membrane. These sacs are usually relatively small and are interposed between structures where there is need for a greater degree of motion than is permitted by fascial clefts. Bursae are located mostly around joints and are inconstant in number and size.

Subcutaneous bursae are found between the skin and underlying deep fascia. An example of this type is present between the skin and the olecranon process at the elbow. Bursae also occur where a muscle slides over a bone, as in the case of the large **submuscular bursa** between the gluteus maximus muscle and the underlying ischial tuberosity. Numerous bursae are located around the knee joint and often communicate with the synovial cavity of the knee.

BURSITIS

Bursitis is inflammation of a bursa. Subcutaneous bursae over bony protuberances are exposed to frequent pressure and trauma. When the trauma is repetitive and prolonged, the bursa becomes swollen, inflamed, and painful. Occupational or sport activities predispose certain bursae to injury. Colorful, descriptive terms are give to these conditions, for example, housemaid's knee (prepatellar bursitis), weaver's bottom (ischial tuberosity bursitis), tailor's ankle (bursitis over the lateral malleolus), and tennis elbow (olecranon bursitis).

Inflammatory processes (bursitis) may increase the secretion of fluid by the serous membrane into the closed bursal sac. The resultant increase of fluid exerts pressure, causes swelling and may be extremely painful.

Tendon Sheaths

Tendon sheaths are more complex structures which are modifications of the simple closed bursal sac. Tendon sheaths are elongated sacs, lined with a synovial (serous) secreting membrane, into which tendons have invaginated. They occur mainly at the ankle, wrist, and along the plantar and palmar surfaces of the digits. They are associated with the long tendons of muscles attaching to bones of the hands or feet that have their muscle bellies located in the forearm or leg.

Inflammation of the synovial sheath may develop into adhesions between the layers of the synovial sheath. If this occurs, movement of the bones to which the tendon is attached becomes severely limited (see tendosynovitis, p 86).

Synovial Cavities

Synovial cavities are present in all freely movable joints of the body. The articulating surfaces of the bones of these joints are surrounded by a cufflike sleeve of connective tissue called the joint capsule, which extends well beyond the articulating surfaces to attach to the periosteum of the bones forming the joint. The internal surface of the fibrous joint capsule is lined by a synovial membrane that secretes fluid into the joint cavity. Along the line of attachment of the connective tissue capsule, the synovial membrane reflects onto the surface of the bone to cover the entire portion of the bone within the joint capsule, except the articular surfaces. The latter are covered by hyaline cartilage. In those cavities having internal ligaments such as the knee, hip, and shoulder, the synovial membrane reflects onto the ligaments so as to exclude them from the synovial cavity. The fluid secreted by the synovial membrane is somewhat viscous in consistency and serves as a lubricant between the articular surfaces.

Serous Cavities

Serous cavities of the body facilitate movement between organs. They are present in the major body cavities. In the thorax, the pericardial cavity surrounds the heart, and a separate pleural cavity surrounds each lung. The peritoneal cavity is located for the most part in the abdominal cavity but extends into the pelvic cavity. Serous cavities (pericardial, pleural, and peritoneal) are lined by serous membranes and are closed cavities, except for the peritoneal cavity in the female. In the female, the peritoneal cavity communicates by way of the uterine tube with the uterine cavity, which in turns opens to the exterior through the vagina.

As organs develop they invaginate into the serous cavities, carrying the serous membrane with them. This membrane becomes intimately adherent (fused) to the external surface of the organ and is designated as the **visceral portion,** for example, visceral pleura, visceral pericardium, or visceral peritoneum. That portion of the serous membrane remaining adherent to the wall of the body cavity is called the **parietal layer.** As organs invaginate into serous cavities they essentially obliterate the lumen of the cavity. Normally, there is a limited space between opposing visceral and parietal layers of a serous cavity. This interval is filled with a minimal amount of fluid secreted by the serous membrane.

At several sites within the abdominal cavity two or more layers of serous membrane become fused to suspend an organ from the body wall. This fusion of layers forms the **visceral ligaments** or **mesenteries** (Fig. 1-9). Visceral ligaments of the peritoneal cavity are of special importance. They provide the only pathway for vessels and nerves to reach organs, since no structures penetrate the serous membrane lining the cavity.

In the abdomen, organs that are not suspended from the body wall by a visceral ligament are described as being retroperitoneal in position. Examples of retroperitoneal organs in the abdominal cavity are the kidneys, duodenum, and the pancreas.

PERITONITIS

Peritonitis is inflammation of the peritoneum. It may be general or localized. It is characterized by an accumulation of a large amount of peritoneal fluid (ascites) containing fibrin and many leucocytes (pus). In the supine patient, the infected fluid tends to collect at two sites, 1)

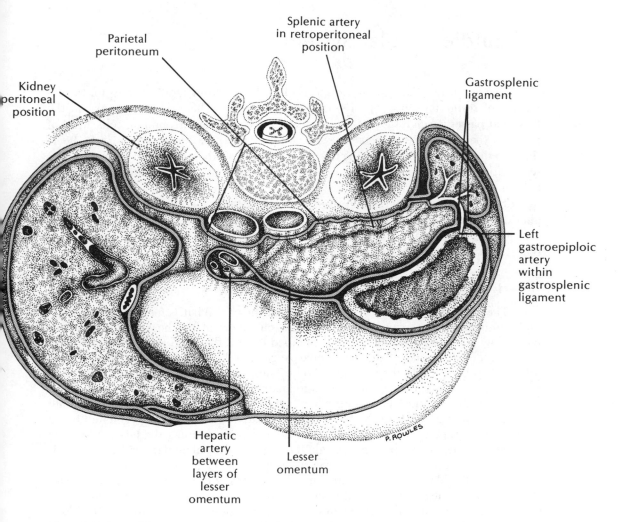

Parietal
peritoneum

Splenic artery
in retroperitoneal
position

Gastrosplenic
ligament

Kidney
peritoneal
position

Left
gastroepiploic
artery
within
gastrosplenic
ligament

Hepatic
artery
between
layers of
lesser
omentum

Lesser
omentum

P. ROWLES

Fig. 1-9. Visceral ligaments (mesenteries).

the pelvic cavity, and 2) in the right posterior subphrenic space. Tapping or draining of excess fluid from the abdomen is accomplished by inserting, under local anesthesia, a trocar and cannula or a needle and plastic tubing through the anterior abdominal wall, and aspirating the peritoneal fluid (paracentesis). When a patient with peritonitis is examined, stretching of the infected parietal peritoneum is very painful. The pain is especially severe when digital pressure over the inflamed area is suddenly released, because this causes the abdominal wall to rebound suddenly. Clinically, this is called rebound tenderness, which localizes the site of inflammation of the parietal peritoneum and often occurs over the infected organ.

Cardiovascular System

The cardiovascular system transports blood to and from capillary beds for the exchange of respiratory gases, nutrients, and metabolites. This exchange occurs in the fluid that passes through the walls of capillaries to bathe the surrounding cells. Most of the fluid that enters tissue spaces is retrieved by the capillaries to return to the heart through the veins. After nourishing the cells, a portion of the tissue fluid drains into the lymphatic vessels as lymph, which follows a different route in returning ultimately to the venous system.

The circulatory system comprises the heart, arteries, capillaries, veins, and lymphatics. However, in dissection, capillary and lymphatic beds, as well as the smaller arteries and veins, are not visible. Therefore, in gross anatomy the functional level of the cardiovascular system is not investigated.

Heart

The clinician often speaks of right and left hearts. What is meant by this seemingly ambiguous expression is that the right chambers of the heart receive venous blood, low in oxygen content, which is pumped toward the lungs to be oxygenated. In contrast, the left cardiac chambers receive oxygenated blood from the lungs and pump it by way of the aorta to the rest of the body. Thus, the expression right and left hearts refers to a functional concept as well as to anatomic divisions.

The heart is a muscular pump whose prime function is to propel the blood into capillary beds in all parts of the body. The **right atrium** receives deoxygenated blood by way of the superior and inferior vena cava, and veins draining the heart. From the right atrium the blood passes through the right **atrioventricular orifice,** which is guarded by the tricuspid valve, into the **right ventricle.** Contraction of the right ventricle forces the blood past the semilunar valve, guarding the **orifice of the pulmonary artery,** into the lungs. Oxygenated blood returns to the heart through the pulmonary veins, which empty into the **left atrium.** Upon contraction of the left atrium, the blood passes through the left atrioventricular orifice, which is guarded by the bicuspid (mitral) valve, into the left ventricle. Contraction of this chamber forces the blood past the cusps of the semilunar valve of the **aortic orifice** for distribution, by way of the aorta, throughout the body.

The functional role of each chamber is reflected in the thickness of its muscular wall. The atrial walls are very thin, having only to propel blood from the atria into the ventricles. The relatively massive wall of the left ventricle is approximately three times as thick as the wall of the right ventricle. This is consistent with its much greater work load since this chamber must develop sufficient pressure to drive the blood to all parts of the body, as opposed to the short distance the right ventricular blood must travel to reach the lungs.

Arteries

Arteries carry blood away from the ventricles of the heart to the beds of the capillary vessels. At each bifurcation, the units of this distributing system decrease in size, from a diameter of over 30 mm at the pulmonary trunk and aorta to arterioles that may be as small as 0.5 mm. In the pulmonary circuit, the arteries carry blood low in oxygen content, whereas in the systemic circulation, they carry highly oxygenated blood.

Arteries are classified as **elastic** or **large arteries, muscular** or **medium-sized arteries,** and **arterioles.** The walls of arteries are much thicker than the walls of veins of corresponding diameter, thus they are less apt to collapse as blood pressure falls during diastole. As branching arteries decrease in diameter, the elastic tissue component of their walls diminishes and is replaced by smooth muscle. At the arteriolar level, all elastic tissue has been essentially replaced.

ANEURYSM

An aneurysm is a thin, weakened section of the wall of an artery or a vein that bulges outward forming a balloonlike sac, or it may cause a permanent dilation of the blood vessel. The aneurysm dilates to a larger and larger size until the vessel wall becomes so thin it bursts and causes massive hemorrhage with shock, severe pain, stroke, or death depending on which vessel is involved. Aneurysms commonly involve the circle of Willis, which can cause stroke and mental impairment, or the thoracic and abdominal aorta. Although atherosclerosis is the most common cause, syphilis, congenital vessel defects, and trauma may also produce life-threatening aneurysms.

The structure of the walls of an artery suggests its function. For example, an elastic artery (aorta, pulmonary trunk) may be stretched lengthwise and distended in diameter to accommodate the large volume of blood that accompanies each ventricular contraction. During ventricular relaxation the elasticity of the walls of large arteries helps to maintain blood pressure and thereby effect a continuous flow through the smaller arteries.

In gross anatomy, most of the study of the cardiovascular system is spent in considering the pattern of arterial distribution. For a particular vessel we study its point of origin, its relationship to other structures along its course, its possible anastomoses with other vessels, and its area of distribution.

Structures of the body usually retain the blood supply they acquired during development so that the pattern of arterial distribution is relatively constant; however, minor variations do occur. The most frequent deviations from the regular pattern include:

1. A reduction in the size of a given artery. If this occurs, an adjacent artery supplying the same general area will usually be increased in size. An example of this may be the uneven size of the two vertebral arteries contributing to the blood supply of the brain.

2. Two vessels that normally arise as direct branches from a larger artery may arise from a common trunk. This occurs frequently with the anterior and posterior humeral circumflex arteries in the upper part of the arm.

3. As an artery bifurcates, some secondary vessels may arise from the parent trunk. An example of this would be the common interosseus artery, normally a

branch of the ulnar artery (formed as the brachial artery bifurcates at the elbow), arising directly from the brachial artery.

In general, the arterial supply to the upper and lower limbs follow an analogous pattern.

Capillaries

While capillaries are much too small (7–10 μ) to be visualized at the dissection table, their role in circulation should be mentioned since exchanges between the blood and tissue fluid occur across their one-cell thick walls.

Capillaries form a complex, anastomotic network between arterioles and venules. Many of their channels parallel one another and not all channels in a network function simultaneously. More metabolically active tissues (glands and muscles) have more extensive capillary beds than do less active tissue (tendons and ligaments). Anastomoses occur between arterioles and venules so that in different physiologic conditions the blood supply to various organs may be changed rapidly. Thus, in vigorous exercise more blood courses through muscles, while during digestion the blood supply to the gastrointestinal tract is increased (Fig. 1-10).

Avascular structures (cartilage, cornea, epidermis) have no capillary beds. Modifications in the vascular communication between arterioles and venules, in the form of sinusoids, occur in some organs (spleen, liver, bone marrow, pituitary). **Sinusoids** are wide, irregular channels, and are partially lined by phagocytic cells.

Veins

Venous channels originate opposite the arteriolar side of capillary networks as small-caliber venules. They coalesce to form veins of increasing diameter that ultimately empty into the atria of the heart. Veins are usually more superficial, have thinner walls, are more numerous, and are of larger caliber than their companion arteries. Blood flow through veins is slower and under much less pressure than through arteries of comparable size.

The pattern of venous tributaries is much more variable than that of corresponding arteries. Veins of the pulmonary circulation drain into the left atrium, while subsystems of venous drainage for the systemic circulation draining the body wall, abdominal viscera, and vertebral column empty into the right atrium. These subsystems include the venous drainage from the gastrointestinal tract that forms the **hepatic portal system** which drains through the portal vein into the liver. The **azygos venous system** drains blood from the wall of the thoracic cavity and provides some drainage of the abdominal wall through the ascending lumbar veins. Extensive venous plexuses surround the vertebral column and the spinal cord as the **vertebral plexus of veins.** Anastomoses between these subsystems occur freely. The anastomoses in the rectal, umbilical, lower esophageal regions, and abdominal mesenteries are of special clinical importance in obstruction of the inferior vena cava flow or in pathologic conditions of the liver, in which venous flow through this organ is impeded.

As in the arterial distributions, the venous return from the upper limb is analogous to the lower limb. Veins that accompany major deep arteries in the upper and

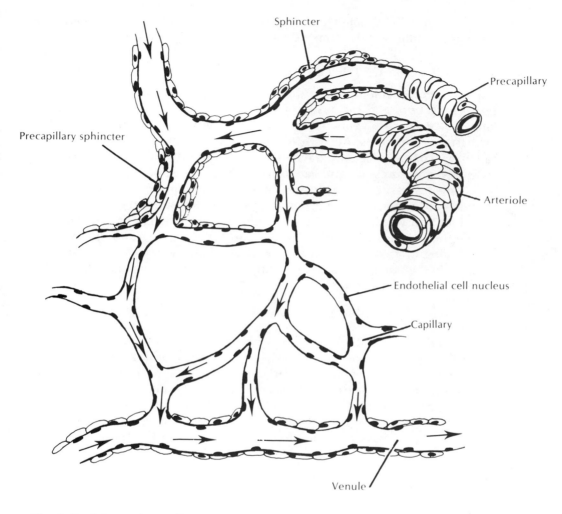

Fig. 1-10. Schematic capillary bed fed by arterioles and drained by venules.

lower extremity usually occur in pairs called **venae comitantes** (*i.e.*, each artery has two companion veins named for the artery they accompany). Veins from the hand and foot coalesce to form two major superficial vessels, the **cephalic** and **basilic** in the upper extremity; veins from the foot form the superficial **greater** and **lesser saphenous** in the lower extremity (these veins provide easy access to the circulatory system for intravenous infusion or drawing of blood).

Many veins are provided with **valves** to insure an unidirectional flow of blood. These are especially important in the limbs as venous blood returns to the heart against the force of gravity. Valves are usually situated in major vessels just distal to their larger tributaries. This arrangement limits retrograde blood flow. Incompetence of valves, as seen especially in the inferior extremity, results in an engorgement and dilatation of veins. If this becomes chronic, varicosities of the veins may occur, which are often removed surgically. Venous return from the extremities is facilitated by muscular contraction that compresses the deep veins and forces the blood toward the heart.

VARICOSE VEINS OF THE LOWER EXTREMITY

One of the most common disorders of the vascular system is the dilation, elongation, and tortuosity of the superficial veins of the lower limb, called varicose veins. The principal cause of this affliction is an increased venous pressure with dilation of the veins, due largely to valvular incompetence and gravity from the upright position. The increased venous pressure and varicosities cause edema of the skin and subcutaneous tissue with decreased blood flow and poor healing. This venous stasis and edema predisposes the skin to develop ulcers following minor skin abrasions or injury. If the deep veins are patent and can handle the venous return from the foot and leg, then surgical removal or stripping of the superficial varicosed veins decreases the venous pressure in the skin and subcutaneous tissue, decreases the edema, and allows the stasis ulcers to heal.

Portal Systems

Portal systems occur in the circulation wherever an extra set of capillaries is interposed in the circuit between blood leaving and blood returning to the heart. A prime example is the hepatic portal circulation. Arterial blood to the gastrointestinal system reaches the capillary beds in the stomach and intestine and is drained by tributaries that ultimately form the portal vein. The portal vein transports blood to the liver where it passes through a second set of capillaries (sinusoids). The hepatic veins which empty into the inferior vena cava are ultimately formed from this second capillary plexus.

PORTAL-SYSTEMIC VENOUS ANASTOMOSIS

Tributaries of the portal vein communicate with systemic veins in several locations, particularly the lower end of the esophagus, rectum, and umbilicus. Ordinarily these channels are collapsed with little blood flow, because blood drains from the abdominal viscera through the portal vein to the liver. If venous flow through the liver is impeded by a blood clot or scarring in the liver (cirrhosis), the portal vein pressure rises markedly. This portal hypertension dilates the veins of the portal system, including some or all of these anastomotic channels. This can cause varicosities of the umbilicus (caput medusae), esophagus (esophageal varices), and rectum (hemorrhoids).

Anastomoses

An **anastomosis** is a union between the distal ends of blood vessels permitting free communication between the involved vessels. Anastomoses may occur between arteries, between veins, or between arterioles and venules. In the latter case, the blood bypasses the capillary network since it is transported directly from arterioles to venules. **Arteriovenous anastomoses** are widely distributed but are found especially in the intestine, kidney, and skin.

Anastomoses may occur between relatively large vessels, for example, in the circle of Willis, located between branches of the internal carotid artery and branches of the basilar artery at the base of the brain, in the arterial arches in the hands and feet, and in the arterial arcades in the intestine. The latter form end-to-end anastomoses providing, in essence, a dual blood supply to all segments of the gut and its accessory organs.

Most anastomoses occur between small vessels or develop from capillary networks or arterioles that supply the same region. Such potential alternate pathways are clinically significant in that they provide for the development of collateral circulation.

COLLATERAL CIRCULATION

Collateral circulation is the mechanism whereby blood may flow to an organ or region after its normal course has been blocked. This is necessary when occlusion of a vessel from a blood clot, foreign body, tumor, or ligation occurs. If the segment distal to the occlusion is to remain viable it must receive a new blood supply. When collateral circulation develops, the blood bypasses the obstruction by an anastomosis, and may even flow in both directions in the anastomotic artery to supply all tissue distal to the occlusion.

Lymphatics

The lymphatic portion of the vascular system consists of a series of lymph channels interrupted by lymph nodes. It differs from the cardiovascular system proper in that it does not have a pumping apparatus, does not form a complete circuit (lymph travels only in one direction), and is a system of vessels with lymph nodes interposed along their course.

Lymph coursing through the vessels is a clear, slightly straw-colored fluid formed by filtration from blood capillaries as tissue fluid and similar in composition to blood plasma. The lymph contains lymphocytes formed largely in the lymph nodes, enzymes, antibodies, and lipids. In intestinal lymphatics, lipids form a milky white, fatty emulsion, the chyle.

Lymphatic vessels begin as an extensive, diffuse network of blind-ending **lymphatic capillaries** that drain tissue fluid from nearly all tissues and organs. These capillaries are lined with endothelium, have variable diameters, and are usually slightly larger than blood capillaries. Lymph capillaries are absent in the central and peripheral nervous systems, and in avascular tissues, such as the cornea, hyaline cartilage, and the epidermis of the skin.

Lymph capillaries coalesce to form large collecting vessels that contain valves giving them a beaded appearance. The smaller vessels form extensive, diffuse anastomosing superficial plexuses in the skin and on the surface of organs, and deep plexuses within organs. While lymphatic vessels largely accompany the venous drainage, the superficial and deep lymphatic plexuses of a given organ usually follow different pathways. In cancer, a metastasis in the pelvic region may either pass by lymphatics superficially to the groin or deeply into the pelvic cavity, depending upon the site of the initial lesion.

Coalescence of larger lymphatic vessels form lymphatic trunks. The largest, the **thoracic duct,** begins at the **cisterna chyli.** The latter is a large irregular lymph sac located at the level of the second lumbar vertebra within the abdominal cavity. From this site the thoracic duct ascends through the thoracic cavity to terminate by emptying into the left subclavian vein near its junction with the left internal jugular vein. Tributaries emptying into the thoracic duct drain the lower half and the left upper quadrant (left side of the head, left upper extremity, and left half of the thorax) of the body. The **right lymphatic duct,** a much shorter vessel, drains the upper right quadrant of the body. It is formed in the vicinity of the right internal jugular vein and the right subclavian vein as they join to form the right brachiocephalic vein. Lymph vessels draining the right side of the head, right upper extremity, and right side of the thorax coalesce to form the right lymphatic duct.

Lymph nodes are flattened, bean-shaped structures varying in size from 0.5 mm to 2 to 3 cm in length. They are interposed along the course of the lymph vessels and act as small filtering stations for the lymph. In addition to screening out foreign particulate matter and detoxifying pathogenic bacteria, they produce lymphocytes and antibodies.

Normally, lymph nodes are only palpable in the axillary (armpit) and inguinal (groin) regions. However, in response to inflammation, they enlarge considerably and can be palpated easily in other parts of the body, such as under the mandible, at the elbow and knee, and along the lateral aspect of the neck.

Large lymph nodes are usually found in aggregrates or chains extending along principal veins and are named from their anatomic location. In Figure 1-11 the major groups of nodes are shown.

METASTASIS

Knowledge of the location of the lymph nodes and the direction of lymph flow is important in the diagnosis and prognosis of spread of carcinoma (metastasis). Cancer cells usually spread by way of the lymphatic system and produce aggregates of tumor cells where they lodge. Such secondary tumor sites are predictable by the direction of lymph flow from the organ primarily involved.

Nervous System

The highly specialized nervous system, together with the endocrine system, provides the remarkable and essential coordination necessary for the well-being of a complex living organism such as man. As the major integrative system of the body the nervous system functions to 1) provide an awareness, through stimuli, of both the internal

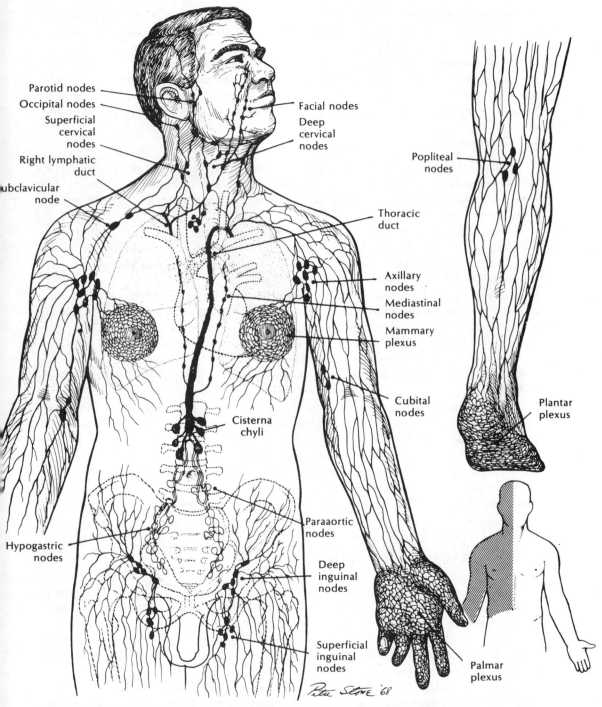

Parotid nodes
Occipital nodes
Superficial cervical nodes
Right lymphatic duct
ubclavicular node

Facial nodes
Deep cervical nodes

Popliteal nodes

Thoracic duct

Axillary nodes
Mediastinal nodes
Mammary plexus

Cubital nodes

Plantar plexus

Cisterna chyli

Paraaortic nodes

Hypogastric nodes

Deep inguinal nodes

Superficial inguinal nodes

Palmar plexus

Fig. 1-11. Principal lymph vessels and nodes. In the figure at the lower right, cross hatching shows the area drained by the right lymphatic duct. Langley LL, Telford IR, Christensen JB: Dynamic Anatomy and Physiology, 5th ed. New York, McGraw-Hill, 1980)

and external environment of the body, 2) make possible voluntary and reflex activities between the various structural elements of the organism, and 3) balance the organism's response to environmental changes.

The complexity of the brain and spinal cord is such that it is usually taught as a separate course, neuroanatomy. Gross anatomy courses are concerned with the distributing nerves which constitute the peripheral nervous system. However, to gain a basic understanding, some aspects of the central nervous system must be appreciated.

Neuron

The functional and anatomic unit of the nervous system is the **neuron** or **nerve cell.** This highly specialized cell is composed of a cell body and one or more protoplasmic processes. Changes in electric potential along the membrane of a process constitutes a nerve impulse.

Functionally, neurons are classified as **afferent** or **sensory** if they transmit impulses from component parts of the body toward centers of integration in the brain and spinal cord, and **efferent** or **motor** if they transmit impulses from the central nervous system toward the distal parts of the body (Fig. 1-12).

The peripheral termination of each sensory nerve cell process is structually modified as a **specialized receptor ending.** Structurally different endings are responsive to different modalities of sensations, such as pain, temperature and pressure. Receptor endings located in the skin, and special sense organs (eye, ear, nose, tongue) provide contact with the external environment. Receptors in muscles, joints, ligaments, and viscera (organs) receive stimuli to provide information on the internal environment of the body.

At **synaptic junctions** (the site where the process of one neuron comes in close proximity to the process or cell body of another) impulses are transmitted along neuronal processes to form conduction pathways. It is the complexity of the conduction pathways and the interplay between the various integrative centers of the central nervous system that necessitate special emphasis on this system of the body.

For descriptive purposes, the nervous system may be divided into a central portion, composed of the brain and the spinal cord, the peripheral portion formed by the nerves of the body, and the autonomic nervous system.

CENTRAL NERVOUS SYSTEM

Cell bodies of neurons in the central nervous system are located in discrete layers or sites collectively referred to as the **gray matter.** In the cerebral and cerebellar hemispheres the cells are situated, for the most part, at the surface of these structures in the cerebral or cerebellar cortex. In contrast to this arrangement, deeply placed clusters of gray matter are called *nuclei.* Functionally, several nuclei may form an **integrative center,** such as the respiratory center.

All gray matter of the spinal cord is located centrally. In a cross section of the spinal cord this accumulation of nerve cell bodies forms a gross configuration of the letter H. Two legs of the H extend dorsally as the **dorsal horns** (or along the length of the cord as the **dorsal columns**). Cell bodies of the dorsal horns are associated with

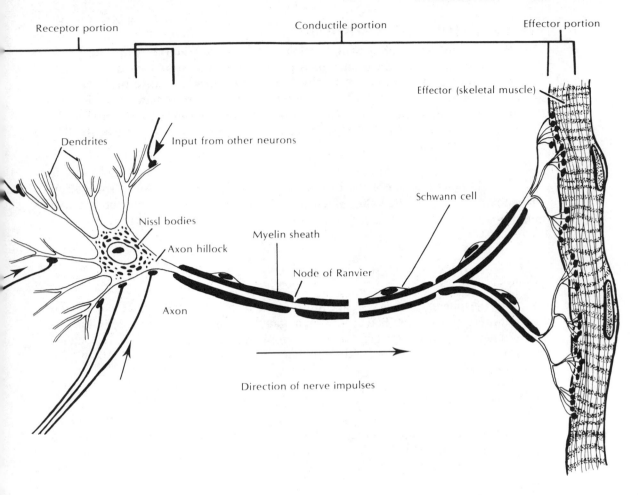

Receptor portion

Conductile portion

Effector portion

Effector (skeletal muscle)

Dendrites

Input from other neurons

Schwann cell

Nissl bodies

Myelin sheath

Axon hillock

Node of Ranvier

Axon

Direction of nerve impulses

Fig. 1-12. Typical large motor neuron showing receptor, conductile, and effector regions. (Copenhover WM, Kelly DE, Wood RL: Bailey's Textbook of Histology, 17th ed. Baltimore, Williams & Wilkins, 1978)

sensory impulses. The other two legs of the H extend ventrally as the **ventral horns** or **ventral columns,** which are composed of motor neuron cell bodies.

Processes of neurons in the central nervous system constitute the **white matter.** Generally they are segregated into sensory and motor components and are gathered together into bundles as **spinal cord** or **brain tracts,** which subserve a specific function, for example, most fibers responsible for transmitting pain and temperature sensations, though present in most peripheral nerves, clump together in the spinal cord as the lateral spinothalamic tract, to transmit this modality as they ascend to higher centers in the brain. Fibers transmitting motor impulses from the brain to the muscles, such as the corticospinal tract, follow entirely separate pathways from the sensory fibers. Additional fiber tracts in the central nervous system connect integrative centers with each other.

RHIZOTOMY AND CORDOTOMY

For relief of intractable pain, cutting of the sensory nerve roots, rhizotomy, or sectioning of the pain pathways in the spinal cord, cordotomy, may be indicated. In the latter procedure the anterolateral columns of the spinal cord are divided surgically. Immediate relief is achieved below the level of the section on the side opposite the incision. Bilateral cordotomy is necessary if the pain is present on both sides of the body.

The supportive tissue of the central nervous system is made up of specialized connective tissue cells called **glia,** which are not involved in the transmission of nerve impulses.

PERIPHERAL NERVOUS SYSTEM

The peripheral nervous system consists of the distributing nerves of the body and small clusters of nerve cell bodies located in **ganglia.** A distinction is made in the peripheral nervous system between the nerves supplying body wall structures and the extremities, the **somatic portion,** and the nerves supplying the viscera of the body, the **autonomic portion.**

Nerves are formed by cell processes of neurons. Nerve cell bodies present in ganglia of the peripheral nervous system are limited in number. All cell bodies in a definitive ganglion are either motor or sensory. The sensory ganglia are located in two sites: either interposed on the dorsal root of a spinal nerve (to be described later), or on those cranial nerves that transmit sensory impulses. All cell bodies of motor neurons in the peripheral nervous system are associated with ganglia of the autonomic portion.

Distributing nerves are also classified as to their origin from the central nervous system. Thirty-one pairs of nerves originating from the spinal cord are called **spinal nerves.** Twelve additional pairs of nerves originate from the brain stem and are called **cranial nerves.** Nerves derived from the spinal cord contain processes that transmit both motor and sensory impulses. Some nerves arising from the brain stem are similar to the spinal nerves in this respect, while others transmit only sensory or only motor impulses.

Supportive connective tissue investments surrounding distributing peripheral nerves are 1) the **endoneurium,** which is a connective tissue ensheathment around an individual nerve fiber; 2) the **perineurium,** which surrounds a bundle of nerve fibers or a nerve fascicle; and 3) the **epineurium,** which is a connective tissue investment of the entire nerve.

Typical Spinal Nerve

To avoid confusion in studying the peripheral nervous system, it is important that the student appreciate the distinction between the roots of origin of a spinal nerve and

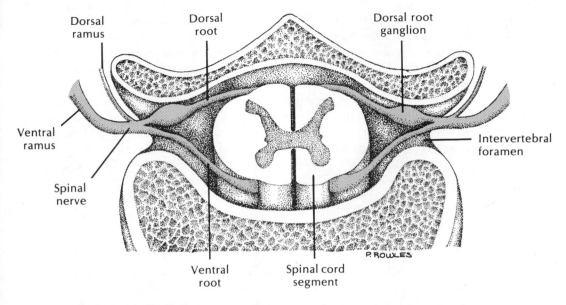

Fig. 1-13. Typical spinal nerve.

the rami or distributing branches of a spinal nerve. A typical spinal nerve is formed by fibers that have their cell bodies limited to a definitive block of the spinal cord termed a **spinal cord segment.** Rootlets arise from the dorsal aspect of each successive segment of the spinal cord and coalesce to form the **dorsal roots.** Similarly, rootlets arising from the ventral aspect of the spinal cord segment form the **ventral root** of a spinal nerve (Fig. 1-13).

Dorsal roots transmit only sensory fibers. The cell bodies of these fibers are located in the ganglia interposed on the dorsal roots called the **dorsal** or **spinal root ganglia.** The ventral root of a spinal nerve transmits only motor fibers. The cell bodies of these fibers are located in the gray matter of the spinal cord, either in the ventral horn if the fibers are destined for skeletal muscle, or in the lateral horn if the fibers (autonomic) supply smooth muscle, cardiac muscle, or glands.

As the dorsal and ventral roots traverse the intervertebral foramen to leave the vertebral canal, they unite to form the definitive **spinal nerve.** A tubular sheath of the meninges (dura, arachnoid, and pia) extends a short distance along the spinal nerve as it exits from the vertebral canal. Distally, this sheath fuses with the epineurium of the spinal nerve.

Shortly after the spinal nerve is formed it divides into two branches, a **dorsal** and a **ventral ramus.** The posterior area of the body is supplied by dorsal rami and the anterior area and the extremities by ventral rami. Except for the second cervical nerve, the ventral rami are much larger branches than dorsal rami and the area they supply is much more extensive than that supplied by the dorsal rami. The latter supply the intrinsic or deep muscles of the back, joints of the vertebral column, and the skin over the back of the trunk, neck, and head. The ventral rami of the second through eleventh thoracic spinal nerves become intercostal nerves (T_{12} is called the subcostal nerve).

These nerves supply the intercostal muscles, the anterolateral muscles of the abdominal wall, and the skin on the anterior aspect of the trunk.

Innervation to the upper and lower limbs is derived from ventral rami of the spinal nerves, which form plexuses. Such a **somatic nerve plexus** is formed as ventral rami branch and coalesce to create a nerve network adjacent to their emergence from the vertebral canal. These networks, formed by the splitting and subsequent uniting of nerve components, vary in their complexity as the process may be repeated several times. The pattern of formation is sufficiently constant in different individuals that each component of a given plexus can be consistently identified. The formation of a plexus results in a mingling of fibers from different spinal cord segments being distributed by a definitive nerve formed from the plexus. For example, the radial nerve, a branch of the brachial plexus, contains fibers from nerve cell bodies located in the fifth, sixth, seventh, and eighth cervical and the first thoracic segments of the spinal cord.

REFERRED PAIN

Referred pain is the localization of visceral pain on a body surface often far removed from the organ involved. Nerves to the painful surface area arise from the same spinal segments as the nerve fibers supplying the viscus in question. For example, cardiac pain is often felt over the anterior chest wall and radiates down the inside of the left axilla and upper extremity. Gallbladder disease may elicit severe pain in the upper right abdominal region and over the right infrascapular region. One explanation for referred pain is that pain fibers from the body wall and those from the viscera share a common neuronal pool in the spinal cord. Bombardment of visceral pain impulses into this common area excites the somatic pain fibers whose response is misinterpreted by the brain as cutaneous pain in the peripheral dermatome supplied by this particular peripheral nerve or nerves.

Skin over the entire body is **supplied segmentally,** that is, each spinal nerve innervates a single constant segment of the skin. With the exception of the first cervical nerve all spinal nerves supply branches to the skin. The skin segment supplied by a given spinal nerve is called a **dermatome.** In the neck and trunk the dermatomes form consecutive bands of skin. In the trunk there is an overlap of adjacent dermatome nerve supply so that to denervate a given dermatome three consecutive nerves must be transected or injected. Most of the skin of the face and scalp is supplied by the trigeminal (fifth cranial) nerve where dermatomes may be assigned to each of its three branches.

SHINGLES

Shingles or herpes zoster is a viral infection of the dorsal (sensory) root ganglion. Small vesicles (blisters) and discoloration of the skin occur over the dermatomal pattern of the skin, supplied by nerves from the involved ganglion. These lesions develop as circular bands of blisters

around the neck and trunk and somewhat vertical bands along the extremities.

The **cervical plexus** is derived from ventral rami of the first through fourth cervical nerves and supplies the skin and muscles of the neck, and the thoracic diaphragm. The latter illustrates the basic concept that migrating (developmental shifting) muscles carry their nerve supply with them; thus, in this case, the diaphragm migrated from the cervical region where it had its embryonic origin.

The **brachial plexus,** derived from the ventral rami of cervical nerves 5 to 8 and the first thoracic nerve, supplies the skin and muscles of the upper extremity, including (with the exception of the trapezius) those muscles of the upper extremity that originate from the chest wall and the back.

The **lumbosacral plexus,** derived from ventral rami of all lumbar and sacral nerves, supplies muscles of the posterior abdominal wall, pelvis, and muscles and skin of the inferior extremity.

As distributing branches of plexuses supply their respective structures, the following generalizations may be made:

1. The nerve supply to a muscle usually joins the vascular supply to form a neurovascular pedicle and enters the deep aspect of the muscle at the proximal end of its fleshy belly.
2. If a nerve pierces a muscle, it usually sends branches to supply the muscle.
3. Nerves supplying a muscle send cutaneous branches to the skin overlying the muscle.
4. Nerves supplying muscles extending over joints also supply the joint.
5. Nerves from plexuses follow a common channel with vessels to enter the extremity. In the upper extremity this is the cervicoaxillary canal, while in the lower extremity the femoral canal leads into the anterior compartment of the thigh, the obturator canal into the medial compartment of the thigh, and the greater sciatic foramen transmits nerves into the posterior compartment of the thigh.

Throughout most of their course, nerves (and vessels) usually traverse fascial clefts between muscles. This relationship is most evident in the thorax and abdomen. Here the intercostal and thoracoabdominal nerves course in the cleft or plane between the second and third layers of muscles of the body wall, the internal intercostal and innermost intercostal muscles in the thorax, or the internal abdominal oblique and the transversus abdominis in the abdominal wall.

Cranial Nerves

Most of the twelve pairs of **cranial nerves** arise directly from the brain stem. They traverse foramina in the floor of the cranial cavity as they course to their area of distribution. They transmit fibers whose cell bodies are located in discrete clumps within the brain called nuclei. They differ from spinal nerves in another respect in that some cranial nerves transmit only sensory fibers, some transmit only motor fibers and others are similar to spinal nerves in that they transmit both motor and sensory fibers.

All cranial nerves transmitting general sensory fibers have ganglia interposed on the nerve, similar to dorsal root ganglia of spinal nerves. Except for the spinal accessory and vagus nerves, the cranial nerves are limited to their distribution to the head and neck.

AUTONOMIC NERVOUS SYSTEM

The autonomic portion of the peripheral nervous system supplies motor innervation to smooth muscle, cardiac muscle, and glands. There is a basic anatomic difference between the impulse pathway of somatic motor fibers to skeletal muscles, and the autonomic pathway to smooth muscle, cardiac muscle, and glands. From the central nervous system motor impulses to skeletal muscle fibers are transmitted along neurons that have their cell bodies in the ventral horn of the spinal cord or in motor nuclei of cranial nerves. Thus, a motor impulse to a skeletal muscle utilizes only one neuron to pass from the central nervous system to motor end plates in muscle fibers.

In contrast to the above, impulses destined for involuntary muscles or glands require two neurons in passing from the central nervous system to the effector organs. The first neuron in this pathway, termed the **preganglionic neuron,** has its cell body in the central nervous system. The second, termed the **postganglionic neuron,** has its cell body in a ganglion outside the central nervous system. Thus, impulses passing along preganglionic fibers activate postganglionic neurons at ganglia. Postganglionic fibers in turn activate the smooth or cardiac muscle fibers and glandular cells.

The **autonomic nervous system** is further subdivided into **sympathetic** (Fig. 1-14). and **parasympathetic portions**. Each of these subdivisions is distinctive in the location of the preganglionic cell bodies, the location of the ganglia, and in their function.

Sympathetic (Thoracolumbar) Division of the Autonomic Nervous System

This portion of the autonomic nervous system is also called the **thoracolumbar portion** because its preganglionic nerve cell bodies are located in the lateral horn (intermediolateral cell column) of the thoracic and upper two lumbar segments of the spinal cord. Processes from these nerve cell bodies leave the central nervous system by way of ventral roots of the twelve thoracic and upper two lumbar spinal nerves. The preganglionic fibers contained in the fourteen pairs of nerves arising from the above spinal cord segments form short filaments called **white rami communicantes.** These extend from the spinal nerve to a ganglion situated adjacent to the body of a respective thoracic or lumbar vertebra. Such a ganglion is part of a chain of ganglia that extends along the entire length of the vertebral column from the base of the skull to the tip of the coccyx.

Once the preganglionic fiber passes into the chain ganglion by a white ramus communicantes, it may do one of three things. It may synapse on a postganglionic nerve cell body in the ganglion at that level. The process of the postganglionic nerve cell body can then return by a **gray ramus communicans** to the spinal nerve, to be dis-

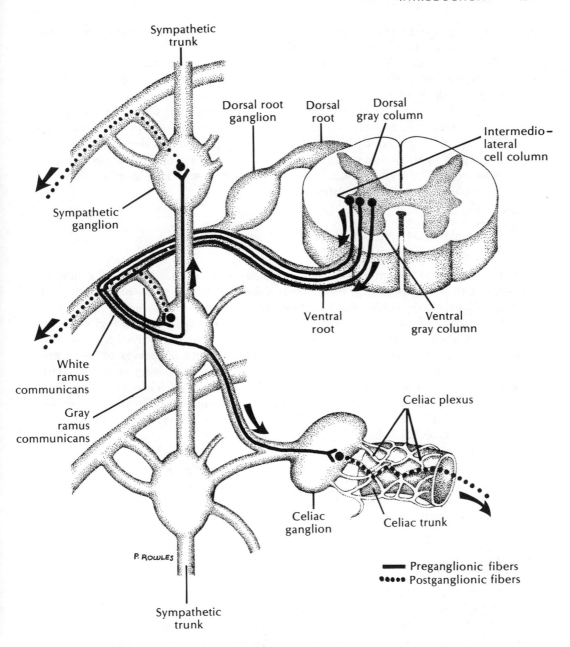

Fig. 1-14. Sympathetic nervous system.

tributed as a component of the spinal nerve. The second alternative is to ascend or descend by communications between adjacent ganglia to a higher or lower level (thereby creating the chain). This is necessary because the preganglionic outflow from the spinal cord is limited to the thoracolumbar level. Gray rami associated with these higher or lower ganglia will then transmit the postganglionic fiber to the spinal nerve at their levels. Thus, all spinal nerves receive gray rami communicantes and, therefore, transmit postganglionic fibers to their areas of distribution.

The third alternative pathway for preganglionic fibers is to traverse the sympathetic chain ganglia without synapsing and to pass as a component of a **splanchnic nerve** to **preaortic ganglia.** These ganglia, containing postganglionic nerve cell bodies, are situated adjacent to major arteries arising from the aorta. The fibers of such a postganglionic nerve cell body contribute to the formation of autonomic plexuses that surround the major branches of the aorta. Extending along the vessel, they innervate the viscera supplied by the arteries they accompany. The major sympathetic plexuses include the carotid plexus, the celiac plexus, the superior mesenteric plexus, the inferior mesenteric plexus, and the hypogastric plexus.

Parasympathetic (Craniosacral) Division of the Autonomic Nervous System

This portion of the autonomic nervous system is also referred to as the **craniosacral portion** since its preganglionic nerve cell bodies are located in the brain stem or in the lateral horn of the second, third, and fourth sacral segments of the spinal cord.

Processes from the preganglionic cell bodies located in the sacral region of the spinal cord leave the cord by way of the ventral root of their respective spinal nerves. From the sacral spinal nerves they form the **pelvic splanchnic nerves** or **nervi erigentes.** Preganglionic fibers carried by the nervi erigentes may unite with sympathetic plexuses along their course toward the postganglionic nerve cell body. The latter cells are widely dispersed on the surface, or in the substance of the organ they innervate. The sacral portion of the parasympathetic system supplies pelvic organs and the descending portion of the colon.

In the brain stem, parasympathetic preganglionic nerve cell bodies are located in the nuclei associated with certain cranial nerves, namely, the oculomotor (III), facial (VII), glossopharyngeal (IX), and vagus nerves (X). Fibers from these cell bodies travel through their respective cranial nerves to parasympathetic ganglia located in the head, except for the vagus nerve.

The Skin

The **skin or** integument is a sturdy, elastic, movable envelope that covers the entire body, blending with the sensitive mucous membranes of the mouth, nose, eyes, and anal and urogenital openings. It consists of two distinct layers, the **epidermis,** an outer epithelial investment of closely packed stratified squamous epithelium, and the **dermis,** a deeper layer of dense irregular collagenous fibers, interlaced with blood vessels, nerves, and lymphatics.

LANGER'S LINES (LINES OF CLEAVAGE)

Langer's lines (lines of cleavage) are faint linear clefts in the skin indicative of the direction of the underlying collagen fibers. If a sharp

rounded object penetrates the skin, it leaves an elongated slit instead of a circular wound. This is because the collageonus fibers of the dermis of the skin are arranged in parallel rows, which are separated along their length during the injury rather than disrupted. The perceptive surgeon will realize that an incision running parallel to the collagen fibers will heal with only a fine scar. However, if the incision is across the rows, the collagen is disrupted, and the wound tends to gape open and to heal in a broad, thick scar. The direction of these rows of collagen "Langer's lines" are quite consistent in most individuals, being longitudinal in the limbs and circular in the neck and trunk. These lines are especially evident on the palmar surfaces of the fingers where they run parallel to the long axis of the digit.

Skin is more, however, than a mere protective covering against injury, loss of body fluids, and temperature changes. It is also an extensive sense organ equipped with exquisite nerve endings that inform us of our external environment, for example, pain, temperature, touch, and pressure. The many functions of the skin are summarized by Whitnall,

". . . . even with our ingenious modern machinery we cannot create a tough but highly elastic fabric that will withstand heat and cold, wet and drought, acid and alkali, microbic invasion, and the wear and tear of three score years and ten, yet effect its own repairs throughout, and even present a seasonable protection of pigment against sun's rays. It is indeed the finest fighting tissue."

The skin has four appendages—hair, nails, sebaceous (oil) glands, and sweat glands. **Hairs** are dead, keratinized cells that protrude from follicles. The follicles are oblique invaginations of epidermis into the dermis and grow from a bulbous, dilated end, the dermal papilla. **Nails** are hardened, flat thickenings of the dead cells from the outermost layer of the epidermis. Growth occurs at the proximal end of the nail, the nail bed.

Sebaceous glands develop as downgrowths from hair follicles into the dermis, and, therefore, they are nearly always associated with hair. Their oily secretion, sebum, makes the skin essentially waterproof. **Sweat glands** are single, tubular glands present over the entire body, except the lips and glans penis. Their excretory ducts open onto the surface of the skin through small pores. Sweat is an effective coolant of the skin. A sedentary person excretes about 500 ml/day, a hard-working manual laborer, about a liter each hour.

2

superior extremity

The superior extremity consists of the shoulder, arm, forearm, and hand. The latter is specially adapted for prehension. The muscles of the shoulder and arm act to place the grasping hand in almost any desired position. The rich nerve supply of the fingertips makes the hand a sensitive tactile organ. The muscles of the hand permit complex activity of the digits, which are moved primarily by muscles in the forearm. The superior extremity articulates with the trunk at the small sternoclavicular joint and is firmly anchored to the chest by several muscles which cover the thorax. Thus, the superficial muscles of the trunk (back and pectoral regions) acting upon this member must be considered in any description of the upper extremity.

Superficial Back and Scapular Region

Surface Anatomy

The most superior structure palpable in the midline of the upper portion of the back is the spinous process of the seventh cervical vertebra, or **vertebra prominens.** Above this level the spinous processes of the cervical vertebrae lie deep to the **ligamentum nuchae,** while inferiorly all vertebral spinous processes are palpable.

The **spine** of the scapula is subcutaneous through most of its extent, although its medial triangular portion is covered by the trapezius muscle. Laterally it is continuous with the **acromion,** which forms the point of the shoulder. The **medial (vertebral) border** and the **inferior** and **superior angles** of the scapula can be felt deep to the superficial musculature of the back. In a muscular individual there is a diamond shaped depression between the two scapulae formed by the lack of muscle fibers in the aponeurosis of the trapezius muscle in this area.

Inferolaterally the **crests of the ilia** project as bony ridges below the waist and are palpable posteriorly to the **posterior superior iliac spines.** Further inferiorly in the midline of the lower portion of the back, the posterior surface of the **sacrum** is subcutaneous, and at its inferior extent the **coccyx** can be felt in the cleft between the buttock. On each side the **lumbar triangle (of Petit),** a small area low in the back, is bounded by the crest of the ilium, the anterior border of the latissimus dorsi, and the posterior extent of the external abdominal oblique muscles.

Cutaneous Innervation

The cutaneous innervation to the shoulder and upper pectoral region is derived from **supraclavicular nerves** from the cervical plexus. Inferior to the first intercostal space, cutaneous branches of the intercostal or thoracoabdominal nerves, continuations of the ventral rami of spinal nerves, supply the anterolateral body wall. The **intercostal nerves** are ventral rami of spinal nerves. They supply segmental bands of skin over the ribs and intercostal spaces, with branches of a single spinal nerve sending overlapping twigs to adjacent skin areas. From these nerves **lateral cutaneous branches** penetrate the skin near the midaxillary line and divide into anterior and posterior branches. The intercostal nerve then continues anteriorly in the intercostal space and terminates as **perforating branches** emerging just lateral to the sternum to divide into **medial** and **lateral cutaneous twigs.**

Skin over the back of the neck is supplied by **dorsal rami** of cervical spinal nerves. Below the shoulder region the dorsal rami of thoracic spinal nerves divide into medial and lateral branches. Above the level of the sixth thoracic vertebra the **medial branch** supplies the skin over the back, while the **lateral branch** is primarily muscular. Below this level the distribution of these branches is reversed.

SUPERFICIAL BACK

Muscles

The muscles of the back are arranged in layers (Fig. 2-1, Table 2-1). Those of the first and second layers, although related topographically to the back, afford attachment of the upper limb to the vertebral column. They are innervated by ventral rami of spinal nerves and are functional muscles of the upper extremity. The first and most superficial layer is composed of the trapezius and the latissimus dorsi muscles.

The **trapezius muscle,** with its companion of the opposite side, forms a large trapezoid over the upper portion of the back. Its anterior border in the cervical region gives the sloping contour to the neck and bulges in the action of shrugging the shoulders. Inferiorly, in the midline it overlaps the superior extent of the origin of the latissimus dorsi. With the scapula drawn forward, the lateral border of the trapezius, the superior border of the latissimus dorsi, and the medial border of the scapula bound the **triangle of auscultation,** an area often utilized in listening to respiratory sounds with the stethoscope. In this position the underlying ribs become essentially subcutaneous.

The **latissimus dorsi muscle** gives the lateral taper to the chest. With the teres major, the latissimus dorsi forms the posterior wall of the axilla, or armpit. Inferiorly, its lateral fibers interdigitate with those of the external abdominal oblique. Its inferior border spirals or turns under in passing to its insertion on the humerus. The latissimus dorsi, acting with the pectoralis major muscle, returns the flexed arm to the anatomic position, as in the action of rowing a boat or climbing a rope.

Deep to the trapezius a sheet of three relatively thin straplike muscles, the **levator scapulae** and the **rhomboidei minor** and **major,** insert sequentially into the posterior lip of the medial border of the scapula. The levator scapulae and the rhomboideus minor are usually fused at their insertion and, therefore, somewhat difficult to differentiate. These muscles elevate the scapula, draw it toward the midline, and assist in lateral rotation. The rhomboideus major draws the inferior angle of the scapula superiorly to depress the lateral angle, which assists in adduction of the arm.

A fourth muscle, the **serratus anterior,** inserts into the anterior lip of the medial border of the scapula. From its origin on the upper eight or nine ribs this muscle follows the contour of the thoracic cage as it passes to its insertion. It acts to hold the scapula onto the rib cage. Loss of its action results in a flaring out of the medial border (winged scapula).

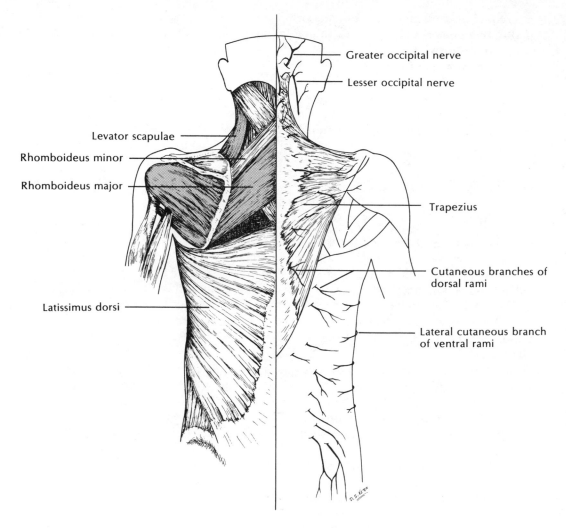

Fig. 2-1. Cutaneous innervation and superficial muscles of the back.

Fasciae

The superficial fascia over the back has no special characteristic features. The deep fascia, in addition to forming muscular envelopes, specializes in the lower back as the thickened thoracolumbar fascia, which will be described with the deep muscles of the back and the abdominal musculature.

Arteries and Nerves

The arterial supply to the trapezius, levator scapulae, and rhomboidei muscles is derived from the transverse cervical artery, a branch of the thyrocervical trunk from

Table 2-1. Muscles of the Superficial Back

Muscle	Origin	Insertion	Action	Nerve
Trapezuis	External occipital protuberance, superior nuchal line, ligamentum nuchae, seventh cervical and all thoracic spinous processes	Anterior border of spine of scapula, acromion, and lateral third of posterior border of clavicle	Adducts and rotates scapula; upper part elevates scapula; lower part depresses scapula	Spinal accessory and twigs from third and fourth cervical nerves
Latissimus dorsi	Spinous processes of all vertebrae below sixth thoracic; lumbodorsal fascia, crest of ilium, and lower three or four ribs	Floor of intertubercular groove of humerus	Lowers, rotates medially, and draws arm posteriorly	Thoracodorsal
Levator scapulae	Transverse processes of first through fourth cervical vertebrae	Posterior lip of medial border of scapula	Elevates scapula and inclines head	Twigs from cervical plexus and dorsal scapular
Rhomboideus major	Spinous processes of second through fifth thoracic vertebrae	Posterior lip of lower half of medial border of scapula	Adducts scapula	Dorsal scapular
Rhomboideus minor	Spinous processes of seventh cervical and first thoracic vertebrae	Root of spine of scapula	Adducts scapula	Dorsal scapular
Serratus anterior	Digitations from lateral surfaces of upper eight ribs	Anterior lip of medial border of scapula	Holds scapula to chest wall; draws scapula anteriorly; and rotates inferior angle laterally	Long thoracic

the first part of the subclavian artery (Fig. 2-2). The **transverse cervical artery** crosses the posterior triangle of the neck to reach the anterior border of the trapezius. It divides into a **superficial branch** ramifying on the deep surface of the trapezius and a **deep branch** passing parallel to the medial border of the scapula, deep to and supplying the levator scapulae and rhomboidei.

Innervation of the trapezius is from the **spinal accessory nerve** together with twigs from the third and fourth cervical nerves; the latter are probably sensory in function. The spinal accessory emerges from the deep surface of the sternocleidomastoid and crosses the posterior triangle of the neck, passing deep to the trapezius to ramify on the deep surface of the muscle. The levator scapulae and rhomboidei are supplied by the **dorsal scapular nerve,** a branch of the brachial plexus. After piercing the scalenus medius muscle, this nerve passes deep to, and supplies the levator scapulae and rhomboidei, passing parallel to the medial border of the scapula with the deep branch of the transverse cervical artery.

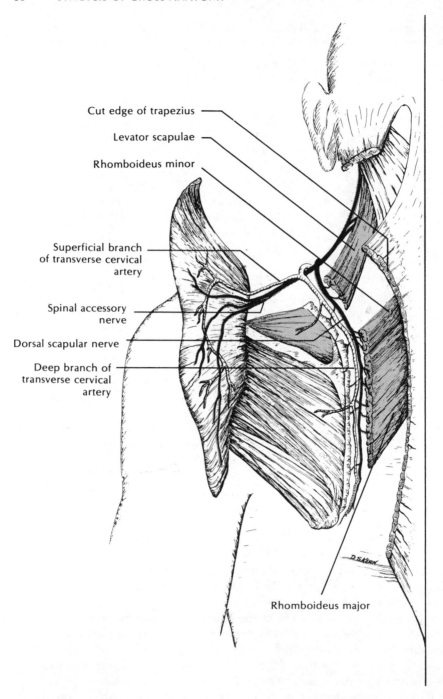

Cut edge of trapezius

Levator scapulae

Rhomboideus minor

Superficial branch
of transverse cervical
artery

Spinal accessory
nerve

Dorsal scapular nerve

Deep branch of
transverse cervical
artery

Rhomboideus major

Fig. 2-2. Innervation and arterial supply to the superficial muscles of the back.

The latissimus dorsi receives its blood supply from the **thoracodorsal artery,** a terminal branch of the subscapular from the third part of the axillary. This vessel passes along the lateral border of the scapula to its distribution in the muscle. The nerve supply of the latissimus dorsi is from the **thoracodorsal nerve,** a branch of the posterior cord of the brachial plexus. It passes anterior to the subscapularis and teres major muscles to descend along the lateral border of the scapula to its termination in the muscle.

The serratus anterior is supplied by the **lateral thoracic artery,** a branch of the second part of the axillary. This vessel arises near, and descends along, the lower border of the pectoralis minor muscle to ramify on the superficial surface of the serratus anterior muscle. Innervation of the latter is by the **long thoracic nerve,** a branch of the brachial plexus, which descends posterior to the brachial plexus to follow the lateral thoracic artery into the muscle.

SCAPULAR REGION

Muscles

Three muscles, the **supraspinatus, infraspinatus,** and **subscapularis,** originate from, and cover, the three shallow fossae of the scapula (Table 2-2). The tendons of these muscles, as well as the tendon of the teres minor, pass over the glenohumeral ar-

Table 2-2. Muscles of the Shoulder

Muscle	Origin	Insertion	Action	Nerve
Supraspinatus	Medial two-thirds of supraspinous fossa	Superior portion of greater tubercle of humerus	Initiates abduction of arm and augments deltoid function of abduction	Suprascapular
Infraspinatus	Medial three-fourths of infraspinous fossa	Midportion of greater tubercle of humerus	Main lateral rotator of arm	Suprascapular
Subscapularis	Medial two-thirds of subscapular fossa	Lesser tubercle of humerus	Principal medial rotator of arm; aids in flexion, extension, adduction, and abduction of arm	Upper and lower subscapular
Deltoideus	Lateral third of anterior border of clavicle, acromion, and posterior border of spine of scapula	Deltoid tuberosity of humerus	Main abductor of arm; aids in flexion, extension, adduction, and medial and lateral rotation of arm	Axillary (circumflex)
Teres major	Posterior surface of inferior angle and lower portion of lateral border of scapula	Medial lip of intertubercular groove of humerus	Adducts and rotates arm medially	Lower subscapular
Teres minor	Upper portion of lateral border of scapula	Inferior portion of greater tubercle of humerus	Rotates arm laterally and acts as weak adductor of arm	Axillary (circumflex)

ticulation deep to the deltoideus muscle to fuse with and strengthen the joint capsule as they insert into the greater and lesser tubercles of the humerus. Their insertions form the **musculotendinous (rotator) cuff** of the shoulder joint. The supraspinatus initiates, then assists the deltoideus in abduction of the arm. The infraspinatus and subscapularis rotate the arm laterally and medially, respectively.

ROTATOR CUFF

The principal stability of the shoulder joint is not provided by the configuration of its articular surfaces or its ligaments, but rather from several stout tendons that cross the joint. These tendons insert on the greater and lesser tubercles of the humerus to form a rotator (musculotendinous) cuff. The cuff greatly strengthens the joint, except in its inferior aspect. Dislocations (subluxations) frequently occur inferiorly.

The superficial **deltoideus muscle** forms the lateral mass of the shoulder. From its extensive origin on the clavicle, the acromion, and spine of the scapula, it acts primarily as an abductor, but segments working independently function also in adduction, extension, flexion, and internal and external rotation of the arm.

The teres major, teres minor, and long head of the triceps have a lineal origin along the lateral border of the scapula, whereas the omohyoideus arises from the scapular notch. Note that with respect to muscular attachments to the scapula, all the muscles attaching to the fossae, including the omohyoid attachment at the scapular notch, are attachments of origin. Muscles attaching to the medial border are all insertions, whereas muscles attaching to the lateral border are all origins, including the supraglenoid tubercular attachment of the long head of the biceps. The trapezius and deltoideus attach parallel to each other, to both the spine of the scapula and the clavicle. They form a U- or V-shaped inner insertion for the trapezius and outer origin for the deltoideus.

Arteries and Nerves

The supraspinatus and the infraspinatus muscles are supplied by branches of the same artery and nerve (Fig. 2-3). The artery, the **suprascapular,** crosses the posterior triangle of the neck to arrive at the scapular notch. Here it passes over the transverse scapular ligament, which bridges the scapular notch, then terminates as **supraspinatus** and **infraspinatus branches** to these respective muscles. The **suprascapular nerve,** a branch of the brachial plexus, also crosses the posterior triangle of the neck to the scapular notch, but passes deep to the transverse scapular ligament, after which its distribution follows the arterial supply.

The subscapularis, teres major, and teres minor muscles receive their major blood supply from a single vessel, the **scapular circumflex,** a terminal branch of the subscapular artery from the third part of the axillary. The **lower subscapular nerve,** a branch of the posterior cord of the brachial plexus, innervates the teres major and

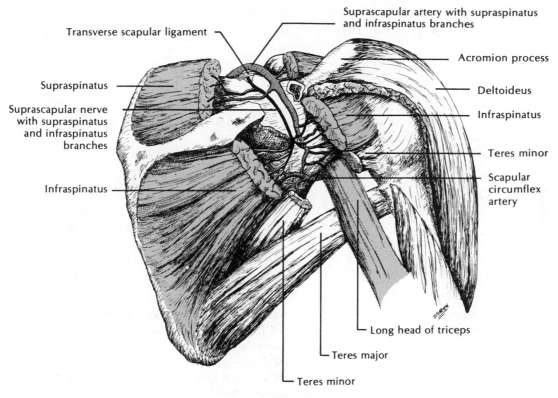

Fig. 2-3. Innervation and arterial supply to the shoulder muscles.

sends twigs to the subscapularis, which receives additional innervation from the **upper subscapular nerve** of the brachial plexus. The teres minor is supplied by a branch of the **axillary nerve** from the posterior cord of the brachial plexus.

Blood is distributed to the deltoideus muscle via the **anterior** and **posterior humeral circumflex arteries** and small deltoid branches of the thoracoacromial. The humeral circumflex arteries, from the third part of the axillary, pass anterior and posterior, respectively, around the neck of the humerus. The larger posterior humeral circumflex passes through the **quadrangular space** with the **axillary nerve** which innervates the deltoideus.

SCAPULAR ANASTOMOSES

If the distal segment of the subclavian artery or the proximal segment of the axillary artery need to be ligated, scapular anastomoses provide a route for a collateral blood supply of the upper extremity. Arteries involved in these anastomatic collateral channels are branches of: 1) the thyrocervical trunk, 2) the subscapular artery; and 3) other smaller arteries, such as, intercostal, pectoral, lateral thoracic, and thoracodorsal.

Pectoral Region

Surface Anatomy

On the anterior surface of the thorax the clavicle demarcates the chest from the neck. It is subcutaneous along its entire length from the manubrium of the sternum to its articulation with the acromion process of the scapula.

FRACTURE OF THE CLAVICLE

The S-shaped clavicle serves as a strut between the point of the shoulder (acromion process) and the sternum that allows the arm to swing away from the body. Because of its position, the clavicle transmits forces from the upper extremity to the trunk. If such forces are excessive, as in falling on one's outstretched arm, the clavicle may be fractured—in fact it is the most frequently broken bone in the body.

A depression in the midline at the base of the neck, the **suprasternal notch,** is bounded inferiorly by the superior border of the manubrium and laterally by the tendons of the sternal heads of the strenocleidomastoid muscles. In the midline, a palpable ridge demarcates the **sternal angle (of Louis)** and corresponds to the junction of the manubrium with the body of the sternum. Laterally, this junction affords articulation for the costal cartilage of the second rib and, therefore, can be utilized for accurate superficial determination of rib number or intercostal space. The **nipple** of the breast in the male usually lies over the fourth intercostal space, about a hand's breadth from the midline, but the level is variable in the female. The inferior and lateral margins of the thoracic cage are easily palpable. In the midline the **xiphoid process** can be felt at the inferior extent of the body of the sternum.

ANGLE OF LOUIS

The angle of Louis (sternal angle) is an important topographical landmark of the anterior thorax. It is a transverse ridge at the junction of the manubrium and body of the sternum that marks the level of articulation of the costal cartilage of the second rib with the sternum. When counting ribs in a physical examination the physician should begin with the second rib at the angle of Louis, since the first rib is behind the clavicle and difficult to palpate.

Muscles

The **pectoralis major,** a large fan-shaped muscle, covers the anterior chest wall from an extensive origin on the clavicle, sternum, and ribs (Table 2-3, Fig. 2-4). Near its insertion into the humerus it forms the anterior wall of the axilla. Its superior border meets the deltoideus to form the **deltopectoral triangle,** a small triangular depression bounded by the anterior border of the deltoid, the superior border of the pectoralis

Table 2-3. Muscles of the Pectoral Region

Muscle	Origin	Insertion	Action	Nerve
Pectoralis major	Clavicular head from medial half of clavicle; sternal head from sternum and costal cartilages; abdominal head from aponeurosis of external abdominal oblique	Lateral lip of intertubercular groove of humerus	Flexes, adducts, and medially rotates arm	Lateral and medial pectorals
Pectoralis minor	Anterior aspect of third, fourth, and fifth ribs	Coracoid process of scapula	Draws scapula inferiorly and elevates ribs	Medial pectoral
Subclavius	Junction of first rib and costal cartilage	Inferior surface of clavicle	Draws clavicle inferiorly and anteriorly	Nerve to subclavius

major, and the midportion of the clavicle. It contains fat, the deltopectoral lymph nodes, the cephalic vein, and the deltoid branch of the thoracoacromial artery.

The **pectoralis minor,** lying immediately deep to the pectoralis major, is a much less extensive muscle and acts as a muscle of forced respiration by elevating the chest wall. As it passes toward its insertion, it lies superficially to the axillary artery dividing it, for descriptive purposes, into three parts. The small **subclavius muscle** lies deep to the clavicle. In fracture of the clavicle this muscle may afford protection for the deeper lying subclavian vessels and the brachial plexus.

Fasciae

Over the pectoral region, the **superficial fascia** contains abundant fat, especially in the female, where it surrounds the mammary gland and gives the gross configuration to the breast.

Deep to the external investing layer of deep fascia in the pectoral region an additional lamina of deep fascia specializes as the **clavipectoral fascia.** This specialization is attached to the clavicle, encloses the subclavius muscle, fuses to span the gap between the clavicle and the pectoralis minor, then separates to enclose the latter. Lateral to the pectoralis minor, the fascia thickens and fuses to the external investing layer of fascia to form the **suspensory ligament of the axilla,** which passes to the floor of the axilla where it blends with the axillary fascia. Between the subclavius and the pectoralis minor this fascial sheet is pierced by branches of the thoracoacromial artery, the medial and lateral pectoral nerves, the cephalic vein, and lymphatic vessels.

Arteries and Nerves

Blood is supplied to both pectoralis major and minor muscles by **pectoral branches** from the thoracoacromial, the **lateral thoracic,** and **perforating branches** of the anterior intercostal arteries. Additional supply to the pectoralis major is derived

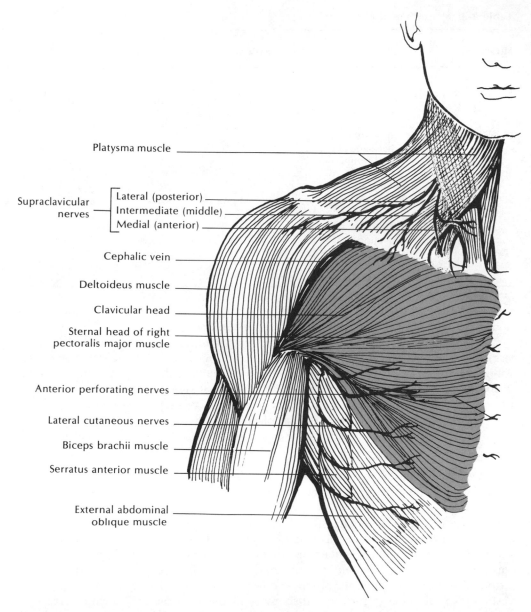

Platysma muscle

Supraclavicular
nerves
⎡ Lateral (posterior)
⎢ Intermediate (middle)
⎣ Medial (anterior)

Cephalic vein

Deltoideus muscle

Clavicular head

Sternal head of right
pectoralis major muscle

Anterior perforating nerves

Lateral cutaneous nerves

Biceps brachii muscle

Serratus anterior muscle

External abdominal
oblique muscle

Fig. 2-4. Superficial muscles of anterior chest wall and arm.

from the **perforating branches** of the internal thoracic artery. The **medial** and **lateral pectoral nerves,** branches from the medial and lateral cords of the brachial plexus, respectively, pierce the clavipectoral fascia to innervate the pectoral muscles. The medial pectoral nerve passes through the pectoralis minor to the pectoralis major to supply both muscles. The lateral pectoral nerve courses medial to the pectoralis minor,

Table 2-4. Nerve Distribution to Scapular and Pectoral Regions

Nerve	Origin	Course	Distribution
Supraclaviculars	Third loop of cervical plexus	From posterior border sternoclei-domastoid penetrate platysma to run in superficial fascia	Skin over shoulder, clavicle and first intercostal space
Intercostal nerves	Ventral rami of spinal nerves	Traverse intercostal spaces; lateral and anterior cutaneous branches penetrate muscles to reach skin	Skin and muscles of respective intercostal spaces
Spinal accessory	Spinal root from upper five segments of spinal cord; cranial root from brain stem	From jugular foramen bisects posterior triangle to reach anterior border of trapezuis	Trapezius and sternocleidomastoid
Dorsal scapular	Ventral ramus of C_5 (upper root of brachial plexus)	Penetrates scalenus medius to run along deep aspect of vertebral border of scapula	Levator scapulae and rhomboids
Suprascapular	Upper trunk of brachial plexus	Crosses posterior triangle to reach scapular notch	Supraspinatus and infraspinatus
Long thoracic	Roots of brachial plexus	Descends along lateral thoracic wall	Serratus anterior
Subscapulars	Posterior cord of brachial plexus	From axilla to anterior aspect of subscapularis	Subscapularis; lower subscapular also supplies teres major
Thoracodorsal	Posterior cord of brachial plexus	Runs along anterior border of latissimus dorsi	Latissimus dorsi
Nerve to subclavius	Upper trunk of brachial plexus	Runs on deep surface of subclavius	Subclavius
Axillary	Posterior cord of brachial plexus	Traverses quadrangular space to reach deep surface of deltoid	Deltoid and twigs to teres minor
Lateral pectoral	Lateral cord of brachial plexus	Penetrates clavipectoral fascia to reach deep surface of pectoral muscles	Pectoralis major and minor
Medial pectoral	Medial cord of brachial plexus	Penetrates clavipectoral fascia to reach deep surface of pectoral muscles	Pectoralis major and minor

usually supplying only the pectoralis major, although there is a communication between the two nerves so in a technical sense, both can be said to supply both muscles.

The blood supply to the subclavius is the **clavicular branch** of the thoracoacromial artery. The **nerve to the subclavius** is a branch of the upper trunk of the brachial plexus as the latter passes deep to the clavicle (Table 2-4).

BREAST

The female **breast**, a modified sweat gland, is located in the superficial fascia of the pectoral region, where it rests upon the deep fascia covering the pectoralis major

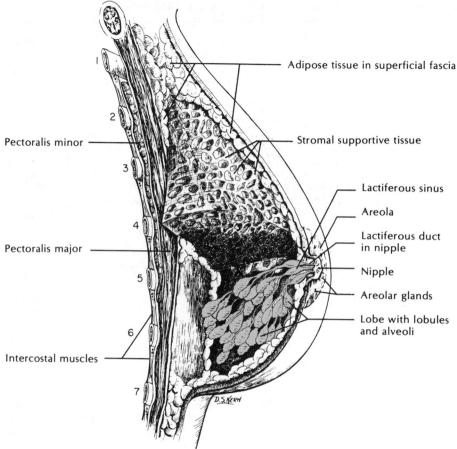

Fig. 2-5. The breast.

(Fig. 2-5). It normally extends between the second and sixth rib and from the lateral border of the sternum to the axilla.

AXILLARY TAIL OF SPENCE

Axillary tail of Spence is a rather constant extension of the breast tissue through an opening in the axillary fascia into the anteromedial part of the axilla. Thus, the axillary tail is not enclosed by the superficial fascia like the rest of the mammary gland. Breast tissue extends along the anterior axillary line beyond where the edge of the breast is normally seen.

It consists of glandular tissue, stroma, and fat. The **mammary gland** has no distinct capsule and is composed of fifteen to twenty **lobes** radiating from the **nipple.** Each lobe has a single **lactiferous duct** that converges toward the **areola.** It dilates near its termination to form a secretory reservoir, the **lactiferous sinus,** then constricts to open individually on the surface of the nipple. The stroma of the mammary gland consists of fibrous connective tissue that loosely envelops the entire gland, extending into the gland to enclose the parenchyma. Condensations of the collagenous fibers form the **suspensory ligaments (of Cooper),** which extend from the skin through the mammary gland to the underlying deep fascia.

PHYSICAL SIGNS OF CANCER OF THE BREAST

Cardinal physical signs of cancer of the breast can be recognized by the patient. Any palpable mass in the breast may be malignant. When cancer cells invade the gland and enlarge, they often attach to the suspensory ligaments of Cooper (retinaculum cutis) and produce shortening of the ligaments causing depression or dimpling of the overlying skin. If the neoplasm attaches to and shortens the lactiferous ducts, the nipple may become retracted or inverted. When advanced cancer of the breast invades the deep fascia of the pectoralis major muscle, contraction of this muscle causes a sudden upper movement of the whole breast.

The arterial supply to the breast is from **pectoral branches of the thoracoacromial,** the **lateral thoracic,** perforating branches of the **internal thoracic,** and the third through the sixth **anterior intercostal** arteries.

Lymphatic drainage from **subareolar** and **circumareolar plexuses** is continuous with the general cutaneous drainage of the thoracic region and may drain toward the neck or the abdomen. **Perilobular** and **interlobular lymphatic plexuses** in the breast proper drain the deeper tissue and communicate with the subareolar plexus. The deeper drainage of the breast is regional. The lateral half of the mammary gland drains to the axillary and pectoral nodes. Lymph drainage from the medial side passes into the thorax to nodes along the internal thoracic artery or may cross the midline to the opposite breast. Inferiorly the lymph may flow toward the abdomen and drain into the nodes in the upper portion of the abdomen.

CANCER OF THE BREAST

Cancer of the breast is the most common metastatic neoplasm in women. It spreads by direct invasion into adjacent tissues, but principally, the cancer cells travel through lymphatics and blood vessels to many parts of the body. The most common pathway is along the lymphatics leading to the axillary lymph nodes. To remove the breast carcinoma with a wide margin of tissue for the direct invasion spread and with the axillary lymph nodes for lymphatic metastases, radical mastectomy is the operation of choice. It includes removal of the entire breast with the underlying fascia, pectoralis major and minor muscles, and all of the lymph nodes of the axilla.

Arm, Forearm, and Hand

Surface Anatomy

The **biceps brachii muscle** makes the anterior, and the **triceps** the posterior, bulge on the arm. Medial and lateral intermuscular septa are extensions of the external investing fascia of the arm that separate the flexor (anterior) and extensor (posterior) compartments. These septa pass deeply to attach to the humerus and are responsible for the **superficial grooves** on either side of the arm. At the elbow, the **medial** and **lateral epicondyles** are easily palpable. The **olecranon process** of the ulna forms the posterior prominence of the elbow. Pressure on the medial side of the olecranon process elicits a tingling sensation ("funny bone"), which demonstrates the superficial position of the ulnar nerve passing along the ulnar groove on the medial epicondyle.

The **cubital fossa** forms a triangular depression anterior to the elbow joint. The **tendon of the biceps** can be palpated within its boundaries. The **lacertus fibrosus** (bicipital aponeurosis), a strong band of fibrous tissue, passes inferomedially from the tendon to the deep fascia of the forearm. The **median cubital vein,** forming a connection between the basilic and cephalic veins, can be seen crossing the fossa superficially. Just distal to the elbow, the increased width of the forearm results from a massing of the bellies of the **flexor muscles** of the wrist and fingers **medially,** and the **extensor group laterally.**

The posterior border of the **ulna** is subcutaneous along its entire length as are its prominent head and styloid process at the wrist. The radius is palpable in its distal half, and its **styloid process** can be felt at the lateral side of the wrist. Extension of the thumb results in a prominent ridge on the dorsum of the wrist formed by the **tendon of the extensor pollicis longus** muscle. With the thumb extended, this tendon, with that of the **abductor pollicis longus,** forms a lateral depression, the "anatomic snuff box," across the bottom of which the radial artery courses. When the wrist is flexed against pressure, its palmar aspect reveals in the midline the **tendon of the palmaris longus,** and about a centimeter laterally, the **tendon of the flexor carpi radialis.**

The palmar aspect of the hand reveals transverse creases at the metacarpophalangeal joints. Note that the webbing of the fingers is distal to these articulations. The **thenar eminence** (ball of the thumb) is formed by the small muscles to the thumb, the **hypothenar eminence** (heel of the hand) by the small muscles to the little finger. The interval between the thumb and the index finger contains the adductor muscle of the thumb anteriorly and the first dorsal interosseous muscle posteriorly. The skin of the palm is thickened, with infiltrated subcutaneous fat, and is firmly bound to deeper structures; over the dorsum of the hand the skin is more delicate and freely movable. Upon maximal extension of the wrist and fingers, the tendons of the **extensor digitorum muscle** form prominent ridges on the dorsum of the hand.

Cutaneous Innervation

The cutaneous innervation to the arm, forearm, and hand is supplied by branches of the brachial plexus with some contribution from the cervical plexus (Figs. 2-6, 2-7,

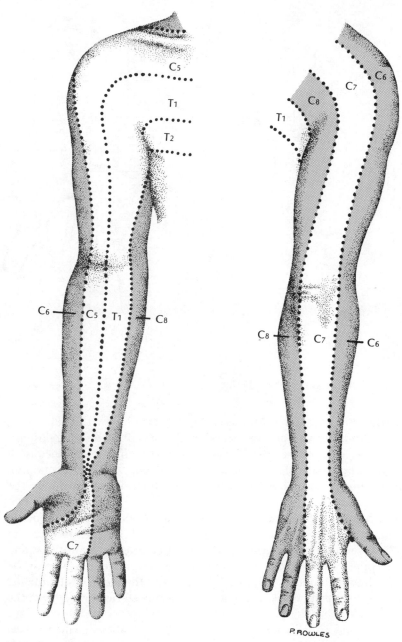

Fig. 2-6. Spinal nerve dermatomes of upper extremity.

2-8). The branches from the cervical plexus are the **supraclavicular nerves,** which supply skin over the upper portion of the deltoideus muscle.

Four cutaneous nerves are distributed to the skin of the arm. The **lateral brachial cutaneous,** a branch of the axillary nerve, supplies skin over the lower half of the deltoideus and the long head of the triceps. The **posterior brachial cutaneous,** from the

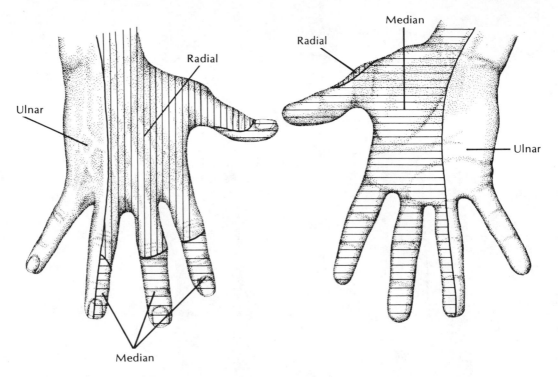

Fig. 2-7. Dermatomes of cutaneous nerves of the hand.

radial nerve, innervates skin on the posterior aspect of the arm below the deltoideus, and the **medial brachial cutaneous,** from the medial cord, is distributed to the postero-medial aspect of the lower third of the arm. The arm receives additional cutaneous innervation from the **intercostobrachial,** the lateral cutaneous branch of the second thoracic nerve, supplying the posteromedial surface of the arm from the axilla to the olecranon process.

Cutaneous branches to the forearm include the **lateral antebrachial cutaneous,** a continuation of the musculocutaneous nerve, giving anterior and posterior branches to the radial half of the forearm; the **posterior antebrachial cutaneous,** from the radial nerve, supplying an upper branch to the distal half of the anterolateral aspect of the arm and a lower branch to the middorsum of the forearm, and the **medial antebrachial cutaneous,** a branch of the medial cord, giving anterior and posterior branches to the medial aspect of the forearm.

The ulnar, median, and radial nerves all contribute to the cutaneous innervation of the hand. The cutaneous branches of the **ulnar nerve** supply both surfaces of the hand and fingers medial to a line passing through the midline of the ring finger. The **median nerve** gives cutaneous branches to the remainder of the palmar surface of the hand and fingers, and to the dorsal surface of the fingers distal to the middle phalanx. Cutaneous branches from the **radial nerve** supply the remainder of the dorsal surface of the hand and fingers, as well as a small area over the lateral aspect of the thenar eminence.

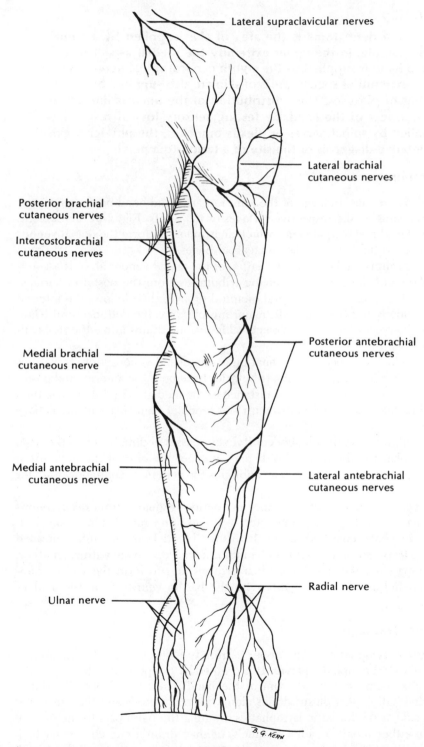

Fig. 2-8. Cutaneous innervation of the dorsal aspect of the superior extremity.

DERMATOME

By definition, a dermatome is the area of skin supplied by a spinal nerve. For example, in the upper extremity the medial aspect of the forearm and hand is supplied by the eighth cervical spinal nerve. However, in the extremities specific areas of skin are also supplied by terminal branches of plexuses. Thus, distribution of the ulnar is limited to the medial aspect of the hand. By testing sensory loss of respective areas supplied by spinal nerves or plexus branches, the physician can make a tentative diagnosis of the site of a lesion to a nerve.

Venous Drainage

The superficial venous drainage of the arm, forearm, and hand begins as **palmar** and **dorsal digital veins** on the respective surfaces of the digits (Fig. 2-9). These veins join to form the **dorsal metacarpal veins,** which anastomose to form the **dorsal venous arch** lying proximal to the heads of the metacarpal bones.

The **cephalic vein** is the lateral continuation of the dorsal venous arch. It ascends along the radial side of the forearm receiving tributaries from the posterior surface. Above the elbow it continues in the lateral bicipital groove, then follows the interval between the deltoideus and the pectoralis major muscles into the deltopectoral triangle. It terminates by perforating the clavipectoral fascia and drains into either the axillary or subclavian vein.

The **basilic vein** is the medial continuation of the dorsal venous arch. It ascends on the ulnar side of the forearm, receiving tributaries from both the anterior and posterior surfaces. In the arm it ascends a short distance in the medial bicipital groove, then penetrates the brachial fascia and unites with the brachial vein to form the axillary vein.

In the antecubital fossa the **median cubital vein** passes obliquely across the fossa connecting the cephalic and basilic veins. It may conduct the bulk of the blood from the cephalic to the basilic vein. The median cubital is the vein most commonly used for venipuncture.

The deep venous drainage of the upper extremity originates from **deep venous arcades** of the hand which parallel the arterial arches. The radial, ulnar, and brachial arteries have **venae comitantes,** which receive blood from the areas supplied by these vessels. Deep and superficial veins communicate extensively with each other. The veins accompanying the radial and ulnar arteries join to form the venae comitantes of the brachial artery, which unite with the basilic vein to form the axillary vein.

Lymphatic Drainage

The **superficial lymphatics** of the upper extremity begin as a meshwork around the fingers as the **digital lymphatic plexus.** It is drained by lymph vessels following the digital arteries that turn onto the dorsum of the hand where they form the **dorsal plexus.** The radial half of this plexus drains along the radial side, and the ulnar half along the ulnar side of the forearm. Lymphatics draining the palm pass to the sides of the hand to join either ulnar or radial channels, or may drain along channels which ascend in the midline of the volar aspect of the forearm. One or two **cubital lymph**

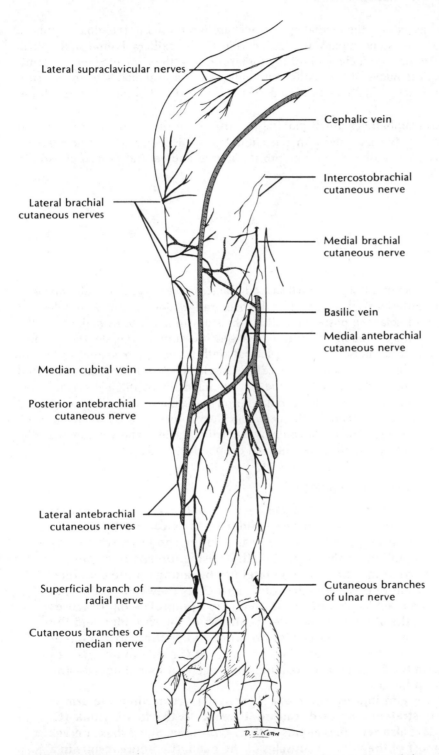

Lateral supraclavicular nerves

Cephalic vein

Intercostobrachial cutaneous nerve

Lateral brachial cutaneous nerves

Medial brachial cutaneous nerve

Basilic vein

Medial antebrachial cutaneous nerve

Median cubital vein

Posterior antebrachial cutaneous nerve

Lateral antebrachial cutaneous nerves

Superficial branch of radial nerve

Cutaneous branches of ulnar nerve

Cutaneous branches of median nerve

D. S. KERN

Fig. 2-9. Superficial venous drainage and cutaneous innervation of the volar aspect of the superior extremity.

nodes, located just above the medial epicondyle, are interposed in the ulnar channels. Their efferent vessels accompany the basilic vein to the **axillary lymph nodes.** The radial and posterior lymph channels of the forearm follow the cephalic vein to terminate in the **apical nodes** of the axilla. Efferents of the apical nodes form the subclavian lymph trunk. A **deltopectoral node** may be interposed before they reach the axilla.

The **deep lymphatic channels** parallel the arteries in the hand and forearm and drain into five or six small nodes in the cubital fossa. Efferent vessels accompany the brachial veins and ascend to drain into the **lateral and central groups of axillary nodes.**

ARM

Axilla

The axilla is a potential pyramidal shaped space. It consists of four walls, an apex, and a base. The **anterior wall** is formed by the pectoralis major and minor muscles and the clavipectoral fascia; the **posterior wall** by the subscapularis, teres major, and the latissimus dorsi muscles; the **medial wall** by the serratus anterior muscle, the first five ribs, and the intercostal muscles; and the **lateral wall** by the upper medial surface of the humerus. The **base** is formed by the axillary fascia. The truncated **apex,** directed supermedially, is bounded anteriorly by the clavicle, medially by the first rib, and posteriorly by the superior border of the scapula. Through the apex the axillary vessels and the brachial plexus, ensheathed by the **cervicoaxillary fascia,** pass between the neck and the upper extremity. The axilla contains the axillary artery and vein, most of the brachial plexus, axillary lymph nodes, fat, and connective tissue.

BRACHIAL PLEXUS LESIONS

The upper arm birth palsy (Erb–Duchenne paralysis) is the most common type of nerve injury during childbirth. It is caused by forcible widening of the angle between the head and the shoulder, which may occur from pulling on the head at birth or using forceps to rotate the fetus in utero. It may be caused in later life by falling on the shoulder. The site of injury is the junction of C_5 and C_6 as they form the upper trunk of the brachial plexus. This is called Erb's point. The injury causes paralysis of the abductors and lateral rotators of the shoulder, and the flexors of the elbow. Weakness is also noted in the adductors and medial rotators of the shoulder. The arm hangs at the side in internal rotation with the forearm pronated and the fingers and wrist flexed—the porter's tip hand.

Lower arm injuries (Klumpke's paralysis) occur when the arm is forcefully stretched upward, causing damage to the lower trunk (C_8 and T_1). It is also seen in scalenus anticus syndrome. Since these nerves supply most of the intrinsic muscles of the hand, the injury results in a claw-hand appearance, similar to an ulnar nerve injury.

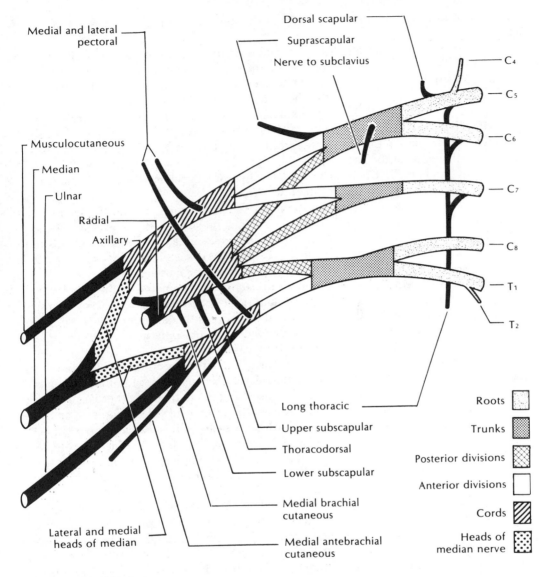

Medial and lateral pectoral

Dorsal scapular

Suprascapular

Nerve to subclavius

C₄

C₅

C₆

Musculocutaneous

Median

Ulnar

Radial

Axillary

C₇

C₈

T₁

T₂

Long thoracic

Upper subscapular

Thoracodorsal

Lower subscapular

Medial brachial cutaneous

Lateral and medial heads of median

Medial antebrachial cutaneous

Roots

Trunks

Posterior divisions

Anterior divisions

Cords

Heads of median nerve

Fig. 2-10. Brachial plexus.

Brachial Plexus

The **brachial plexus** is a network of nerves derived from the **ventral rami of the fifth through eighth cervical** (C₅ through C₈) and **first thoracic** (T₁) **nerves** (Fig. 2-10). It receives contributions from the fourth cervical (C₄) and second thoracic (T₂) nerves. The brachial plexus supplies muscular, sensory, and sympathetic fibers to the upper extremity. The plexiform arrangement permits intermingling of nerve components from several segments of the spinal cord to form composite nerves that supply individual structures. It is composed of five roots, three trunks, six divisions, three cords, and sixteen named branches. It is located partly in the neck, under the clavicle, and in the axilla.

The **roots** of the brachial plexus are continuations of the ventral rami of its component spinal nerves, which emerge between the scalenus anterior and medius muscles in line with similar, but more cranially situated roots constituting the cervical plexus. The roots of C_5 and C_6 unite to form the **upper trunk,** C_7 becomes the **middle trunk,** and C_8 and T_1 constitute the **lower trunk.** Each trunk divides distally into **anterior** and **posterior divisions,** which anastomose to form lateral, medial, and posterior cords, so named for their relation to the second part of the axillary artery. The **posterior cord** is formed by the union of the posterior divisions of all three trunks; the anterior divisions of the upper and middle trunks form the **lateral cord;** and the **medial cord** is the continuation of the anterior division of the lower trunk.

Two branches are derived from the roots, the dorsal scapular and the long thoracic. The **dorsal scapular** (C_5) pierces the scalenus medius to course parallel to the medial border of the scapula and deep to the levator scapulae and the rhomboids to supply these muscles. The **long thoracic** (C_5, C_6, and C_7) descends posterior to the roots of the plexus to ramify on and supply the serratus anterior muscle.

Two nerves are also derived from the upper trunk. The **nerve to the subclavius** (C_5 and C_6) descends anterior to the plexus and posterior to the clavicle to supply the subclavius muscle. The **suprascapular** (C_5 and C_6) crosses the posterior triangle of the neck to the lesser scapular notch, where it divides into **supraspinatus and infraspinatus branches** to supply muscles in their respective scapular fossae

Two nerves originate from the lateral cord. The **lateral pectoral** (C_5, C_6, and C_7) pierces the clavipectoral fascia to supply the pectoralis major. The **musculocutaneous** (C_5, C_6, and C_7) enters the coracobrachialis, supplying it and the two other flexor muscles in the arm, then continues into the skin on the lateral side of the forearm as the **lateral antebrachial cutaneous nerves.** The **lateral head of the median nerve** is also derived from the lateral cord. It joins with the **medial head** from the medial cord to form the **median nerve** (C_5, C_6, C_7, C_8 and T_1), which supplies flexor muscles in the forearm, most of the short muscles of the thumb, and skin of the lateral two-thirds of the palm of the hand and the fingers.

In addition to the medial head of the median nerve, the medial cord gives rise to four nerves. The **medial pectoral** (C_8 and T_1) passes through and supplies the pectoralis minor as well as the overlying pectoralis major. The **medial brachial cutaneous** (C_8 and T_1) supplies skin over the medial and posterior aspect of the distal third of the arm. The **medial antebrachial cutaneous** (C_8 and T_1) innervates skin of the medial and posterior aspect of the forearm. The largest branch, the **ulnar** (C_8 and T_1) is distributed to some of the flexors in the forearm, the intrinsic muscles of the hand, the skin of the medial side of the hand, all the skin of the little finger, and the skin of the medial half of the ring finger.

The posterior cord gives origin to the **upper subscapular nerve** (C_5 and C_6), which passes posteriorly to enter the subscapularis muscle; the **lower subscapular** (C_5 and C_6), which descends to supply the subscapularis and terminates in the teres major, and the **thoracodorsal** (C_5, C_6, and C_7), which descends anterior to the subscapularis and teres major to terminate in the latissimus dorsi. Other branches include the **axillary** (C_5 and C_6), which passes posteriorly through the quadrangular space and innervates the deltoideus and teres minor, and the **radial** (C_5, C_6, C_7, C_8, and T_1), which courses posteriorly to the axillary artery to innervate the extensors of the arm and forearm. It also supplies skin on the posterior aspect of the arm , forearm, lateral two-thirds of the dorsum of the hand, and the dorsum of the lateral three and one-half digits over the proximal and intermediate phalanges.

Axillary Artery

The **axillary artery,** a continuation of the **subclavian,** extends from the lateral border of the first rib to the lower border of the teres major, where it becomes the **brachial artery** (Fig. 2-11). For descriptive purposes, it is subdivided into **three parts** by

Fig. 2-11. Axillary artery.

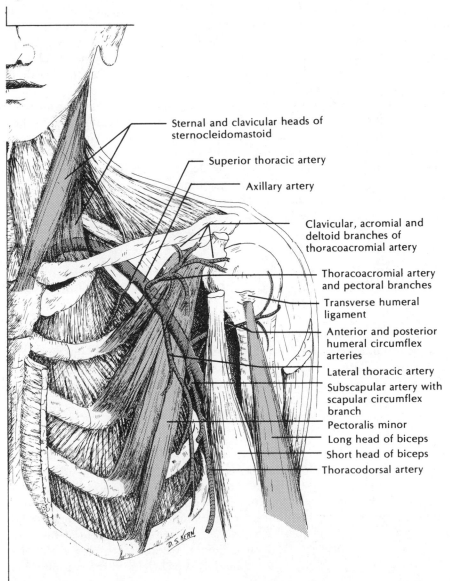

Sternal and clavicular heads of sternocleidomastoid

Superior thoracic artery

Axillary artery

Clavicular, acromial and deltoid branches of thoracoacromial artery

Thoracoacromial artery and pectoral branches

Transverse humeral ligament

Anterior and posterior humeral circumflex arteries

Lateral thoracic artery

Subscapular artery with scapular circumflex branch

Pectoralis minor

Long head of biceps

Short head of biceps

Thoracodorsal artery

D.S. KERN

the overlying pectoralis minor muscle. The first portion lies proximal, the second deep, and the third distal to the muscle.

Six branches originate from the axillary artery, one branch from the first segment, two from the second, and three from the third. The single branch from the **first part,** the **superior thoracic** (highest thoracic), is distributed to the first intercostal space.

The branches from the **second portion** are the thoracoacromial artery and the lateral thoracic. The **thoracoacromial artery** pierces the clavipectoral fascia to divide into four branches. The **acromial branch** passes laterally over the coracoid process, gives twigs to the deep surface of the deltoideus, and finally ramifies about the acromial process. The **deltoid branch** courses in the deltopectoral triangle to supply the deltoideus and pectoralis major. The **pectoral branches** pass between and supply both pectoral muscles and, in the female, distribute deep branches to the mammary gland, and the **clavicular branch** supplies the sternoclavicular joint and the subclavius muscle. The **lateral thoracic artery** arises near and descends along the lower border of the pectoralis minor, supplying the two pectoral muscles and the serratus anterior. In the female it sends deep branches to the mammary gland.

Branches from the **third part** of the axillary artery include the smaller **anterior** and the larger **posterior humeral circumflex arteries,** which encircle the neck of the humerus to supply the deltoideus and anastomose with the ascending branch of the profunda brachii artery. The largest branch of the axillary, the **subscapular artery,** arises opposite the lower border of the subscapularis muscle. It divides into the **scapular circumflex artery,** which passes through the triangular space to supply muscles on the dorsum of the scapula, and the **thoracodorsal branch,** which continues along the lateral border of the scapula to supply the latissimus dorsi.

COLLATERAL CIRCULATION

Collateral circulation of the axillary and brachial arteries will depend upon the level of ligation of the artery. For example, if the ligation of the axillary artery is distal to the humeral circumflex and subscapular branches, the blood flow in the limb is re-established through an anastomosis between these branches and the profunda brachii artery. If the ligation of the brachial artery is distal to the profunda brachii and the superior ulnar collateral arteries, these vessels will establish circulation distally, with the inferior ulnar collateral, radial, ulnar, and interosseous recurrent arteries.

Muscles of the Arm (Table 2-5)

The four muscles of the arm, three flexors, and one extensor, are located in flexor (anterior) and extensor (posterior) compartments delineated by the lateral and medial intermuscular septa (Fig. 2-12). In the flexor compartment the fusiform **biceps brachii** originates by two heads, a short head (in common with the medially situated **coracobrachialis muscle**) from the coracoid process, and a long head from the supraglenoid tubercle of the scapula. Lying in the intertubercular groove, the tendon of the long head passes deep to the transverse humeral ligament, where it acquires a synovial sheath as it traverses the joint cavity of the shoulder to its origin on the scapula. The

Table 2-5. Muscles of the Arm

Muscle	Origin	Insertion	Action	Nerve
Biceps brachii	Long head, supraglenoid tubercle; short head, tip of coracoid process	Tuberosity of radius and antebrachial fascia via the bicipital aponeurosis	Flexes forearm and arm; supinates hand	Musculocutaneous
Coracobrachialis	Tip of coracoid process	Middle third of medial surface of humerus	Flexes and adducts arm	Musculocutaneous
Brachialis	Distal two-thirds of anterior surface of humerus	Coronoid process and tuberosity of ulna	Flexes forearm	Musculocutaneous and small branch of radial
Triceps brachii	Long head, infraglenoid tubercle; lateral head, posterior surface and lateral border of humerus; medial head, posterior surface of distal half of humerus	Posterior aspect of olecranon process of ulna	Extends forearm; long head aids in extension and adduction of arm	Radial
Anconeus	Lateral epicondyle of humerus	Lateral aspect of olecranon and upper fourth of posterior surface of ulna	Acts as weak extensor of forearm	Radial

Fig. 2-12. Compartments of arm.

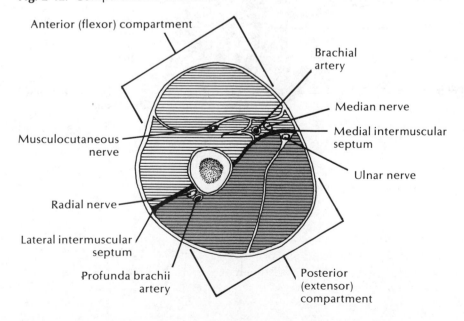

Anterior (flexor) compartment

Brachial artery

Median nerve

Medial intermuscular septum

Ulnar nerve

Musculocutaneous nerve

Radial nerve

Lateral intermuscular septum

Profunda brachii artery

Posterior (extensor) compartment

biceps is a powerful supinator as well as the main flexor of the forearm. The **brachialis,** the pure flexor of the forearm, covers the lower anterior half of the humerus and the capsule of the elbow joint, forming a bed for the more superficially placed biceps.

The **triceps brachii** fills the posterior compartment. Its long and lateral heads obscure the more deeply placed medial head. Both biceps and triceps cross the elbow and shoulder joints, and hence, act upon both. The small triangular **anconeus muscle** is located superficially at the lateral aspect of the elbow.

Arteries and Nerves in the Arm (Fig. 2-13)

The **brachial artery,** the continuation of the axillary, begins at the lower border of the teres major, passes obliquely from a medial to a midline position in the arm, and

Fig. 2-13. Innervation and arterial supply to the muscles on the posterior aspect of the arm.

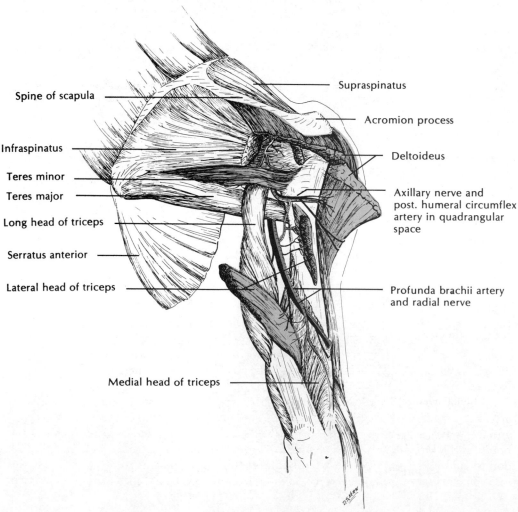

Spine of scapula

Infraspinatus

Teres minor

Teres major

Long head of triceps

Serratus anterior

Lateral head of triceps

Medial head of triceps

Supraspinatus

Acromion process

Deltoideus

Axillary nerve and post. humeral circumflex artery in quadrangular space

Profunda brachii artery and radial nerve

terminates in the cubital fossa by dividing into the radial and ulnar arteries. It may be palpated throughout its entire course. Laterally it lies on the muscular septum and medial head of the triceps; distally it is located on the brachialis muscle medial to the tendon of the biceps and lateral to the median nerve.

ARTERIAL BLOOD PRESSURE

Arterial blood pressure is routinely determined by using a sphyg-momanometer (an inflatable cuff), pressure recorder, a manometer, and a stethoscope. The cuff is placed around the midarm and inflated with air, which compresses the brachial artery against the humerus, thus occluding the vessel. A stethoscope is placed over the artery below the cuff and the air is slowly released. The first sound of blood flowing through the artery and the decreasing pressure is noted on the manometer as the systolic pressure. As the cuff is further deflated a murmur develops. The pressure just before this sound disappears is the diastolic pressure.

Four named branches arise from the brachial artery. The largest, the **profunda brachii** (deep brachial), accompanies the radial nerve in the radiospiral groove of the humerus. Posteriorly this vessel divides into an ascending branch that courses proximally between the lateral and long heads of the triceps, and a descending branch that runs with the medial head of the triceps to the posterior aspect of the elbow. The small **nutrient artery** of the humerus arises about the middle of the arm to enter the nutrient canal on the anteromedial aspect of the bone. The **superior ulnar collateral branch** begins about the middle of the arm, pierces the medial intermuscular septum, accompanies the ulnar nerve, and sends branches to either side of the medial epicondyle. The **inferior ulnar collateral artery** arises about an inch proximal to the medial epicondyle. It divides into a posterior branch that pierces the intermuscular septum to descend deep to the triceps, and an anterior branch that passes inferiorly between the biceps and the brachialis muscles.

The **median nerve** arises from the brachial plexus lateral to the axillary artery, descends, crossing the brachial artery in the midarm, then proceeds into the cubital fossa where it then lies medial to the artery. The **ulnar nerve,** the terminal branch of the medial cord, courses distally medial to the brachial artery. In the midarm it pierces the medial intermuscular septum. At the elbow the nerve passes between the olecranon process and the medial epicondyle where it is superficial and easily palpable. Impingement upon the nerve at this point gives rise to the tingling sensation interpreted as arising from the "crazy bone" of the elbow. The median and ulnar nerves have no branches in the arm.

The **musculocutaneous nerve,** from the lateral cord, courses lateral to the axillary artery, pierces the coracobrachialis muscle, and continues distally between the biceps and the brachialis to innervate all three muscles. Crossing to the lateral side of the arm between the biceps and the brachialis, it pierces the deep fascia above the biceps tendon to continue distally as the **lateral antebrachial cutaneous nerve.**

The **axillary** (circumflex) **nerve,** arising from the posterior cord at the lower border of the subscapularis muscle, passes posteriorly through the quadrangular space with the posterior humeral circumflex artery to supply the teres minor and ramify on the deep surface of the deltoideus muscle. It gives a branch, the **lateral brachial cutaneous,** which supplies skin over the deltoideus.

The **radial nerve,** the largest branch of the brachial plexus and the continuation of the posterior cord, initially lies posterior to the axillary artery. It passes distally between the teres major and the long head of the triceps and spirals around the posterior aspect of the humerus in the radiospiral groove between the lateral and medial heads of the triceps. In the groove it is accompanied by the profunda brachii artery. It pierces the lateral intermuscular septum and follows the interval between the brachialis and brachioradialis muscles into the cubital fossa. Along its course it supplies the triceps and, just proximal to the elbow, sends branches to the brachialis, brachioradialis, extensor carpi radialis longus, and anconeus muscles.

FRACTURES OF THE HUMERUS

Fractures of the humerus are often serious because of injury to closely related nerves and blood vessels. The surgical neck is a common fracture site, which often involves damage to the axillary nerve, thus limiting abduction of the arm. A fracture in the middle third of the humerus may involve the radial nerve where it lies in the radiospiral groove. Such damage causes wrist drop since the nerve supply to the extensor muscles of the hand is lost.

Cubital Fossa

The **cubital fossa** is a triangular depression at the anterior aspect of the elbow joint (Fig. 2-14). Its base is formed by a line passing through the epicondyles of the humerus. Its apex is directed distally, with the brachioradialis muscle forming the lateral side and the pronator teres muscle the medial side. The floor of the fossa is formed by the brachialis and the supinator muscles. The fossa is covered by deep and superficial fasciae and skin. The **median cubital vein** (utilized in drawing blood and for intravenous injection) crosses the fossa obliquely, superficial to the deep fascia. The tendon of the biceps brachii descends through the middle of the fossa to insert onto the radial tuberosity. It sends a secondary insertion as a fibrous expansion, the **bicipital aponeurosis,** into the deep fascia over the flexor muscles. The brachial artery passes through the fossa medial to the biceps tendon to bifurcate into the radial and ulnar arteries. The median nerve descends between the brachioradialis and brachialis muscle to enter the fossa medial to the brachial artery and the radial nerve.

THE MEDIAN CUBITAL VEIN

The median cubital vein, the most common site for venipuncture, links the cephalic and basilic veins in the antecubital fossa. The tough, bicipital aponeurosis separates the deeper median nerve and the brachial artery from this superficial vein. This is an important relationship

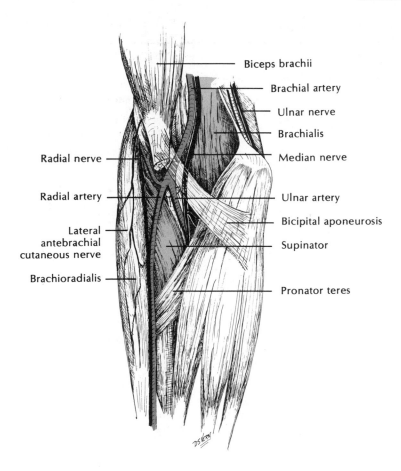

Biceps brachii

Brachial artery

Ulnar nerve

Brachialis

Radial nerve

Median nerve

Radial artery

Ulnar artery

Lateral
antebrachial
cutaneous nerve

Bicipital aponeurosis

Supinator

Brachioradialis

Pronator teres

Fig. 2-14. Cubital fossa.

as the fibrous band protects against the accidental injection of drugs into the artery, as well as possible damage or irritation to the nerve (Table 2-6).

FOREARM

Muscles (Table 2-7)

The flexor muscles of the forearm are subdivided into superficial and deep groups. The superficial group, the **palmaris longus, pronator teres, flexor carpi radialis,** and **flexor carpi ulnaris,** form the muscle mass at the medial side of the proximal forearm. The superficial group and the **flexor digitorum superficialis** originate, in part, from a common tendon attached to the medial epicondyle of the humerus. In the deep group, the **flexor pollicis longus** and **flexor digitorum profundus** pass superficially to the more distally placed **pronator quadratus.** With the exception of the flexor carpi ul-

Table 2-6. Nerve Distribution to the Arm

Nerve	Origin	Course	Distribution
Lateral brachial cutaneous	Axillary	From deep to deltoid, winds around posterior border of muscle to reach the skin	Skin over the deltoid
Lateral brachial cutaneous (lower)	Radial	Pierces lateral intermuscular septum to reach skin of lower portion of arm	Skin over lateral and posterior aspect of arm below deltoid
Posterior brachial cutaneous	Radial	From axilla penetrates deep fascia to reach skin	Skin on posterior aspect of proximal third of arm
Medial brachial cutaneous	Medial cord of brachial plexus	From axilla penetrates deep fascia to reach skin	Skin on medial aspect of distal third of arm
Intercostobrachial cutaneous	Lateral cutaneous branch of second intercostal nerve	From intercostal space passes through axilla to reach skin	Skin on medial aspect of proximal portion of arm
Axillary	Posterior cord of brachial plexus	Passes through quadrangular space to reach deep surface of deltoid	Deltoid and twig to teres minor
Musculocutaneous	Lateral cord of brachial plexus	Pierces coracobrachialis and emerges between biceps and brachialis to continue as lateral antebrachial cutaneous	Flexor muscles in the arm—coracobrachialis, biceps and brachialis
Radial	Posterior cord of brachial plexus	Follows radiospiral groove between lateral and medial heads of triceps; pierces lateral intermuscular septum and continues to forearm	In arm, supplies ceps, brachioradialis, extensor carpi radialis longus, anconeus and twigs to brachialis

naris and the ulnar half of the flexor digitorum profundus, which are supplied by the **ulnar nerve,** all the flexors are supplied by the **median nerve.** The flexors are separated from the extensor muscles by the **interosseous membrane** passing between the radius and ulna.

The extensor muscles are also divided into groups. The superficial group, consisting of the **brachioradialis, extensor carpi radialis longus** and **brevis, extensor digitorum, extensor digiti minimi,** and the **extensor carpi ulnaris,** forms the muscle mass at the lateral aspect of the proximal portion of the forearm. Two of these muscles, the brachioradialis and the extensor carpi radialis longus, extend above the elbow joint, taking origin from the lateral supracondylar ridge of the humerus; the remainder arise, in part, by a common tendon attached to the lateral epicondyle of the humerus. The **brachioradialis,** while grouped with the extensors and innervated by the radial nerve, is functionally a flexor of the forearm. The deep extensor group (the **abductor pollicis longus,** the **extensor pollicis brevis** and **longus,** and the **extensor indicis**) are distal to the **supinator** and lie parallel to each other in the above order. They extend obliquely

Table 2-7. Muscles of the Forearm

Muscle	Origin	Insertion	Action	Nerve
Palmaris longus	Medial epicondyle of humerus	Flexor retinaculum and palmar aponeurosis	Flexes hand	Median
Pronator teres	Humeral head, medial epicondyle; ulnar head, coronoid process of ulna	Middle of lateral surface of radius	Pronates hand	Median
Flexor carpi radialis	Medial epicondyle of humerus	Bases of second and third metacarpals	Flexes hand and elbow; slightly pronates and abducts hand	Median
Flexor carpi ulnaris	Humeral head, medial epicondyle; ulnar head, medial border of olecranon process and posterior border of ulna	Pisiform, hook of hamate, and base of fifth metacarpal	Flexes and adducts hand	Ulnar
Flexor digitorum superficialis	Humeral head, medial epicondyle; ulnar head, coronoid process; radial head, anterior border of radius	Palmar surface and sides of middle phalanges of fingers	Flexes middle phalanx; continued action flexes proximal phalanx and hand; aids in flexion of elbow	Median
Flexor digitorum profundus	Medial and anterior surface of ulna and adjacent interosseus membrane	Bases of distal phalanges of fingers	Flexes terminal phalanx; continued action flexes proximal phalanges and hand	Ulnar and median
Flexor pollicis longus	Anterior surface of radius, adjacent interosseus membrane, and coronoid process of ulna	Base of distal phalanx of thumb	Flexes thumb	Median
Pronator quadratus	Anterior surface of distal fourth of ulna	Anterior surface of distal fourth of radius	Pronates hand	Median
Brachioradialis	Lateral supracondylar ridge of humerus and lateral intermuscular septum	Lateral side of base of styloid process of radius	Flexes forearm	Radial
Extensor carpi radialis longus	Lateral supracondylar ridge of humerus	Posterior surface of base of second metacarpal	Extends and abducts hand	Radial
Extensor carpi radialis brevis	Lateral epicondyle of humerus	Posterior surface of base of third metacarpal	Extends and abducts hand	Radial
Extensor digitorum communis	Lateral epicondyle of humerus	Into extensor expansion on fingers	Extends fingers and hand	Radial
Extensor carpi ulnaris	Lateral epicondyle and posterior border of ulna	Base of fifth metacarpal	Extends and adducts hand	Radial

Table 2-7. Continued

Muscle	Origin	Insertion	Action	Nerve
Extensor digit minimi	From extensor digitorum communis and interosseus membrane	Extensor expansion on proximal phalanx of fifth digit	Extends fifth digit	Radial
Supinator	Lateral epicondyle of humerus, ligaments of elbow joint, and supinator crest and fossa of ulna	Lateral surface of upper third of radius	Supinates hand	Radial
Abductor pollicis longus	Posterior surface of ulna and middle third of posterior surface of radius	Base of first metacarpal	Abducts thumb and hand	Radial
Extensor pollicis longus	Middle third of posterior surface of ulnar and adjacent interosseus membrane	Base of distal phalanx of thumb	Extends distal phalanx and abducts hand	Radial
Extensor pollicis brevis	Posterior surface of middle third of radius	Base of proximal phalanx of thumb	Extends and abducts hand	Radial
Extensor indicis	Posterior surface of ulna	Extensor expansion on index finger	Extends index finger	Radial

across the interosseous space. All the extensors are supplied by the deep branch (posterior interosseous) of the radial nerve.

TENNIS ELBOW

The medical term for this condition is lateral epicondylitis, which means an inflammation of the lateral epicondyle of the humerus or the tissues surrounding it. Recall that most of the extensors of the fingers and wrist originate from a common tendon that attaches to the lateral epicondyle. Repeated strenuous contraction of these muscles, as occurs in the backhand stroke in tennis, causes a strain on the tendon, and results in exquisite tenderness and pain around the epicondyle. Rest eliminates the causative factor, and recovery usually follows.

Arteries and Nerves

One of the terminal branches of the brachial artery, the **radial artery,** courses distally in the forearm between the brachioradialis and the pronator teres, and passes to the anterior aspect of the radius to become superficial at the wrist where the pulse is usually taken (Fig. 2-15). It gives a **radial recurrent branch** in the cubital fossa which courses proximally anterior to the lateral epicondyle to anastomose with the profunda brachii artery. Muscular branches of the radial artery are distributed to the superficial extensor muscles. At the wrist it gives **palmar** and **dorsal carpal branches,** which aid in the formation of the **carpal arches** which supply the wrist and carpal joints.

The larger of the two terminal branches of the brachial, the **ulnar artery,** passes deep to the pronator teres and the superficial flexors to lie on the surface of the flexor digitorum profundus, where it is overlapped distally by the flexor carpi ulnaris. After

Fig. 2-15. Anterior aspect of right arm and forearm showing deep muscles and arteries.

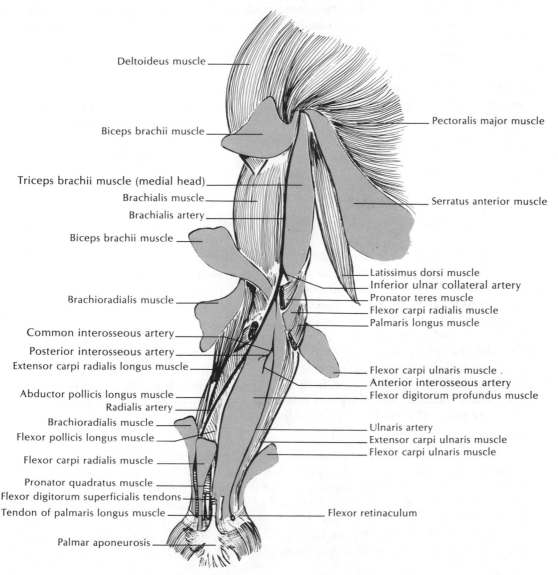

giving **palmar** and **dorsal carpal** branches to the **carpal arches,** it passes into the hand deep to the palmar carpal ligament. At the origin of the ulnar artery, **anterior** and **posterior ulnar recurrent branches** arise and course proximally to pass respectively anterior and posterior to the medial epicondyle, and anastomose with superior and inferior ulnar collateral branches of the brachial artery. The largest ulnar branch, the **common interosseous,** arises about 3 cm distal to the bifurcation of the brachial. A **recurrent branch** of the common interosseous artery turns proximally, deep to the supinator, to join the anastomosis around the elbow joint. The common interosseous artery terminates by dividing into the **anterior** and **posterior interosseous branches,** which pass to their respective sides of the interosseous membrane to supply the deep flexor and deep extensor muscles. The **anterior interosseous artery,** extending further distally than the posterior branch, pierces the interosseous membrane in the distal third of the forearm to supply the muscles on the dorsum of the forearm before it anastomoses with the carpal arches.

The **median nerve** leaves the cubital fossa by passing between the heads of the pronator teres to gain a position deep to the flexor digitorum superficialis, where it continues distally between this muscle and the flexor digitorum profundus. In the forearm it supplies the pronator teres, the pronator quadratus, and all the flexors, except the flexor carpi ulnaris. It shares the innervation of the flexor digitorum profundus with the ulnar nerve. At the wrist the median nerve becomes superficial to lie at the ulnar side of the tendon of the palmaris longus muscle, then passes deep to the flexor retinaculum to enter the hand.

MEDIAN NERVE INJURIES

At the wrist in median nerve injury, sensation is lost over the lateral two-thirds of the palm, the palmar surface of the thumb, index, and middle fingers, the lateral half of the ring finger, and the dorsal surface of the distal phalanges of these digits. Paralysis occurs in the muscles of the thenar eminence, and the ball of the thumb flattens out as the muscles atrophy. Furthermore, the thumb cannot be placed in opposition to the other fingers and its pinching action is lost.

If the injury is at or above the elbow, additional paralysis is noted, for example, loss of pronation, and loss of the digital flexors, except for the little and ring fingers, which have an additional innervation from the ulnar nerve. Therefore, in making a fist the thumb will be extended and adducted; the index and middle fingers will be in extension, and the ring and little fingers flexed (papal sign).

From its position between the olecranon process and the medial epicondyle, the **ulnar nerve** enters the forearm by passing between the heads of the flexor carpi ulnaris. It continues distally in the forearm deep to this muscle, lying on the flexor digitorum profundus. In the distal half of the forearm it parallels the course of the ulnar artery to pass to the lateral side of the tendon of the flexor carpi ulnaris and enters the hand deep to the palmar carpal ligament. In the forearm it supplies the flexor carpi ulnaris and the ulnar half of the flexor digitorum profundus.

ULNAR NERVE INJURY

Ulnar nerve injury may result in extensive motor and sensory loss to the hand. After complete interruption of the ulnar nerve above the elbow, the following changes in the hand are noted: 1) spaces between the tendons become sunken, called "guttering," due to atrophy of the interossei, third and fourth lumbricales, and the adductor pollicis; 2) partial (ulnar) *claw hand* involves hyperextension of the fourth and fifth digits at the metacarpophalangeal joints and flexion of the interphalangeal joints; 3) flattening of the hypothenar eminence from atrophy of the palmaris brevis muscle and the intrinsic muscles of the little finger; 4) persistent abduction of the little finger due to loss of action of the fourth palmar interosseous with unopposed action of extensor digiti quinti.

Clinical tests for the ulnar nerve damage include: 1) difficulty in making a fist since the fourth and fifth fingers cannot flex at the distal interphalangeal joint; 2) adduction of thumb is lost—when the patient attempts to grasp a piece of paper between the thumb and the index finger he accomplishes it by strong flexion of the thumb at the interphalangeal joint (Froment's sign); 3) abduction and adduction of the fingers are weakened and difficult to perform.

After entering the cubital fossa, the **radial nerve** divides into superficial and deep branches. The **superficial branch** passes distally, deep to the brachioradialis, to the dorsum of the wrist, where it divides into medial and lateral branches that supply cutaneous innervation to the dorsum of the hand. The **deep branch** of the radial (posterior interosseous) nerve pierces the supinator muscle to be distributed to all the extensors muscles within the forearm.

RADIAL NERVE INJURY

In a complete radial nerve injury there is loss of extension with "wrist drop." The patient is unable to extend the wrist against gravity. If the lesion is in the distal third of the arm, extension of the elbow by the triceps, weak supination by the brachioradials, and weak extension by the extensor carpi radialis longus will be preserved. In lesions at both levels sensory loss will occur on the lateral aspect of the dorsum of the hand (Table 2-8).

HAND

Muscles (Table 2-9)

The intrinsic muscles of the hand are divided into three groups: the thenar, the hypothenar, and the interossei and lumbricales (Fig. 2-16).

The **thenar eminence** (ball of the thumb) is formed by the laterally placed super-

Table 2-8. Nerve Distribution to Forearm

Nerve	Origin	Course	Distribution
Medial ante-brachial cutane-ous	Medial cord of brachial plexus	Follows brachial artery and in ante-cubital fossa pierces deep fascia to course in superficial fascia along me-dial aspect of forearm	Skin over medi aspect of forear
Lateral ante-brachial cutane-ous	Continuation of musculocu-taneous nerve	At antecubital fossa pierces deep fas-cia and courses in superficial fascia along lateral aspect of forearm	Skin of lateral a pect of forearm
Posterior ante-brachial cutane-ous	Radial nerve	Pierces lateral intermuscular septum in distal portion of arm to course in superficial fascia on posterior aspect of forearm	Skin on posteri aspect of forear
Ulnar	Medial cord of brachial plexus	Follows brachial artery and at elbow passes along ulnar groove of medial epicondyle; passes through forearm between flexor carpi ulnaris and flexor digitorum profundus	Flexor carpi uln and ulnar half flexor digitorum profundus
Median	Lateral and medial cords of brachial plexus	Follows brachial artery; at cubital fossa passes between heads of pro-nator teres to course through forearm deep to the flexor digitorum superfi-cialis	Supplies all flex in forearm not plied by ulnar
Deep branch of radial (posterior interosseous)	Radial in proximal portion of forearm	Penetrates supinator to reach poste-rior compartment	Supplies all ext sors in forearm supplied by rad nerve proper

ficial **abductor pollicis brevis,** the intermediate **opponens pollicis,** and the more deeply placed **flexor pollicis brevis.** The latter is divided into a deep and superficial portion by the tendon of the flexor pollicis longus. The **hypothenar eminence** (heel of the hand) is composed of the medially situated **abductor digiti minimi,** the more lateral **flexor digiti minimi brevis** lying superficial to and usually blending with the **opponens digiti min-imi.**

The seven interossei are disposed in two layers and are composed of three palmar and four dorsal muscles. These muscles are located in the spaces between the metacar-pal bones. The unipennate **palmar interossei** arise from single metacarpals, while the bipennate **dorsal interossei** arise from two adjacent metacarpals. All the interossei in-sert into bases of proximal phalanges. The dorsal interossei act as abductors and the palmar interossei acts as adductors of the digits. Abduction and adduction of the digits is relative to a line passing through the central axis of the middle finger. The rela-tively extensive **adductor pollicis** muscle is in the same plane as the palmar interossei muscle and arises by two heads. It is sometimes referred to as the fourth palmar inter-osseous.

The four **lumbricales** consist of muscular slips, which originate from the four tendons of the flexor digitorum profundus. They course to the radial side of the meta-carpophalangeal joint to insert into the extensor expansion. Their action is to flex the

Table 2-9. Muscles of the Hand

Muscle	Origin	Insertion	Action	Nerve
Palmaris brevis	Medial aspect of flexor retinaculum	Skin of palm	Wrinkles skin of palm	Ulnar
Abductor pollicis brevis	Flexor retinaculum, scaphoid, and trapezium	Lateral side of base of proximal phalanx of thumb	Abducts thumb; assists in flexion of proximal phalanx	Median
Flexor pollicis brevis	Flexor retinaculum and trapezium	Base of proximal phalanx of thumb with sesamoid interposed	Flexes thumb; assists in apposition	Median
Opponens pollicis	Flexor retinaculum and trapezium	Entire length of lateral border of first metacarpal	Draws first metacarpal toward center of palm	Median
Abductor digiti minimi	Pisiform and tendon of flexor carpi ulnaris	Medial side of base of proximal phalanx of fifth digit	Abducts fifth digit	Ulnar
Flexor digiti minimi brevis	Flexor retinaculum and hook of hamate	Medial side of base of proximal phalanx of fifth digit	Flexes proximal phalanx of fifth digit	Ulnar
Opponens digiti minimi	Flexor retinaculum and hook of hamate	Medial border of fifth metacarpal	Draws fifth metacarpal forward in cupping of hand	Ulnar
Abductor pollicis	Oblique head from capitate and bases of second and third metacarpals; transverse head from anterior surface of third metacarpal	Medial side of base of proximal phalanx of thumb	Adducts thumb; assists in apposition	Ulnar
Lumbricales (4)	Tendons of flexor digitorum profundus	Extensor expansion distal to metacarpophalangeal joint	Flex metacarpophalangeal and interphalangeal joints	Two lateral muscles by median; two medial by ulnar
Dorsal interossei (4)	By two heads from adjacent sides of metacarpal bones	Lateral sides of bases of proximal phalanges of index and middle fingers; medial sides of bases of middle and ring fingers	Abduct index, middle, and ring fingers; aid in extension of interphalangeal and flexion of metacarpophalangeal joints	Ulnar
Palmar interossei (3)	Medial side of second metacarpal; lateral sides of fourth and fifth metacarpals	Base of proximal phalanx in line with its origin	Adduct index, ring, and little fingers; aid in extension of interphalangeal and flexion of metacarpophalangeal joints	Ulnar

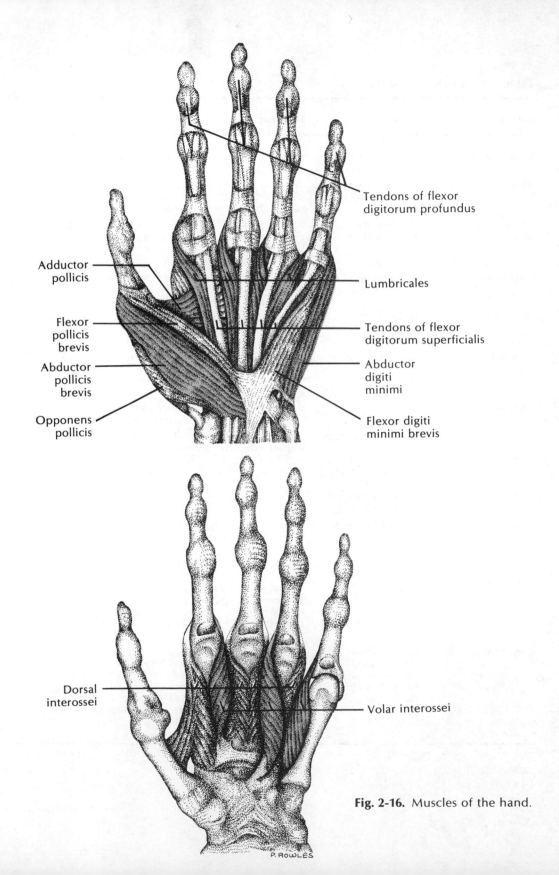

Adductor pollicis

Flexor pollicis brevis

Abductor pollicis brevis

Opponens pollicis

Tendons of flexor digitorum profundus

Lumbricales

Tendons of flexor digitorum superficialis

Abductor digiti minimi

Flexor digiti minimi brevis

Dorsal interossei

Volar interossei

Fig. 2-16. Muscles of the hand.

P. ROWLES

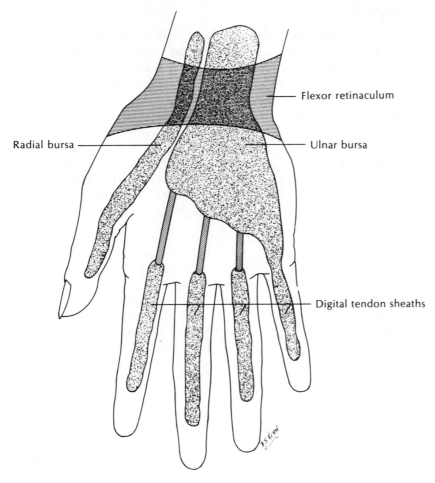

Radial bursa

Flexor retinaculum

Ulnar bursa

Digital tendon sheaths

Fig. 2-17. Radial and ulnar bursae.

metacarpophalangeal joints and extend the interphalangeal joints, thus placing the fingers at right angles to the palm of the hand.

Synovial Tendon Sheaths

At the wrist and in the hand, lubricating synovial sheaths surround tendons as the latter extend from the forearm to the hand (Fig. 2-17). The **ulnar bursa** is a large synovial sac invaginated on the radial side by tendons of the flexor digitorum profundus and superficialis muscles. It extends from approximately 2 to 3 cm above the flexor retinaculum to the middle of the palm. Around the tendons of the fifth finger it continues to the distal phalanx. Separate from the ulnar bursa, the **digital synovial sheaths** invest the digital parts of the deep and superficial flexor tendons to the second, third, and fourth digits. These sheaths extend from the heads of the metacarpals to the insertions of the long flexor tendons.

TENDOSYNOVITIS

Tendosynovitis is an infection of a synovial tendon sheath. Digital sheath infections are common and are often caused by a nonsterile penetrating wound from a thorn, needle, knife, or splinter, which carries pathogenic organisms into the closed synovial digital sheath. Cardinal signs of such an infection are exquisite tenderness over the sheath, sustained flexion of the infected digit, and upon extension, severe pain over proximal end of finger. The closed chamber of the digital sheath swells with edema fluid and pus and the increasing pressure of the sheath can cause exquisite pain and necrosis of the flexor tendons. Tendosynovitis of the digital sheath is an emergency because the sheath must be opened and drained early to prevent ischemic necrosis of the tendons and a serious loss of hand function.

The **radial bursa** surrounds the flexor pollicis longus tendon, extending about 2–3 cm proximal to the flexor retinaculum to the insertion of this tendon. A separate synovial sheath surrounds the tendon of the flexor carpi radialis from a point proximal to the flexor retinaculum to its insertion into the base of the second metacarpal bone.

Synovial sheaths surround **extensor tendons** within six osseofibrous compartments or tunnels located deep to the extensor retinaculum. The sheaths, enclosing the tendons of the extensor carpi radialis longus and brevis and the extensor carpi ulnaris, extend from the proximal border of the extensor retinaculum to the insertions of these tendons. Sheaths of the digital extensors, including the extensors of the thumb and the abductor pollicis longus, terminate at about the middle third of the metacarpal bones.

Arteries and Nerves

After passing deep to the volar carpal ligament, the **ulnar artery** enters the hand by passing to the radial side of the pisiform bone to give muscular branches to the hand (Fig. 2-18). It terminates by dividing into a **deep branch** that joins with a similar branch of the radial artery to form the **deep palmar arch,** and a **superficial branch** joining with a corresponding branch of the radial to complete the **superficial palmar arch.** The latter, predominately from the ulnar artery, lies transversely in the palm immediately under the palmar aponeurosis in line with the fully extended thumb. It gives a **proper digital branch** to the medial side of the little finger and three **common palmar digital branches,** which subsequently divide into **proper digital branches** to supply contiguous sides of the four fingers.

Passing from the anterior surface of the radius medial to the styloid process, the **radial artery** enters the hand deep to the tendons of the abductor pollicis longus and the extensor pollicis longus and brevis muscles. It pierces the first dorsal interosseous to lie between this muscle and the adductor pollicis. Branches of the radial artery include the **superficial palmar,** which with a similar branch of the ulnar, completes the **superficial palmar arch.** The **princeps pollicis branch** and the **indicis proprius branch** both arise as the radial artery pierces the first dorsal interosseous muscle. They supply, respectively, both sides of the thumb, and the radial side of the index finger. The **deep branch** of the radial artery forms most of the **deep palmar arch,** which lies on the carpal

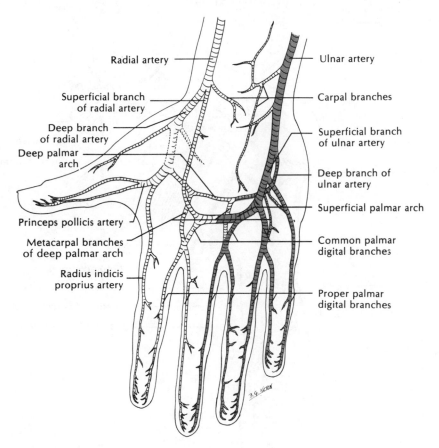

Radial artery

Superficial branch
of radial artery

Deep branch
of radial artery

Deep palmar
arch

Princeps pollicis artery

Metacarpal branches
of deep palmar arch

Radius indicis
proprius artery

Ulnar artery

Carpal branches

Superficial branch
of ulnar artery

Deep branch of
ulnar artery

Superficial palmar arch

Common palmar
digital branches

Proper palmar
digital branches

Fig. 2-18. Arterial supply of the hand.

bones a thumb's breadth proximal to the superficial arch. Branches of the deep arch are the **palmar metacarpal vessels** which join the **common palmar digital arteries** of the superficial arch to supply the fingers. Figure 2-19 illustrates arterial supply of the upper extremity.

On entering the hand deep to the flexor retinaculum, the **median nerve** gives a **motor (lateral) branch** to the muscles of the thenar eminence, namely, the opponens pollicis, the abductor pollicis brevis, the superficial head of the flexor pollicis brevis, and the two lateral lumbricales. Terminally the nerve supplies cutaneous innervation to the central area of the palm, palmar surface of the thumb, index, middle, and lateral half of the ring fingers, and all of the skin on the distal phalanges of these digits (Fig. 2-20).

CARPAL TUNNEL SYNDROME

Carpal tunnel syndrome is a painful compression of the median nerve by the transverse carpal ligament, which forms the volar aspect of the

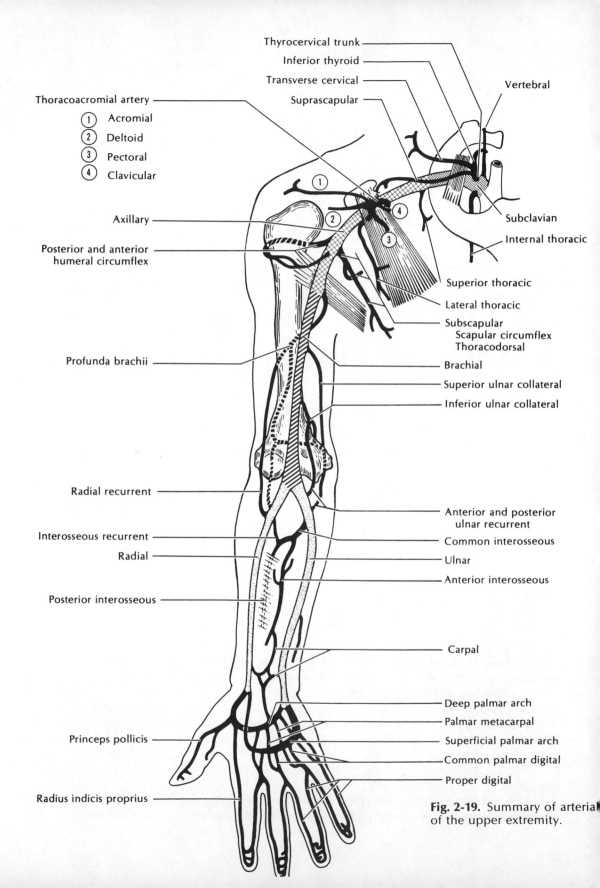

Thyrocervical trunk

Inferior thyroid

Transverse cervical

Suprascapular

Vertebral

Thoracoacromial artery

① Acromial

② Deltoid

③ Pectoral

④ Clavicular

Axillary

Posterior and anterior
humeral circumflex

Subclavian

Internal thoracic

Superior thoracic

Lateral thoracic

Subscapular
Scapular circumflex
Thoracodorsal

Profunda brachii

Brachial

Superior ulnar collateral

Inferior ulnar collateral

Radial recurrent

Anterior and posterior
ulnar recurrent

Common interosseous

Interosseous recurrent

Ulnar

Radial

Anterior interosseous

Posterior interosseous

Carpal

Deep palmar arch

Palmar metacarpal

Princeps pollicis

Superficial palmar arch

Common palmar digital

Proper digital

Radius indicis proprius

Fig. 2-19. Summary of arterial
of the upper extremity.

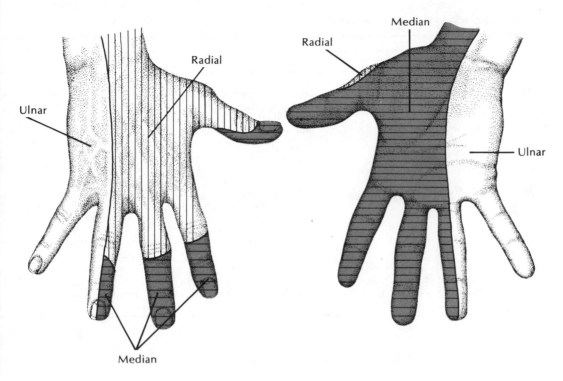

Fig. 2-20. Dermatomes of hand.

carpal tunnel. Median nerve compression at the wrist causes a burning sensation, or numbness felt over the lateral three and one-half digits, and weakness or atrophy of the thenar muscles. The compression may be due to inflammation from chronic irritation of the transverse carpal ligament and is often found in patients with rheumatoid arthritis. Dividing the transverse carpal ligament decompresses the carpal tunnel and the median nerve and after several months, usually results in almost complete return of median nerve function.

At the wrist the **ulnar nerve** divides into a **superficial branch** that supplies the palmaris brevis and the cutaneous innervation to both surfaces of the hand medial to a line passing through the midline of the ring finger, and a **deep branch** that supplies all the intrinsic muscles of the hand except those supplied, as listed above, by the median nerve. The deep branch pierces the opponens muscle and passes around the hamulus of the hamate bone to cross the palm in company with the deep palmar arch.

The **radial nerve** supplies no muscles in the hand. Its terminal distribution is the cutaneous innervation to a small area on the palmar surface of the thenar eminence and the dorsum of the hand lateral to a line passing through the midline of the ring finger, except for skin over the distal phalanges (Table 2-10).

Table 2-10. Nerve Distribution to the Hand

Nerve	Origin	Course	Distribution
Median	Lateral and medial cords of brachial plexus	Enters hand by passing deep to flexor retinaculum; distal to palm divides into motor branch and palmar digital (cutaneous) branches	Motor branch supplies thena muscles, lateral lumbricales opponens, and short flexor a abductor of thumb; cutaneo branches supply lateral ⅔ of palm, palmar aspect, and all skin of distal phalanges of la eral 3½ digits
Ulnar	Medial cord of brachial plexus	Enters hand by passing deep to the palmar carpal ligament; distal to pisiform bone divides into superficial and deep branches of hand	Superficial branch supplies s medial to a line bisecting the ring finger; deep branch supplies all intrinsic muscles of hand not supplied by the me dian nerve
Radial	Superficial branch—the continuation of radial after deep branch is given off at elbow	In forearm, courses deep to brachioradialis; at wrist, turns onto dorsum of the hand	Skin on dorsum of hand, exc for distal phalanges, lateral t line bisecting ring finger

FASCIAE

The external **investing layer** of deep fascia completely ensheathes the upper extremity. In the arm it sends distinct **lateral** and **medial intermuscular septa** to attach to the humerus, which separate the musculature of the arm into extensor and flexor compartments. At the wrist this layer of deep fascia condenses and thickens to form a distinct band surrounding the wrist as the flexor and extensor retinacula (Fig. 2-21). The **extensor retinaculum** has deep attachments to the lateral side of the dorsum of the radius, the pisiform and triquetrum bones, and the styloid process of the ulna. It sends septa to attach to ridges on the dorsum of the radius and forms **six osseofibrous compartments** for the extensor tendons passing from the forearm into the hand. The six compartments transmit tendons, from the radial to the ulnar side respectively, as follows: the tendons of the abductor pollicis longus and extensor pollicis brevis traverse the **first compartment,** located on the lateral border of the styloid process of the radius; the **second compartment,** situated on the radial side of the tubercle of the radius, contains the tendons of the extensor carpi radialis longus and brevis muscles; the **third,** on the ulnar side of the tubercle of the radius, conveys the tendon of the extensor pollicis longus muscle. The **fourth** and largest compartment, covering the ulnar third of the dorsum of the radius, transmits the four tendons of the extensor digitorum communis and the extensor indicis proprius; the **fifth,** located over the distal radioulnar articulation, contains the tendon of the extensor digiti minimi; and the **sixth,** located on the head of the ulnar bone, affords passage for the tendon of the extensor carpi ulnaris.

The **flexor retinaculum** (transverse carpal ligament) stretches across the concavity formed by the articulated carpal bones. It attaches to the tuberosity of the scaphoid

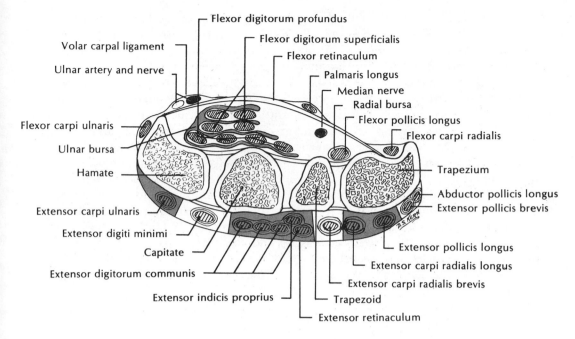

Fig. 2-21. Relationships of flexor and extensor retinacula.

and trapezium bones laterally, and the hamate and pisiform medially. A **large** and a **small osseofibrous tunnel** transmit the tendons of the flexor muscles and the median nerve. The smaller compartment contains the tendon of the flexor carpi radialis, while the larger transmits the tendons of the flexor digitorum superficialis and profundus. The tendons of the superficialis to the third and fourth digits lie superficially, those to the second and fifth digits lie intermediately, and the deeply placed tendons of the profundus lie side by side in the compartment. Superficial to the flexor retinaculum the palmaris longus muscle passes distally in the midline, and the ulnar nerve and artery course to its ulnar side between the retinaculum and the **palmar carpal ligament.**

DUPUYTREN'S CONTRACTURE

Dupuytren's contracture involves an insidious fibrosis of the palmar fascia, which forms subcutaneous nodules that attach first to the flexor sheath of the ring finger and then progressively involves the little, middle and index finger in that order. The progressive retraction of the flexor tendons pulls the fingers into continuous flexion which is like the sign of "papal benediction" with the ring and little fingers flexed into the palm and the thumb, index and middle fingers extended. With increasing disability of the hand from these flexed fingers, surgical excision of this overgrowth of palmar fibrous tissue may be necessary to restore the hand to normal function.

Deep fascia is also specialized on the palmar aspect of the hand to form compartments. The **thenar compartment** is formed by fascia that surrounds the small muscles of the thumb. Structures within this space include the abductor, flexor, and opponens muscles of the thumb, the muscular branch of the median nerve, and the superficial palmar branch of the radial artery. The **hypothenar compartment** is formed by fascia that surrounds the small muscles of the little finger. It contains the abductor, flexor brevis, and opponens muscles of the fifth digit, and the deep branches of the ulnar artery and nerve. The **central,** or intermediate, **compartment** (midpalmar space) is bounded superficially by the **palmar aponeurosis,** medially and laterally by fascia covering hypothenar and thenar muscles, and dorsally by the interosseous adductor compartment. The central space is incompletely divided by septa separating flexor tendons and associated lumbricale muscles. It contains the long flexor tendons, the lumbricales, median and ulnar nerves, and the superficial palmar arch. The **interosseous adductor compartment** is bounded by dorsal and palmar interosseous fasciae and encloses the interossei muscles and the metacarpal bones. It contains the interossei muscles, metacarpal bones, the adductor of the thumb, the deep palmar arch, the deep branch of the ulnar nerve, and the arterial arch on the dorsum of the hand.

SURGICAL REPAIR OF FLEXOR TENDONS

Surgical repair of flexor tendons in the midpalmar space is often unsuccessful because of the limited blood supply to the tendons and the great difficulty in reestablishing an efficient sliding movement of the tendons within the damaged synovial sheaths. Because fully functional repair of these sheaths is almost impossible, the midpalmar space has been called the "no man's land" of tendon surgery.

Joints of the Superior Extremity

SHOULDER JOINT

The multiaxial, **ball-and-socket** shoulder joint, consisting of the **head of the humerus** articulating with the much smaller, relatively flat **glenoid fossa** of the scapula, has the greatest freedom of movement of any joint of the body (Fig. 2-22). The bony configurations result in a mobile but inherently unstable joint. Such anatomic instability is compensated for by the presence of an **articular** (rotator) **cuff of muscles and tendons,** which holds the humeral head in place and reinforces the joint capsule. A fibrocartilaginous rim around the glenoid cavity, the **glenoidal labrum,** deepens the articular fossa. The **articular capsule** attaches proximally to the margins of the glenoid fossa and distally at the anatomic neck of the humerus (Fig. 2-23). The joint cavity, lined by synovial membrane, is traversed by the **tendon of the long head of the biceps** ensheathed by a synovial membrane. The tendon continues distally in the **intertuber-**

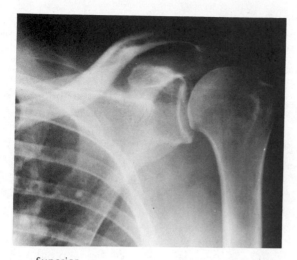

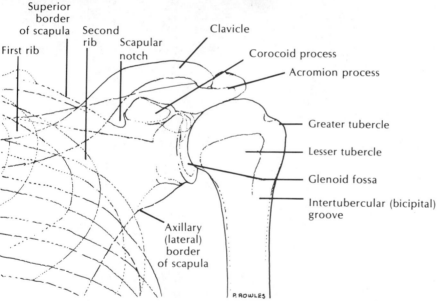

Superior
border
of scapula　Second　　　　Clavicle
　　　　　rib　Scapular
First rib　　　　notch　　　　Corocoid process
　　　　　　　　　　　　　Acromion process

Greater tubercle

Lesser tubercle

Glenoid fossa

Intertubercular (bicipital)
groove

Axillary
(lateral)
border
of scapula

P. ROWLES

Fig. 2-22. Anteroposterior radiograph and schematic of right shoulder joint.

cular groove, which is bridged by the **transverse humeral ligament** to hold the tendon in place. The **glenohumeral ligaments** are three bands of tissue that blend with and strengthen the articular capsule. The **coracohumeral ligament** passes from the coracoid process to blend with the upper posterior part of the capsule and attaches to the anatomic neck and greater tubercle of the humerus. The **coracoclavicular ligament** forms the strongest union between the scapula and the clavicle. Passing from the coracoid process of the scapula, its lateral **trapezoid portion** attaches to the trapezoid ridge of the clavicle, while medially the **conoid portion** passes to the conoid tubercle of the clavicle.

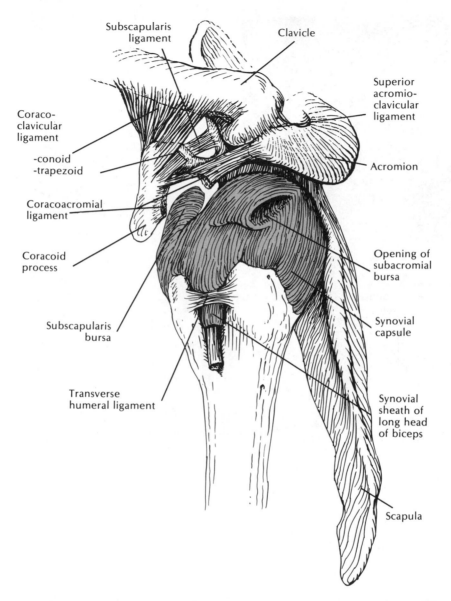

Fig. 2-23. Synovial capsule of shoulder joint with important bursae and ligaments.

DISLOCATION OF THE SHOULDER JOINT

Dislocation of the shoulder joint occurs frequently. The joint stability is largely dependent on the strong tendons (rotator cuff) of the scapular muscles. Because the rotator cuff is not present at the inferior aspect of the joint capsule, most dislocations occur in this area and are directed downward and forward. Such a dislocation may damage the

axillary nerve. Superior dislocation seldom occurs because of the additional protection afforded by the acromial and coracoid processes that form an arch or cradle for the head of the humerus.

ELBOW JOINT

The elbow is a double **hinge (ginglymus) joint** (Fig. 2-24 A and B). The spool-shaped **trochlea** of the humerus articulates with the **trochlear notch** of the ulna, and the **capitulum** of the humerus articulates with the **head of the radius.** A single articular capsule encloses the joint. Anteriorly and posteriorly the capsule is relatively thin and membranous but is reinforced laterally and medially by strong **radial** and **ulnar collateral ligaments.** The synovial membrane lining the joint capsule is extensive and passes distally under the **annular ligament** to permit free rotation of the radius.

Radioulnar joints form the proximal and distal articulations between the radius and ulna, which permit rotation of the radius in pronation and supination of the forearm. The **superior radioulnar** articulation is enclosed in the articular capsule of the elbow joint and shares its synovial membrane. The **radial notch** of the ulna and the annular ligament of the radius form a ring in which the radial head rotates. The **annular ligament** forms four-fifths of the ring and attaches to the margins of the radial notch cupping the head of the radius and blending with the articular capsule and the **radial collateral ligament.** The **quadrate ligament** forms a loose band extending from the distal border of the radial notch to the medial surface on the neck of the radius.

The **interosseous membrane,** a strong tendinous sheet connecting the shafts of the radius and ulna, separates the extensor from the flexor compartment of the forearm and gives attachment to deep muscles of both groups. It extends proximally to within 2 to 3 cm of the tuberosity of the radius and distally to the inferior radioulnar articulation.

WRIST JOINT

COLLES' FRACTURE

When one falls on the outstretched arm, the radius bears the brunt of forces transmitted through the hand. If a fracture occurs in such a fall, it is usually a transverse break of the lower end of the radius about 3 cm proximal to the radiocarpal joint. In this type of injury, called Colles' fracture, the hand is displaced backward and upward. A tentative diagnosis can be made by the characteristic deformity of the wrist, resembling an overturned eating fork.

The **inferior radioulnar articulation** is a **pivot-type joint** lying between the head of the ulna and the ulnar notch of the radius. An **articular disc** of fibrocartilage, attaching to the medial edge of the distal end of the radius and the internal surface of the

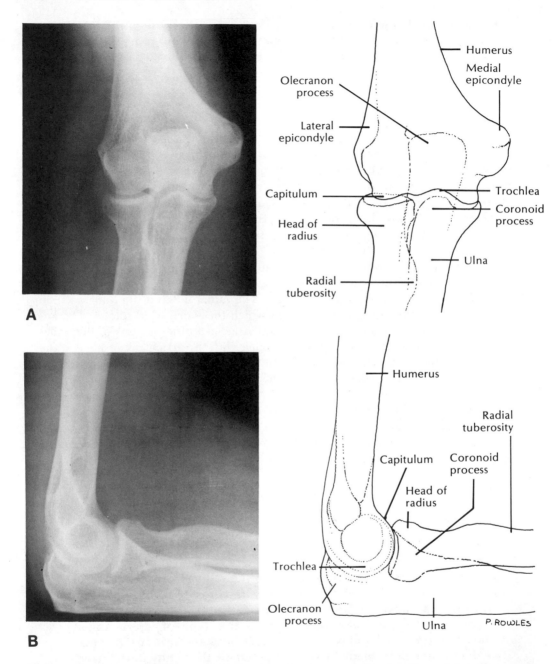

Fig. 2-24. A. Anteroposterior radiograph and schematic of right elbow joint. **B.** Lateral radiograph and schematic of right elbow joint.

styloid process of the ulna, is interposed between the ulna and the proximal row of carpal bones. The articular capsule is relatively weak.

A **condyloid joint** is present between the distal end of the radius, the articular disc, and the proximal row of carpal bones, namely, the **scaphoid** (navicular), **lunate,** and **triquetrum** (triangularis). The articular capsule enclosing the joint is strengthened by **dorsal** and palmar radiocarpal ligaments and the **radial** and **ulnar collateral ligaments.**

CARPAL JOINTS

The intercarpal articulations are **gliding joints** between the two transverse rows of carpal bones, with **dorsal** and **palmar intercarpal ligaments** passing transversely, and interosseous **intercarpal ligaments** linking the lateral borders of the proximal row to the distal row of bones (Fig. 2-25). The **midcarpal joint** between the proximal and distal rows allows a considerable range of movement. The central portion of this articulation forms a limited ball-and-socket type of joint, with the **scaphoid** and **lunate** forming the socket proximally and the **capitate** and **hamate** forming the ball distally. The lateral portions are **gliding joints** between the **trapezium** and **trapezoid** articulating with the **scaphoid** laterally and the **hamate** with the **triquetrum** medially. **Collateral ligaments** of the radial and ulnar borders connect the scaphoid with the trapezium and the triquetrum with the hamate. The **pisiform** articulates with the **triquetrum,** which is surrounded by a separate articular capsule.

Fig. 2-25. Posteroanterior radiograph and schematic of the left hand.

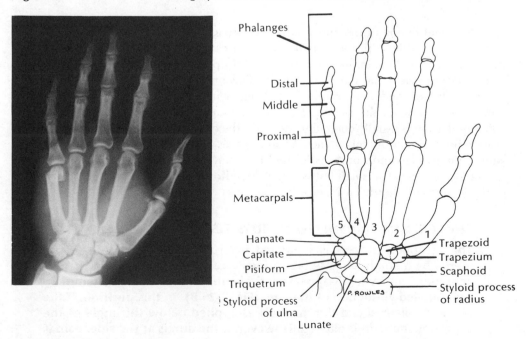

JOINTS OF THE HAND

The **carpometacarpal joint** of the thumb forms a **saddle-shaped articulation** between the **trapezium** and the **first metacarpal** and is surrounded by an articular capsule. The remaining carpometacarpal joints have a common synovial cavity between the **intercarpal** and **intermetacarpal joints** linked by **dorsal, palmar,** and **interosseous ligaments.**

The **intermetacarpal joints** formed by contiguous sides of the bases of **second through fifth metacarpals** are joined by **dorsal, palmar,** and **interosseous ligaments.** The **deep transverse metacarpal ligaments** connect the heads of the **second, third,** and **fourth metacarpals** on the palmar aspect and thereby limit the abduction of these bones.

The condyloid **metacarpophalangeal joints** are formed by the rounded **head of the metacarpal bones** articulating with the concavities of the **proximal phalanges.** These joints are linked by articular capsules that are reinforced dorsally by the extensor tendons. **Palmar ligaments** bridge the joints on the palmar aspect of the hand and are continuous laterally with the strong cordlike **collateral ligaments,** which attach proximally to the tubercle and distally to the lateral aspect of the base of the phalanx.

The **interphalangeal joints** are **hinge-type joints,** structurally the same as the metacarpophalangeal articulations, with a **palmar ligament** and two **collateral ligaments** that are reinforced dorsally by the extensor expansion.

BURSAE

Synovial bursae of the upper extremity are associated with the joints and sometimes communicate with the articular synovial cavity. The **subacromial (subdeltoid) bursa,** about 2 to 3 cm in diameter, is located deep to the deltoid between the tendon of the supraspinatus and the joint capsule. This bursa sends an extension deep to the acromial process and the coracoacromial ligament and may communicate with the joint cavity. Between the tendon of the subscapularis and the neck of the scapula the **subscapular bursa** usually communicates with the cavity of the shoulder joint. The **olecranon bursa** lies in the subcutaneous tissue spaces over the olecranon process, and the **subtendinous olecranon bursa** is located between the tendon of the triceps and the olecranon process. The small but constant **bicipitoradial bursa** is between the tendon of the biceps and the radial tuberosity.

SUBACROMIAL (SUBDELTOID) BURSITIS TEST

The acromion process is separated from the tuberosities of the humerus by a single large bursa variously called the subdeltoid or subacromial. This cavity disappears entirely undercover of the acromion in right angled abduction of the arm. (Fig. 2-26 B). In this position, if the bursa is inflamed and firm pressure is applied below the angle of the acromion, no pain is elicited. However, if the arm is at the side, pain is

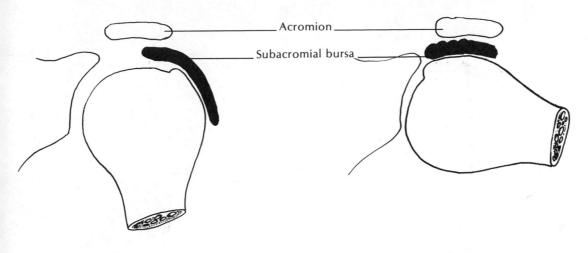

A. Arm at side   B. Arm abducted to right angle

Fig. 2-26. Subacromial bursitis test.

felt on palpation because the bursa has emerged from under shelter of the acromion (Fig. 2-26 A). This observation (Dawborn's sign) aids in distinguishing subacromial bursitis from other lesions of the shoulder joint.

3

thorax

Thoracic Cage

The thoracic cage, formed by the sternum, the ribs, and the thoracic vertebrae, gives protection for the lungs and heart and affords attachment for muscles of the thorax, upper extremity, back, and diaphragm.

Skeleton of the Thorax

The **sternum** is a flat bone consisting of three parts, the manubrium, the body and the xiphoid process. The **manubrium,** united with the body by fibrocartilage at the **sternal angle (of Louis),** presents superiorly a **suprasternal notch** and articulates laterally with the clavicle and the first costal cartilage. At its junction with the body of the sternum, lateral demifacets are present for articulation with the second costal cartilage. The **body,** forming the bulk of the sternum, articulates with the second through seventh costal cartilages and is united with the xiphoid process inferiorly at the **xiphisternal junction.** The **xiphoid process,** the smallest part of the sternum, is thin, flattened, and elongated, and presents a demifacet for the seventh costal cartilage. In youth the xiphoid process consists of hyaline cartilage, which is gradually replaced with bone.

There are twelve pairs of ribs. The first seven are **true ribs** in that they articulate by way of their costal cartilages with the sternum. The remaining five pairs are called **false ribs** because either their attachment anteriorly is through the costal cartilage of the rib above, rather than directly to the sternum; or in the case of the eleventh and twelfth ribs, they have no anterior attachment and are called **floating ribs.** Ribs one through seven increase progressively in length, while the remaining ribs decrease. The first rib has the greatest curvature and thereafter the curvature diminishes.

A **typical rib** (third through ninth) can be divided into three parts: head, neck, and shaft (Fig. 3-1). The **head** is wedge-shaped with a crest at the apex presenting two demifacets that articulate with the numerically corresponding thoracic vertebra and the vertebra above. At the **junction of the neck and shaft,** a **tubercle** on the posterior surface articulates with the transverse process of the numerically corresponding vertebra. The long, thin **shaft,** or body, is curved, and twisted on its long axis. It forms the **angle** of the rib at its point of maximal curvature. The **costal groove** on the inferior border of the rib gives protection to intercostal nerves and vessels.

THORACENTESIS

Pus in the thorax, an excess of fluid in pleural effusion, or bleeding into the pleural cavity may necessitate draining this material from the thorax. In this procedure, called thoracentesis, insertion of a large bore needle close to the superior border of a rib avoids injury to the vessels and nerves in the costal groove.

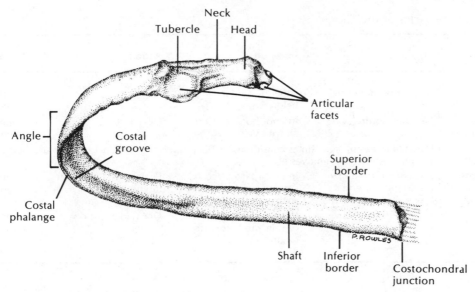

Fig. 3-1. A typical rib.

The first, second, tenth, eleventh, and twelfth ribs are classified as **atypical.** The **first rib** is short, broad, and flat. It has a long neck and only one facet on its head. There is a prominent tubercle on this rib for articulation with the transverse process of the first thoracic vertebra. The tubercle is separated by a shallow groove, formed by the subclavian artery and the brachial plexus, from the **scalene tubercle.** The latter affords attachment for the anterior scalenus muscle. The **second rib** is appreciably longer but similar to the first. Its shaft is not twisted but is still strongly curved, and the costal groove is indistinct. Its roughened external surface gives partial origin for the serratus anterior muscle. The tenth through twelfth ribs are atypical in that their heads have only single facets for articulation with the bodies of corresponding vertebrae. In addition, the **eleventh rib** is short, presents a slight angle, but has no neck or tubercle, while the **twelfth rib** is very short and lacks neck, tubercle, angle, and costal groove.

The **costal cartilages** are bars of hyaline cartilage that prolong the ribs anteriorly. The upper seven cartilages articulate with the sternum, the next three with the costal cartilages immediately above; the last two terminate in abdominal wall musculature.

The ribs and intercostal spaces constitute the thoracic wall. Each **intercostal space** is filled by muscles between which arteries, veins, and nerves course. The intercostal muscles elevate the ribs and receive their nerve supply from corresponding intercostal or thoracoabdominal nerves.

Intercostal Muscles

The intercostal muscles are disposed in three layers (Table 3-1, Fig. 3-2). The most superficial, the **external intercostal muscle,** partially fills the intercostal space from the vertebral column to the costochondral junction, where its muscle fibers are

Table 3-1. Muscles of the Thorax

Muscle	Origin	Insertion	Action	Nerve
External intercostals	Inferior border of rib	Superior border of rib below origin	Elevate rib	Segmentally intercostals
Internal intercostals	Superior border of rib	Inferior border of rib above origin	Elevate rib	Segmentally intercostals
Innermost intercostals	Variable in extent, sometimes considered as deep portion of internal intercostals, being separ. by intercostal nerves and vessels			
Transversus thoracis	Posterior surface of xiphoid process and lower third of sternum	Inner surface of costal cartilage of second to sixth ribs	Draws costal cartilages inferiorly	Segmentally intercostals
Subcostales	Inner surface of lower ribs near their angle	Inner surface of second or third rib below rib of origin	Draw adjacent ribs together	Segmentally intercostals

lost but its fascia continues to the sternum as the **anterior intercostal membrane.** The **internal intercostal muscle** forms the middle layer of muscle. Its muscular fibers extend from the sternum anteriorly to the angle of the ribs posteriorly, where only its fascia continues to the vertebral column as the **posterior intercostal membrane.** The third muscular layer is discontinuous and is described as three individual muscles—anteriorly the **transversus thoracis** (sternocostalis), laterally the **innermost intercostals** (sometimes considered as a splitting of the internal intercostal), posteriorly the **subcostalis.** The nerves, arteries, and veins within the intercostal space course between the second and third layers of muscles.

Intercostal Arteries and Nerves

Arteries passing in the intercostal space supply intercostal muscles, send twigs to the overlying pectoral muscles, mammary gland, and skin, and, at the lower intercostal spaces give branches to the diaphragm (Fig. 3-3). The **internal thoracic** (internal mammary) **artery,** arising in the root of the neck as a branch of the first part of the subclavian, descends vertically at the side of the sternum to divide behind the sixth intercostal space into **musculophrenic** and **superior epigastric branches.** The **anterior intercostal arteries** supplying the first six intercostal spaces are branches of the internal thoracic artery. The remaining spaces are supplied by similar branches of the musculophrenic artery. The **posterior intercostal arteries** to the first two intercostal spaces are branches of the **superior intercostal,** a branch of the costocervical trunk from the subclavian. The remaining intercostal spaces receive their posterior (aortic) intercostal branches directly from the descending aorta. Additional branches from the posterior intercostals pass posteriorly to supply the deep muscles of the back, the vertebral column, and the spinal cord. The **subcostal artery,** the most inferior paired branch of the thoracic aorta arising in line with the posterior intercostals, passes inferiorly to supply structures in the abdominal wall, as do the lower two posterior intercostal vessels.

The **intercostal nerves** are ventral rami of thoracic spinal nerves. The **typical in-**

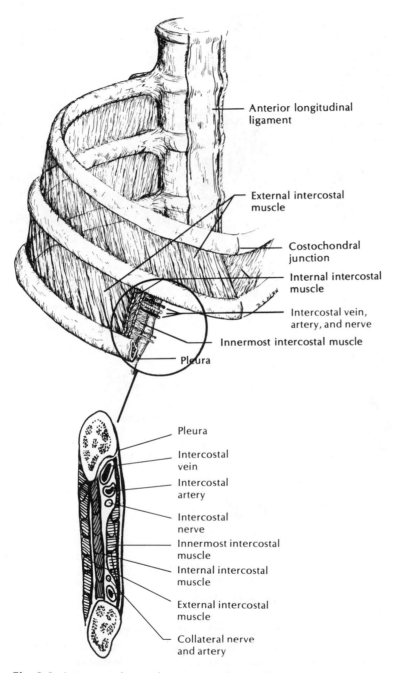

Fig. 3-2. Intercostal muscles. (Lower figure after Crafts, Roger C: A Textbook of Human Anatomy, 2nd ed. New York, John Wiley & Sons, 1979)

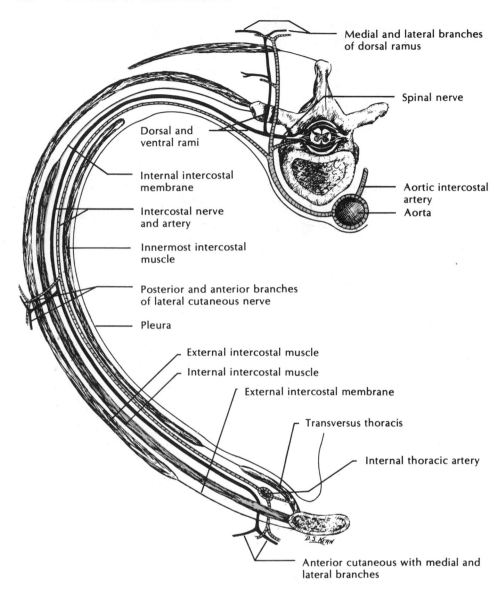

Fig. 3-3. Typical intercostal nerve and intercostal arteries.

tercostal nerves (third through sixth thoracic) course along the posterior intercostal membrane to reach the **costal groove** on the inferior border of the rib between the internal and innermost intercostal muscles and inferior to the intercostal vessels. They supply all the intercostal muscles and, in the midaxillary line give off **lateral cutaneous nerves,** which send **anterior** and **posterior branches** to supply the skin of the lateral chest wall. The intercostal nerves then proceed anteriorly to penetrate the anterior body wall just lateral to the sternum as the **anterior cutaneous nerves,** which divide into **medial** and **lateral branches** and supply the skin on the anterior aspect of the thorax.

Atypical intercostal nerves are also ventral rami, but have several distinctive characteristics. The **first thoracic nerve** is short and thick and divides unevenly into two branches. The small inferior branch supplies structures in the first intercostal space, while the much larger superior branch joins the brachial plexus. The cutaneous innervation over the first intercostal space is not supplied by this nerve but by the **medial** and **intermediate supraclavicular nerves.** The lateral cutaneous branch of the second thoracic nerve distributes with the brachial plexus as the **intercostobrachial branch** to supply cutaneous innervation to the floor of the axilla as well as to the posteromedial aspect of the arm.

At the point where the costal cartilages course upward, the ventral rami of the seventh through eleventh thoracic (**thoracoabdominal**) nerves pass obliquely anteroinferiorly. They pass between the internal abdominal oblique and transverse abdominis muscles and supply the anterolateral abdominal wall musculature and the overlying skin. The ventral ramus of the twelfth thoracic nerve is the **subcostal nerve.** It passes anterior to the quadratus lumborum muscle, then to the abdominal wall, where its distribution is similar to that of the lower intercostal nerves. The dermatome flanking the umbilicus is supplied by the tenth thoracic nerve.

Pleural Cavity

PLEURA

The thorax contains three serous cavities, the two laterally situated pleural sacs and the midline pericardial cavity. The two **pleural cavities** are completely closed, separate, and lined by a serous membrane, the **pleura.** This membrane is a continuous sheet in each cavity but is subdivided for descriptive purposes into parietal and visceral portions. The **parietal pleura** applied to the thoracic cage is designated as the **costal parietal pleura;** the portion fused to the diaphragm, the **diaphragmatic parietal pleura,** and the portion adjacent to the mediastinum, the **mediastinal parietal pleura.** The **visceral pleura** is intimately adherent to the lung, covering its entire surface and continuing deeply into its fissures. In the adult the lung lies free in the cavity, except at the root of the lung, where the visceral pleura reflects to become the parietal pleura. The **pulmonary ligament** extends inferiorly from the root of the lung as a fusion of two layers of mediastinal pleura. Near the diaphragm the ligament ends in a free falciform border. Anteriorly the costal parietal pleura reflects to become the mediastinal parietal pleura forming the **costomediastinal recess;** inferiorly it is reflected onto the diaphragm to form the **costodiaphragmatic (phrenicocostal) recess.** The lung does not extend into these recesses except in maximal inspiration.

Projection of the **pleural reflection** onto the surface of the chest is related to certain ribs or costal cartilages. The anterior border of the parietal pleura passes inferiorly from the **cupola** (pleura over the apex of the lung) to the sternoclavicular joint, continues obliquely across the manubrium, and proceeds inferiorly along the midline of the sternum from the level of the second to the fourth costal cartilage. Here the margin of the left pleural reflection moves laterally, then inferiorly at the cardiac notch to reach the sixth intercostal space. On the right side, the pleural reflection continues inferiorly in the midline, and at the sixth costal cartilage both angle laterally to reach the midclavicular line at the level of the eighth costal cartilage, both reflections crossing the tenth rib at the midaxillary line and the twelfth rib at the midscapular line.

PNEUMOTHORAX

Because of the extension of the pleural cavity and lung into the posterior triangle of the neck, a puncture wound from a knife, needle or bullet superior to the clavicle could penetrate the pleura and the lung and produce a pneumothorax (air in pleural cavity).

LUNGS

The lungs are organs of respiration. **Inspiration** is an active process that results from a decrease in the intrathoracic pressure as the thoracic cavities increase their volume by the activity of the chest muscles and the diaphragm. Normal **expiration** is passive. Relaxation of the chest muscles results in a resilient recoil of the ribs and lungs with a concomitant intrathoracic volume decrease and pressure increase. The most extensive movement of the lung occurs in the lateral, anterior, and inferior directions. Each lung is accurately adapted to the space in which it lies and, when hardened *in situ*, bears impressions of structures in contact with its surfaces.

The **shape of the lung,** described as a bisected cone, presents an apex, a base, costal and mediastinal surfaces, and anterior, inferior, and posterior borders (Fig. 3-4). The rounded **apex** lying behind the middle third of the clavicle and rising about 3 to 4 cm above the first rib, is grooved by the subclavian artery and separated from structures in the neck by the **suprapleural membrane,** or **Sibson's fascia.** The concave semilunar **base** is adapted to the upper surface of the diaphragm. The convex and extensive **costal surface** follows the contour of the rib cage. The **mediastinal surface** is adjacent to, and marked by, structures in the mediastinum. It presents a large centrally depressed area, the **hilus,** through which bronchi, nerves, vessels, and lymphatics enter or leave the lung.

Due to the high position of the right lobe of the liver the **right lung** is shorter than the left. However, it has a greater volume than the left lung because of the displacement of the left lung by the position of the heart. It is related superiorly to the superior vena cava, the brachiocephalic and azygos veins, and, anterior to these structures, the ascending aorta. Posterior to the hilus, the right lung is grooved by the esophagus and more superiorly by the arch of the aorta and the trachea.

The **left lung** displays a deep depression adjacent to the apex and left surface of the heart and is deeply grooved above and behind the hilus by the arch of the aorta and the descending aorta. Superiorly the left lung is grooved by the subclavian and common carotid arteries. The **anterior borders** of both lungs are comparatively short and thin and extend into the costomediastinal recesses. On the left lung the anterior border is displaced laterally by the heart to form the cardiac notch. The rounded, thick **posterior borders** form an indistinct confluence of costal and mediastinal surfaces occupying a deep hollow on either side of the vertebral column. Along its **inferior border** the me-

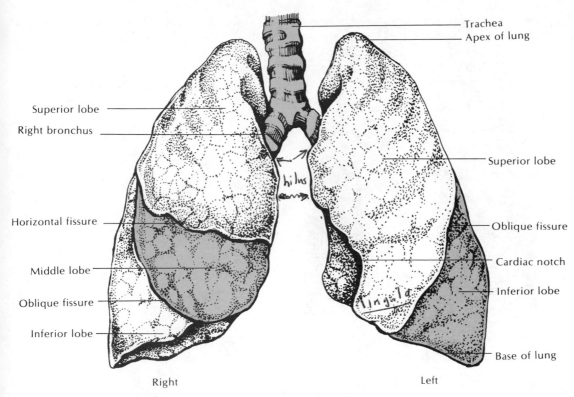

Trachea

Apex of lung

Superior lobe

Right bronchus

Superior lobe

hilus

Horizontal fissure

Oblique fissure

Cardiac notch

Middle lobe

lingula

Inferior lobe

Oblique fissure

Inferior lobe

Base of lung

Right

Left

Fig. 3-4. Anterior view of trachea, bronchi, and lungs. (After Hollinshead, W Henry: Textbook of Anatomy, 3rd ed. Hagerstown, Harper & Row, 1974)

diastinal portion is blunt and related to the lower border of the pericardium. Anteriorly, laterally, and posteriorly this border is thin and sharp and extends into the costodiaphragmatic recess.

Each lung is partially transected by an **oblique fissure** from its surface to within a short distance of the hilus. On the left lung the fissure extends from the posterior border approximately 6 to 7 cm below the summit, runs obliquely and inferiorly to cross the fifth rib in the midaxillary line, and ends anteriorly opposite the sixth costal cartilage. On the right lung the oblique fissure begins slightly lower. The oblique fissure in each lung separates the **upper** (superoanterior) **lobe** from the **lower** (posteroinferior) **lobe.** The right lung is further subdivided by the **horizontal fissure,** which extends from the anterior border of the lung at the fourth intercostal space to follow the upper border of the fifth rib to the oblique fissure to form the wedge-shaped **middle lobe.** In the left lung the **lingula** is homologous to the middle lobe of the right lung and consists of a small tonguelike appendage between the cardiac notch and the oblique fissure.

Air Conduction System

The **trachea** is a wide tube about 10 cm in length. It is kept patent by a series of sixteen to twenty U-shaped horizontal cartilaginous bars embedded in its wall. Posteriorly the open cartilages of the tube are closed by the **trachealis muscle** and fibrous

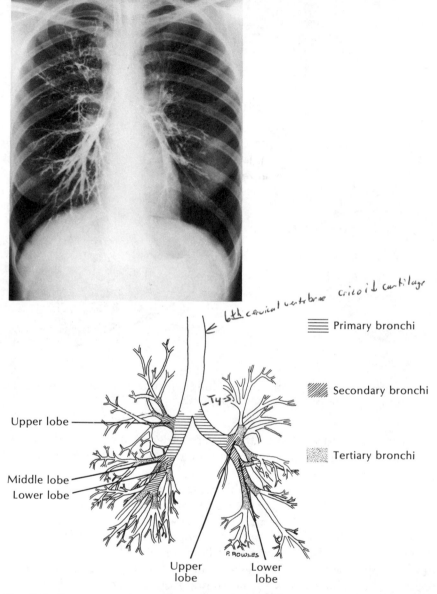

Fig. 3-5. Posteroanterior radiograph of bronchial tree (refer to Fig. 3-6 for tertiary bronchi).

tissue. The trachea begins at the cricoid cartilage of the larynx, at the level of the sixth cervical vertebra. It terminates between the fourth and fifth thoracic vertebrae by dividing into right and left primary bronchi (Fig. 3-5). In inspiration the respiratory shift carries the bifurcation as low as the seventh vertebra. The trachea is related posteriorly to the esophagus and inferior laryngeal nerves; anteriorly to the left innominate vein, left common carotid artery, and the arch of the aorta. Laterally on the right it is related to the pleura, vagus nerve, and arch of the azygos vein, and on the left to the common carotid and subclavian arteries, phrenic and vagus nerves, and the arch of the aorta.

TRACHEOTOMY

Tracheotomy is a life-saving procedure often needed to relieve an upper respiratory obstruction from inhalation of a foreign object, which can cause spasm of the laryngeal muscles and airway obstruction. A tracheotomy can be performed by making a midline skin incision from just above the cricoid cartilage to the jugular notch of the manubrium. The incision is deepened so that the third and fourth tracheal rings can be incised. A metal or plastic tracheal tube is then inserted into the trachea and breathing is reestablished. Structures to be avoided include the inferior thyroid veins, the jugular venous arch, and the isthmus of the thyroid. To avoid these structures and some complications from the tracheal tube, some physicians prefer to open the trachea just inferior to the cricoid cartilage.

The **right primary bronchus,** a more direct continuation of the trachea, is shorter, straighter, and larger than the left bronchus; therefore, foreign bodies are more apt to pass into the right than into the left lung. The right primary bronchus enters the lung at the hilus where it divides into **secondary bronchi** passing to the upper, middle, and lower lobes. The superior lobe bronchus passes above the pulmonary artery as the eparterial bronchus. Secondary bronchi give rise to **tertiary bronchi** named for the **bronchopulmonary segments** they supply. The secondary bronchus to the right upper lobe divides into apical, posterior, and anterior tertiary bronchi. The secondary bronchus to the right middle lobe divides into lateral and medial tertiary bronchi and the secondary bronchus to the right lower lobe divides into superior, medial basal, anterior basal, lateral basal, and posterior basal tertiary bronchi. Thus, ten bronchopulmonary segments are present in the right lung (Fig. 3-6).

The **left primary bronchus** is smaller in caliber but roughly twice as long as the right. It passes initially superior to the pulmonary artery, but divides inferior to the artery into upper and lower lobe **secondary bronchi.** The secondary bronchus to the upper lobe subdivides into apicoposterior, anterior, superior, and inferior tertiary bronchi. The lower lobe bronchus subdivides into superior, anteromedial basal, lateral basal, and posterior basal tertiary bronchi. This results in eight bronchopulmonary segments in the left lung.

The **root of the lung** is formed by structures entering or leaving the lung. It consists of the bronchi, the pulmonary artery (carrying venous blood), pulmonary veins (carrying arterial blood), bronchial arteries supplying the bronchial tree, lymph vessels and nodes, and the pulmonary plexus of nerves. These structures are held together by connective tissue and are surrounded by visceral pleura reflecting from the surface of the lung to become the parietal pleura. The superior vena cava and right atrium are anterior to the root of the right lung, and the azygos vein arches above it. The root of the left lung lies anterior to the descending aorta and inferior to the arch of the aorta. The phrenic nerves, pericardiophrenic arteries and veins, and the anterior pulmonary

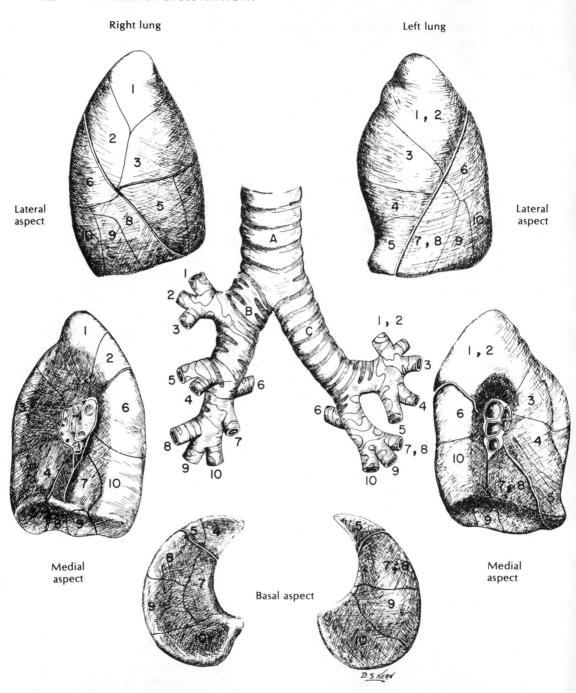

Fig. 3-6. Lung and tracheobronchial tree. **A.** Trachea. **B.** Right primary bronchus. **C.** Left primary bronchus. Tertiary bronchi and bronchopulmonary segments include: (*right lung*) 1. apical, 2. posterior, 3. anterior, 4. medial, 5. lateral, 6. superior, 7. medial basal, 8. anterior basal, 9. lateral basal, 10. posterior basal; and (*left lung*) 1. and 2. apicoposterior, 3. anterior, 4. superior, 5. inferior, 6. superior, 7. and 8. anteromedial basal, 9. lateral basal, 10. posterior basal.

plexuses are located anterior to both roots. The posterior pulmonary plexuses and the vagus nerves are posterior to both roots.

BRONCHOSCOPY

The trachea, main bronchi, and the first segmental bronchi can be viewed directly through a bronchoscope. Such an instrument also enables the physician to remove foreign objects from these passageways. These objects usually fall into the right bronchus because it has a larger caliber and is more in line with the vertical trachea. Spraying a radiopaque oil into the bronchi outlines the deeper bronchi and the lung proper on a chest x-ray. This helps localize bronchial lesions or obstructions. The procedure is called a bronchiogram.

Blood Vessels, Lymphatics, and Nerves

The **right** and **left pulmonary arteries** are branches of the **pulmonary trunk,** which, in turn, stems from the right ventricle. The **right branch,** longer and of larger caliber than the left, passes posterior to the ascending aorta and superior vena cava and anterior to the right bronchus. It divides into two branches, one to the upper lobe and a second to the middle and lower lobes. The **left pulmonary artery** passes horizontally anterior to the descending aorta and the left bronchus, and divides into branches to the upper and lower lobes. Further subdivisions of the pulmonary arteries correspond to the divisions of the bronchial tree.

Pulmonary veins begin from alveolar capillaries and coalesce at the periphery of the lung lobule. They maintain a peripheral relation in the bronchopulmonary segments running in intersegmental connective tissue. They drain adjacent rather than single segments. On the right side the middle lobe vein joins the lower lobe vein at the hilus, usually resulting in two veins from each lung draining into the left atrium.

The **bronchial arteries** are small vessels coursing along the posterior aspect of the bronchi to supply the bronchi and bronchioles to the level of the respiratory bronchiole. Classically, two **left bronchial arteries** arise from the anterior aspect of the descending aorta, while a single **right bronchial artery** arises from the first aortic intercostal or the upper left bronchial artery.

The **lymphatics** of the lungs are disposed as a **superficial plexus** lying immediately deep to the visceral pleura, and a **deep plexus** arising in the submucosa of the bronchi and peribronchial connective tissue. The efferent lymph vessels of the superficial plexus follow the surface of the lung to the bronchopulmonary trunks; those of the deep plexus follow the pulmonary vessels to the hilus. The **nodes** of the lung are disposed in five groups: the tracheal or paratracheal nodes along the trachea, the superior tracheobronchial nodes in the angle between the trachea and bronchi, the inferior tracheobronchial nodes in the angle between the bronchi, the bronchopulmonary nodes at the hilus of each lung, and the pulmonary nodes in the substance of the lung (Fig. 3-7).

Innervation to the lung is supplied by the vagus and thoracic sympathetic nerves distributed through the **anterior** and **posterior pulmonary nerve plexuses,** located on

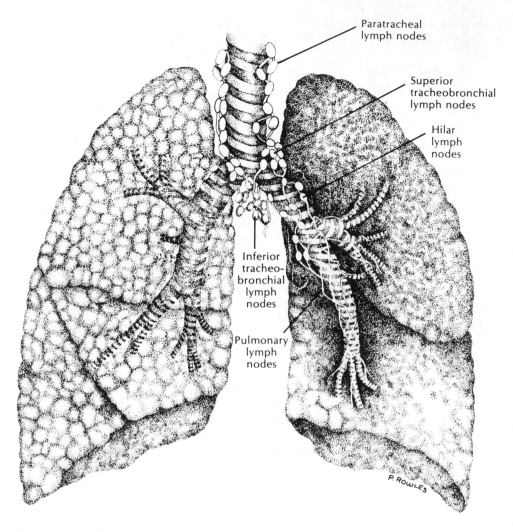

Fig. 3-7. Lymphatics of lungs.

corresponding surfaces of the root of the lung. The cell bodies of the preganglionic sympathetic neurons are located in the upper three to five thoracic segments of the spinal cord.

ASTHMA

Efferent (motor) vagal stimuli cause smooth muscle contraction, thus narrowing the lumens of bronchi and bronchioles. Efferent sympathetic fibers are bronchiodilatory, therefore epinephrine (adrenalin) injections will relieve the bronchial spasms in asthma by blocking the vagal stimuli.

Pericardial Cavity

The **pericardial cavity,** a conical fibroserous sac, surrounds the heart and proximal portions of the great cardiac vessels. The **pericardium** is disposed into three layers: an external strong **fibrous layer,** which superiorly and inferiorly blends with the adventitia of the great vessels, a **parietal layer** of serous pericardium adherent to the inner surface of the fibrous layer, and a **visceral layer** of serous pericardium that reflects onto and is intimately applied to the heart and great vessels. The proximal portions of the aorta, the pulmonary trunk, and the superior vena cava are surrounded by visceral pericardium. Near the median plane, the base of the pericardial cavity rests on, and fuses with, the central tendon of the diaphragm. The right posteroinferior aspect of the base is pierced by the inferior vena cava and posteriorly at the junction of the upper part of the posterior and lateral surfaces by the pulmonary veins.

CARDIAC TAMPONADE

Inflammation of the pericardium (pericarditis) with an effusion of fluid or extensive bleeding into the pericardial sac, if untreated, is a life threatening condition. The tough, fibrous, inelastic pericardium will not stretch to accommodate the excess fluid. This results in a compression of the heart (cardiac tamponade) and circulatory failure.

Anteriorly the pericardium is separated from the sternum and the second to sixth costal cartilages by the lungs and pleura, except in the region of the left fourth through sixth intercostal spaces, where the left lung is deficient at the **cardiac notch.** In this area, known as the "bare area," the pericardium comes in contact with the anterior chest wall. **Superior** and **inferior sternopericardial ligaments** attach the pericardium to the sternum. Laterally the pericardium is adjacent to the mediastinal parietal pleura, with the phrenic nerve and pericardiophrenic vessels interposed between them. Posteriorly it is related to the descending aorta, the esophagus, and the bronchi.

The **transverse sinus** of the pericardium is a space within the pericardial cavity situated between the aorta and pulmonary trunk, and the superior vena cava. It is formed by the reflection of the serous pericardium. The **oblique sinus** of the pericardium, a boxlike diverticulum of the pericardial cavity within the irregularities of the pericardial reflection, is bounded by the pericardial reflection over the right pulmonary veins and the inferior vena cava on one side and the pericardial reflection over the left pulmonary veins on the other.

Heart (Fig. 3-8)

Anteriorly the heart presents a **sternocostal surface** formed superiorly by the right and left auricular appendages and by the infundibulum (Fig. 3-9). Inferiorly two-thirds of this aspect of the heart is formed by the right ventricle and one-third by the

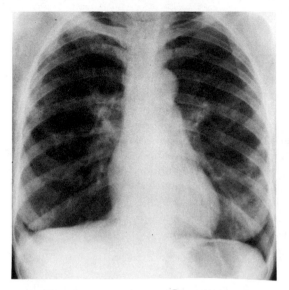

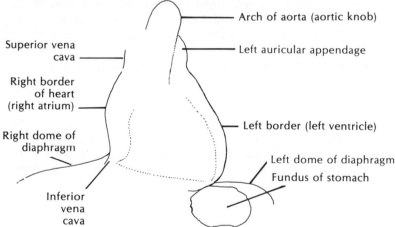

Arch of aorta (aortic knob)

Superior vena cava

Left auricular appendage

Right border of heart (right atrium)

Left border (left ventricle)

Right dome of diaphragm

Left dome of diaphragm

Fundus of stomach

Inferior vena cava

Fig. 3-8. Posteroanterior radiograph of heart.

left ventricle, with the **apex** of the heart being formed entirely by the left ventricle. The **lower border** of the sternocostal surface presents a sharp margin (margo acutis) formed almost entirely by the right ventricle. The **right border,** formed by the right atrium, is continuous superiorly with the superior vena cava and inferiorly with the inferior vena cava. The more convex **left border** is formed mostly by the left ventricle. The **diaphragmatic surface** of the heart rests on the diaphragm and is formed entirely by the ventricles, with the greatest contribution from the left ventricle. The **base** of the heart, or **posterior surface,** is formed almost entirely by the left atrium. The **atrioventricular groove** completely encircles the heart and separates the atria, which lie superiorly and to the right, from the ventricles, which lie inferiorly and to the left.

The nutrient **coronary arteries** of the heart may be considered as greatly enlarged

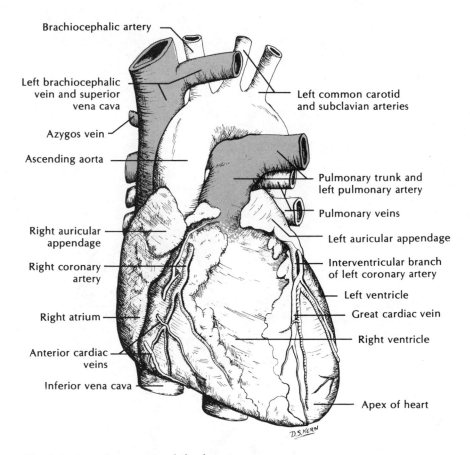

Brachiocephalic artery

Left brachiocephalic
vein and superior
vena cava

Azygos vein

Ascending aorta

Right auricular
appendage

Right coronary
artery

Right atrium

Anterior cardiac
veins

Inferior vena cava

Left common carotid
and subclavian arteries

Pulmonary trunk and
left pulmonary artery

Pulmonary veins

Left auricular appendage

Interventricular branch
of left coronary artery

Left ventricle

Great cardiac vein

Right ventricle

Apex of heart

D.S.KERN

Fig. 3-9. Anterior aspect of the heart.

vasa vasorum (Fig. 3-10). They originate from the **aortic sinuses** (dilations of the aorta) immediately above the aortic cusps. The **right coronary artery** arises from the anterior aortic sinus, courses forward between the pulmonary trunk and the right auricular appendage, then passes inferiorly to the atrioventricular groove, in which it runs posteriorly to anastomose with the left coronary. It gives small branches to the root of the aorta, pulmonary trunk, and the walls of the right atrium and ventricle. It supplies both the sinoatrial and the atrioventricular nodes. A **marginal branch** of the right coronary passes from right to left along the lower border and anterior aspect of the heart. A **posterior interventricular** (posterior descending) **branch** extends anteriorly on the diaphragmatic surface in the interventricular groove. The large **left coronary,** originating from the left posterior aortic sinus, passes a short distance to the left, behind the pulmonary trunk, then between the pulmonary trunk and the left auricular appendage, where it divides into an anterior interventricular (anterior descending) branch and the circumflex branch. The **anterior interventricular branch** descends in the interventricular groove on the anterior aspect of the heart toward the apex, where it anastomoses with the posterior interventricular branch. The **circumflex artery** continues in the

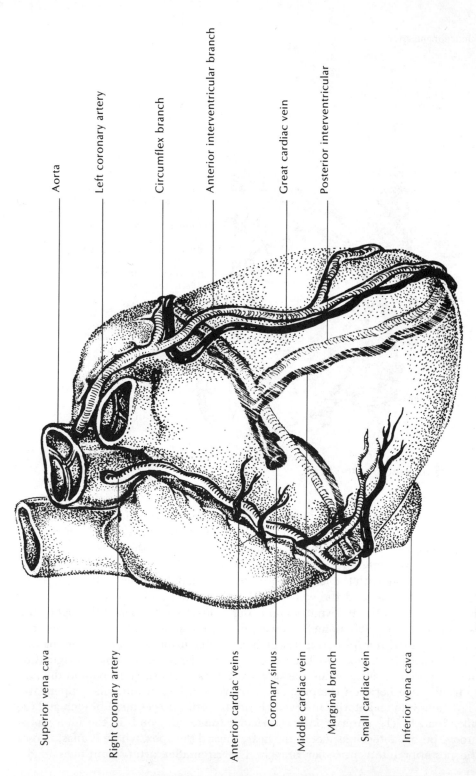

Superior vena cava

Right coronary artery

Anterior cardiac veins

Coronary sinus

Middle cardiac vein

Marginal branch

Small cardiac vein

Inferior vena cava

Aorta

Left coronary artery

Circumflex branch

Anterior interventricular branch

Great cardiac vein

Posterior interventricular

Fig. 3-10. Vasculature of the heart.

atrioventricular groove to the lower border of the base of the heart to anastomose with the right coronary.

HEART ATTACKS

An insufficient blood supply to the heart muscle (myocardial ischemia) may precipitate a heart attack, which causes the patient severe chest pains (angina pectoris) over the area of the heart (precordium). These severe chest pains are an example of referred pain because the pain usually extends from the chest wall, to the left shoulder and down the left arm. A common cause for the reduced blood supply is a marked reduction of the diameter of the lumen of one or more of the coronary arteries, caused by an accumulation of atherosclerotic plaques within the inner wall of the artery. Another cause is by occlusion of a coronary artery by a blood clot (thrombus).

If the circulatory deficiency is of sufficient duration, local tissue necrosis results, causing permanent loss of muscle fibers (myocardial infarction). If the damage covers a substantial area of the heart wall, the heart cannot contract and cardiac arrest results.

The largest vein draining the heart is the **coronary sinus.** It begins at the upper end of the anterior interventricular groove as a continuation of the **great cardiac vein,** the companion vein of the anterior descending branch of the left coronary artery. It runs posteriorly from left to right in the atrioventricular groove. The **middle cardiac vein,** accompanying the posterior descending branch of the right coronary artery, empties into the coronary sinus near its termination. The **small cardiac vein** curves around the lower margin of the right border of the heart to also empty into the coronary sinus at its termination. In its course it follows the marginal branch of the right coronary artery. The small **oblique vein of the left atrium** descends over the posterior surface of this chamber to drain into the midportion of the coronary sinus. **Valves** in the coronary sinus are usually present at the junction of the great and small cardiac veins, and at its termination. The **anterior cardiac veins** are small vessels on the right atrial and ventricular surfaces that empty directly into the right atrium near the atrioventricular groove. A small part of the blood is collected directly from the heart musculature by small veins, the **venae cordis minimae,** which arise within the muscle of the heart wall and drain directly into all chambers of the heart.

Nerves

Modifications of the intrinsic rhythm of heart muscle is produced through autonomic nerve plexuses. **Parasympathetic innervation,** from the vagus nerve, slows the rate and reduces the force of the heart beat. **Sympathetic innervation,** from the thoracic and cervical ganglia has the opposite effects. The **superficial cardiac plexus** lies in the concavity of the aortic arch, proximal to the ligamentum arteriosum, and receives the inferior cervical cardiac branch of the left vagus and the superior cervical cardiac branch of the left sympathetic trunk. The more extensive **deep cardiac plexus** is located on the base of the heart posterior to the aortic arch and anterior to the tracheal bifurcation. From the cervical sympathetic ganglia the plexus receives all the cardiac branches on the right side and the middle and inferior cardiac branches on the left. From the right vagus the deep cardiac plexus receives the superior and inferior cervical branches, and from the left vagus, the superior cervical branch. In the thorax additional vagal and sympathetic branches pass directly to the deep cardiac plexus.

Coronary plexuses, surrounding and accompanying the coronary arteries, are extensions from the cardiac plexus. Vagal fibers within the coronary plexus produce vasoconstriction, while sympathetic fibers cause vasodilatation of the arterioles. **Visceral afferent fibers** from the heart and coronary arteries end almost entirely in the first four thoracic segments of the spinal cord.

Conduction (Purkinje) System

The orderly sequence in which ventricular contraction follows atrial contraction and proceeds from the apex to the base of the heart is due to a specialized conduction system composed of two aggregates of nodal tissue and a band of specialized cardiac muscle (Fig. 3-11).

The **sinoatrial node** is a small collection of specialized myocardial tissue at the

Fig. 3-11. Conduction system.

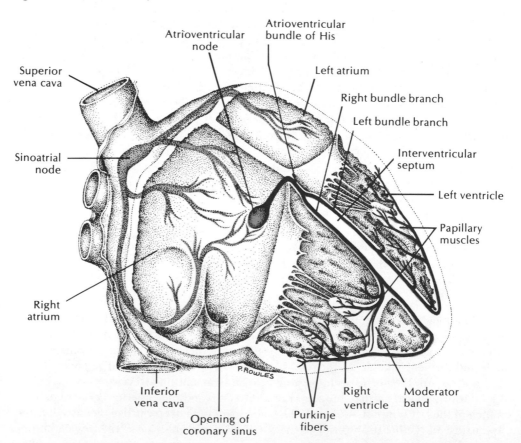

junction of the right atrium and the superior vena cava, at the superior end of the sulcus terminalis. This node, the **pacemaker,** appears to initiate the heart beat, and the impulse is propagated over the atria.

ARTIFICIAL PACEMAKER OF THE HEART

An artificial pacemaker of the heart is a battery-operated device that supplies electric impulses to regulate the heart beat in patients suffering from arrhythmias or bradycardia. The pacemaker is implanted subcutaneously over the upper anterior chest. A special electrode catheter is passed intravenously into the cephalic, external jugular, or internal jugular vein through the superior vena cava and into the right ventricle. The tip of the catheter is positioned in the endocardium, so that a small voltage through the wire from the pacemaker will cause ventricular contraction.

The **atrioventricular node** is a nodule of similar tissue located in the septal wall of the right atrium, immediately above the opening of the coronary sinus. Activity at this node is probably initiated by stimuli reaching it through the atrial musculature. The impulse is then directed toward the apical portions of the ventricles by way of the **atrioventricular bundle of His.** This structure is a strand of specialized myocardium (Purkinje fibers) passing from the atrioventricular node into the interventricular septum, where it divides into a right and left bundle within the septal musculature.

HEART OR AV BLOCK

Heart or AV block is an interruption of nervous impulses through the atrioventricular bundle (of His) causing the ventricles to beat at a rhythm slower than the atria. Such a loss of synchronization causes the ventricles to contract whether filled or not. Thus, the efficiency of the pumping action of the heart is drastically reduced and cardiac arrest may follow.

The **right bundle branch** follows the endocardium along the septum of the right ventricle giving branches to the septum, then extends across the ventricular cavity, traversing the **moderator band** to reach the anterior papillary muscle of the right ventricle. The **left bundle branch** follows the endocardium of the left ventricle to the apex, where it is distributed to the anterior and posterior papillary muscles and the ventricular myocardium.

VENTRICULAR FIBRILLATION

Ventricular fibrillation follows a complete functional breakdown in the conduction system. The impulse follows no regular pattern but seems to be stimulating all regions of the ventricle simultaneously. Such a loss of coordination of the ventricular contraction causes a loss of

blood flow to the myocardium and low blood pressure. To stop ventricular fibrillation or to defibrillate the heart, an electric shock is delivered to the heart through external paddles. This stops the fibrillating ventricles. A normal heart beat will usually follow defibrillation.

Interior of the Heart

The interior of the heart is divided into right and left atria, each with an auricular appendage, and right and left ventricles (Fig. 3-12).

The **right atrium** is described as having two parts, the sinus venarum and the atrial portion. The sinus venarum, a large quadrangular cavity with smooth walls, lies between the two venae cavae and is continuous inferiorly with the right ventricle. The auricle (proper) is a small conical muscular pouch that projects from the upper anterior part of the sinus venarum to overlap the root of the aorta. The **crista terminalis** is a vertical ridge extending from the orifice of the superior vena cava to the opening of the inferior vena cava. The **musculi pectinati** are parallel muscular ridges at right angles to the crista terminalis and extend into the auricular appendage.

Openings into the right atrium include: 1) the superior vena cava, devoid of valves, opening into the superoposterior part of the chamber; 2) the inferior vena cava, with a thin rudimentary valve along the anterior margin between its orifice and the atrioventricular orifice, opening into the inferoposterior part; 3) the opening of the coronary sinus, guarded by a crescentic valve directly posterior to the atrioventricular orifice; and 4) the right atrioventricular aperture, closed by the tricuspid valve in the

Fig. 3-12. Internal aspect of the right atrium.

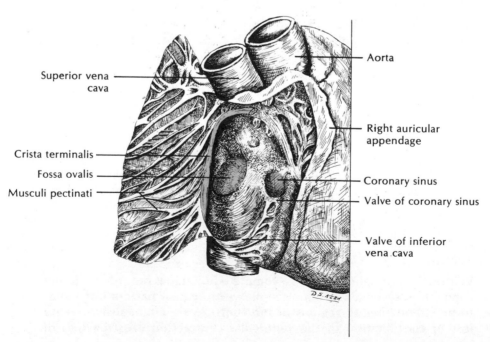

anteroinferior part of the atrium, opening into the right ventricle. The rudimentary valve of the inferior vena cava is continuous on its left with the **annulus ovalis** of the interatrial septum, which surrounds a wide shallow depression, the **fossa ovalis.** The upper end of the fossa is open in fetal circulation as the **foramen ovale** and permits passage of blood from the right to the left atrium.

AUSCULTATION OF THE HEART SOUNDS

To interpret cardiac valve sounds, one must be able to locate the sounds accurately as they are projected onto the anterior chest wall. The distance between the heart and the skin of the anterior thorax affects the intensity of the heart sounds—softer or lower intensity, with increased distances; and louder or higher intensity with shorter distance. Using a stethoscope, the areas of greatest audibility for the various valves are: 1) left atrioventricular (mitral or bicuspid), over the apex of the heart in the fifth intercostal space in the midclavicular line; 2) right atrioventricular (tricuspid), midline over the inferior surface of the sternum, or the right fifth intercostal space at the sternal margin; 3) aortic semilunar, over the right second intercostal space at the sternal margin; and 4) pulmonary semilunar, over the left second intercostal space at the sternal margin.

The **right ventricular cavity** is triangular in shape. Its funnel-shaped superior portion, the **infundibulum,** or **conus arteriosus,** leads into the pulmonary trunk. The infundibulum has smooth internal walls, while the remainder of the cavity possesses projecting muscular bundles, the **trabeculae carneae. Papillary muscles** are conical projections from the wall into the cavity that are connected at their apices to the tricuspid valve by the **chordae tendineae.** The **moderator band,** a well-marked trabeculum, projects from the interventricular septum to the base of the anterior papillary muscle and may act to prevent over-distention of the chamber. The **right atrioventricular orifice,** about 2 to 3 cm in diameter, is guarded by a valve with three triangular cusps, the **tricuspid valve** (Fig. 3-13). The anterior cusp intervenes between the atrioventricular orifice and the infundibulum, the medial cusp lies in relation to the septal wall, and the inferior cusp is adjacent to the inferior wall. The bases of the cusps are attached continuously to a fibrous ring, the **anulus fibrosus,** a deeply lying structure at the periphery of the atrioventricular orifice. The apex of each cusp extends into the ventricle and is attached by chordae tendineae to papillary muscles. The **pulmonary orifice,** at the apex of the infundibulum, is surrounded by a thinner fibrous ring to which the bases of three cuplike **pulmonary semilunar cusps** are attached. The center of the free margin of each semilunar cusp is thickened to form a **nodule** (corpus Arantii). The **lunulae** of the valves are the thinner crescentic regions at either side of the nodules. When the valve is closed, the three nodules meet at the center of the orifice, effecting complete occlusion.

The **left atrium** forms most of the base of the heart and consists of the atrium proper and its auricular appendage. Four **pulmonary veins,** two on each side, enter the chamber at its superolateral aspect. The internal surface of the wall of the left atrium is

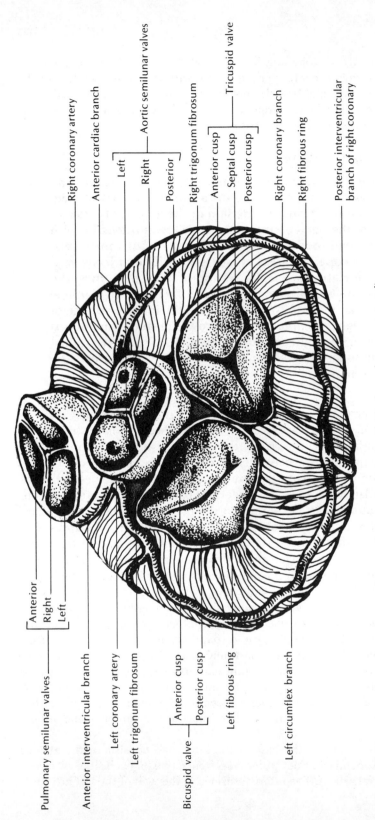

Fig. 3-13. Superior view of heart valves and coronary arteries (atria have been removed).

Pulmonary semilunar valves
Anterior
Right
Left

Anterior interventricular branch

Left coronary artery

Left trigonum fibrosum

Bicuspid valve
Anterior cusp
Posterior cusp

Left fibrous ring

Left circumflex branch

Right coronary artery

Anterior cardiac branch

Aortic semilunar valves
Left
Right
Posterior

Right trigonum fibrosum

Tricuspid valve
Anterior cusp
Septal cusp
Posterior cusp

Right coronary branch

Right fibrous ring

Posterior interventricular branch of right coronary

generally devoid of muscular ridges except in the auricular appendage. Anteroinferiorly the left atrioventricular orifice, guarded by the bicuspid valve, opens into the left ventricle.

CARDIAC MALFORMATIONS

Cardiac malformations usually arise from developmental defects in the formation of the heart valves, septa, or both. An example of an atrioseptal defect is a postnatal patent foramen ovale. Such an abnormal opening allows venous blood from the right atrium to mix freely with the well-oxygenated blood in the left atrium. A ventricular septal defect usually occurs in the uppermost (membranous) portion of the septum. It is more serious than an atrial septal defect because the blood flow from the left to the right ventricle is much greater and heart failure occurs much sooner.

Valvular abnormalities may be either a stenosis (constriction) or an incompetence (leakage) of valves. In the latter the valve does not close completely, which allows a backflow (leakage) of blood and causes a sound called a heart murmur. Most valvular and septal defects are now amenable to surgical correction.

The cone-shaped cavity of the **left ventricle** is longer and narrower than the right and has much thicker walls. The internal surface is covered with a dense meshwork of muscular ridges, **trabeculae carneae,** which are finer and more numerous than those of the right ventricle. The **papillary muscles,** much larger and stronger than in the right ventricle, are usually two in number. The interventricular septum and the upper portion of the anterior wall are relatively free of muscular ridges. The **left atrioventricular orifice,** slightly smaller than the right, is closed by the **bicuspid (mitral) valve.** The cusps are set obliquely, with the larger anterior cusp to the right and the smaller posterior cusp to the left. The bases of the cusps are attached to the **anulus fibrosus** surrounding the left atrioventricular orifice. Their apices project into the cavity with **chordae tendineae** extending from each cusp to both papillary muscles. The **aortic vestibule** is that portion of the left ventricle immediately below the aortic orifice. Its walls, composed mainly of fibrous tissue, are quiescent during ventricular contraction so that outflow of blood from the left ventricle is not impeded. In the upper right posterior part of the cavity the **aortic orifice** is surrounded by an **anulus fibrosus** for the attachment of the bases of the three **semilunar aortic cusps** guarding the opening. This valve is similar to the previously described pulmonary valve. The musculomembranous **interventricular septum** separates the ventricles as well as separating the left ventricle from the inferiormost part of the right atrium. The muscular portion is thickest near the apex and gradually thins to become membranous near the aortic opening.

TETRALOGY OF FALLOT

Tetralogy of Fallot is a combination of septal and valvular defects. It has four (tetra-) consistent malformations: 1) stenosis of the pulmo-

nary trunk, 2) ventricular septal defect (VSD), 3) overriding of the aorta over the VSD, and 4) hypertrophy of the right ventricle. Although it is one of the most crippling of the cardiac abnormalities, it was the first to be alleviated by surgery. Dr. Alfred Blalock of Johns Hopkins University devised an operation to anastomose the right subclavian artery to the right pulmonary artery and thus provide adequate circulation through the lungs and correct the severe hypoxemia, which caused all infants with tetrology to be cyanotic or "blue" babies.

The **fibrous skeleton** of the heart consists of fibrous rings, the annuli fibrosi, surrounding the atrioventricular, aortic, and pulmonary orifices; intervening fibrous connections, namely, the right and left trigona fibrosa, and the conus tendon. The **right trigonum fibrosum,** situated between the two atrioventricular orifices, sends a strong dense expansion into the membranous portion of the interventricular septum. The **left trigonum fibrosum** lies between the left atrioventricular orifice and the root of the aorta, and the **conus tendon** extends between the root of the aorta and the pulmonary artery. In addition to giving attachment to the bases of the valves of the heart, this fibrous skeleton also serves for the attachment of the various muscle bundles composing the myocardium.

Mediastinum

The **mediastinum** is the midline area between the two pleural cavities and contains all the thoracic structures except the lungs (Fig. 3-14). The heart occupies a central subdivision of this area, the middle mediastinum. The rest of the mediastinum is subdivided in relation to the heart (or pericardial cavity) into superior, anterior, and posterior portions (Fig. 3-15). The anterior, middle, and posterior mediastina are sometimes referred to collectively as the **inferior mediastinum.**

The **superior mediastinum** is the area above the fibrous pericardium. It is demarcated from the inferior mediastinum by a line passing from the sternal angle to the intervetebral disc between the fourth and fifth thoracic vertebrae. It contains all structures passing between the neck and the thorax, including the remnant of the thymus, brachiocephalic veins and superior vena cava, aortic arch and its three branches, thoracic duct, trachea, esophagus, phrenic, vagus, and cardiac nerves, and the sympathetic trunk. The **anterior mediastinum** is a limited area anterior to the pericardium and contains connective tissue with some remains of the thymus, fat, and lymph nodes. The **middle mediastinum,** centrally located and limited by the fibrous pericardium, contains the heart and its eight great vessels, namely, aorta, pulmonary artery, superior and inferior venae cavae, and the four pulmonary veins. Laterally, to either side of the pericardium, the phrenic nerve and the pericardiophrenic vessels course between the parietal pericardium and the mediastinal parietal pleura. The **posterior mediastinum** is posterior to the pericardium and the diaphragm. It contains the descending (thoracic)

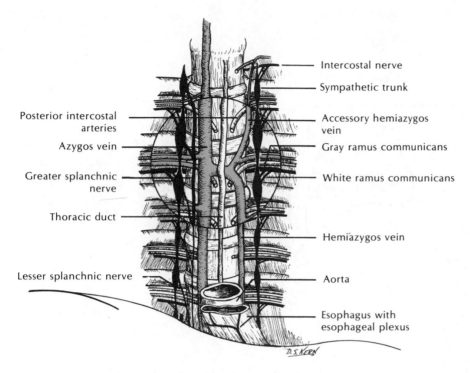

Posterior intercostal arteries

Azygos vein

Greater splanchnic nerve

Thoracic duct

Lesser splanchnic nerve

Intercostal nerve

Sympathetic trunk

Accessory hemiazygos vein

Gray ramus communicans

White ramus communicans

Hemiazygos vein

Aorta

Esophagus with esophageal plexus

D.S.KERN

Fig. 3-14. Structures within the posterior mediastinum.

aorta, posterior intercostal vessels, azygos and hemiazygos veins, thoracic duct, vagus and splanchnic nerves, sympathetic trunk, and esophagus.

Aorta

In the thorax the **aorta** is divided, for descriptive purposes, into the ascending, arch, and descending parts. Originating at the aortic orifice, the **ascending aorta** passes superiorly to the level of the second costal cartilage, where it becomes the arch. At its origin the lumen is not uniform, owing to the three aortic sinuses, which are opposite the cusps of the aortic semilunar valves. The right and left coronary arteries originate from two of these dilatations. The **arch of the aorta** begins at the level of the second costal cartilage, ascends to the level of the middle of the manubrium, and then arches to the left and anteroposteriorly to reach the lower border of the fourth thoracic vertebra where it becomes the descending or thoracic aorta. The **brachiocephalic, left common carotid,** and **left subclavian arteries** arise from the arch. The **ligamentum arteriosum,** the remnant of the ductus arteriosus, passes from the inferior aspect of the arch to the left pulmonary artery.

AORTIC ANEURYSM

The aorta has a higher incidence of aneurysms than any other artery, probably due to its curved shape and large size. In the thorax, aneu-

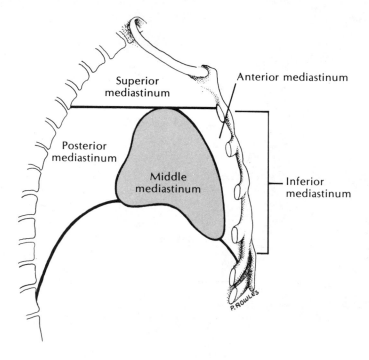

Fig. 3-15. Mediastinum.

rysms usually occur in the ascending or descending aorta but seldom in the aortic arch. Symptoms of thoracic aortic aneurysm usually depend on pressure exerted on adjacent structures. For example, pressure on the inferior (recurrent) laryngeal nerve causes hoarseness and a brassy cough. Pressure on the esophagus may cause dysphagia (difficulty in swallowing). Dyspnea (difficulty in breathing) may follow pressure on the trachea, root of the lung, or phrenic nerve.

The **left inferior laryngeal (recurrent branch)** of the vagus nerve loops around the arch adjacent to the ligamentum arteriosum to pass to the tracheoesophageal groove, where it ascends to reach the larynx. The **descending aorta** continues from the arch to pass through the posterior mediastinum, inclining to the right and then anteriorly to gain a position anterior to the vertebral column. It passes into the abdomen through the aortic opening in the diaphragm, opposite the lower border of the twelfth thoracic vertebra, to become the abdominal aorta. From the anterior aspect its branches include **two left bronchial** and several small branches to the esophagus, mediastinum, diaphragm, and pericardium. From the posterior aspect, nine pairs of **posterior** (aortic) **intercostal** and the two **subcostal arteries** arise. The posterior intercostal arteries divide to send one branch anteriorly into the intercostal space and a second branch posteriorly to the deep muscles of the back, the vertebral column, and the spinal cord. The subcostal arteries follow the lower border of the twelfth rib to supply the abdominal wall musculature.

Azygos System of Veins

The **azygos system of veins** drains most of the blood from the thoracic wall and posterior wall of the abdomen. The **azygos vein** is formed by the junction of the right subcostal and the right ascending lumbar veins and ascends through the posterior and superior mediastina. It receives blood from the posterior intercostal veins as high as the level of the third intercostal space. Arching over the root of the right lung, it empties into the superior vena cava. During development it may deviate to the right, cleaving the developing right lung to demarcate an **azygos lobule.** The **hemiazygos vein,** formed by the junction of the left subcostal and the left ascending lumbar veins, receives the posterior intercostal veins draining the ninth, tenth, and eleventh left intercostal spaces. It crosses the midline to empty into the azygos vein. The **accessory hemiazygos** usually begins at the fourth intercostal vein on the left side, descends to the eighth or ninth space receiving the intervening intercostal veins to empty into the azygos or hemiazygos. The **superior intercostal veins** drain the second and third intercostal spaces. The right vessel joins the azygos, and the left vein drains into the left brachiocephalic vein. Each of the **highest intercostal veins** drain directly into the brachiocephalic vein on the same side.

Sympathetic Trunk, Phrenic and Vagus Nerves

The **sympathetic trunk,** or vertebral chain, is a series of ganglia (collections of nerve cell bodies) connected by interganglionic nervous tissue (Fig. 3-16). It extends from the **superior cervical ganglion** at the base of the skull to the **ganglion impar** at the coccyx. The cell bodies of the preganglionic fibers passing to the chain ganglia are located in the **intermediolateral cell column** of gray matter of the spinal cord from the **first thoracic to the second lumbar spinal cord segment.** The **preganglionic fibers** leave the central nervous system by way of the ventral root of a spinal nerve to pass to the ganglia by way of the **white rami communicantes.** The cell bodies of the **postganglionic neurons,** located within the chain ganglia, send their fibers by way of **gray rami communicantes** back to the spinal nerve to be distributed with its peripheral branches. Preganglionic fibers may also form **splanchnic nerves** by passing through the chain ganglia without synapsing and continue to para- or preaortic ganglia, where they synapse with the postganglionic cell bodies. In the thorax the **greater splanchnic nerve** is formed by filaments from the fifth through ninth ganglia, the **lesser splanchnic** from the tenth and eleventh, and the **least splanchnic** from the twelfth ganglion. These nerves descend anteromedially and pass through the diaphragm to terminate at **preaortic ganglia** lying adjacent to the major branches of the abdominal aorta. Preganglionic fibers may also pass up or down the chain to synapse in ganglia at higher or lower levels, where gray rami communicantes carry the postganglionic fibers to the spinal nerves.

The **phrenic nerve,** which arises from the fourth cervical spinal nerve (with additional twigs from the third and fifth cervical spinal nerves), descends obliquely across the scalenus anterior muscle, passing anterior to the first part of the subclavian artery to enter the thorax. It descends through the thorax between the mediastinal parietal pleura and the fibrous pericardium to supply the diaphragm.

After traversing the neck, the **vagus nerves** enter the thorax anterior to the sub-

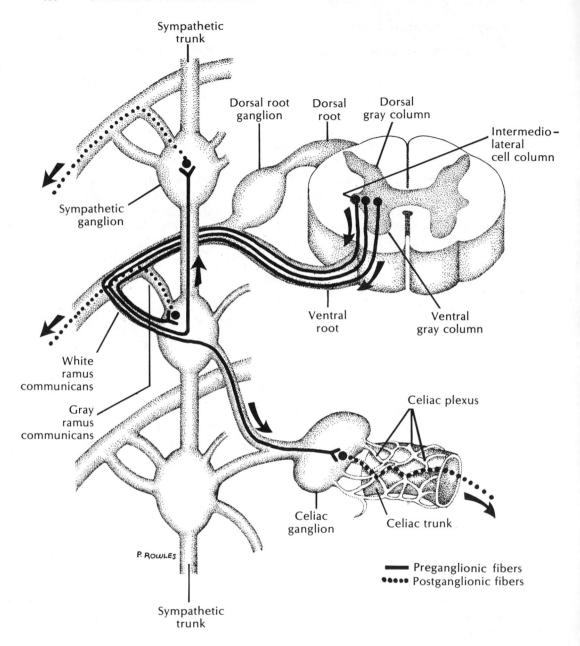

Fig. 3-16. Sympathetic trunk.

clavian arteries. The right vagus gives an **inferior laryngeal** (recurrent) branch that loops around the corresponding subclavian artery, while the inferior laryngeal nerve from the left vagus loops around the arch of the aorta. Both vagi pass posterior to the root of the lung, giving branches to the **pulmonary plexus,** and thoracic cardiac branches to the **cardiac plexus,** after which they become plexiform around the esophagus. As the **esophageal plexus,** the vagi continue through the thorax to pass through

Table 3-2. Nerve Distribution to Thorax

Nerve	Origin	Course	Distribution
Supraclaviculars	Third loop of cervical plexus	From midpoint posterior border of sternocleidomastoid pierces platysma to course in superficial fascia	Skin over shoulder, lower neck, and first two intercostal spaces
Phrenic	Third loop of cervical plexus	Passes on superficial aspect of scalenus anterior to thoracic inlet; in thorax, lies between mediastinal pleura and pericardium	Central portion of diaphragm
Intercostals	Ventral rami of thoracic nerves; first intercostal twig from T_1 as it passes to brachial plexus; ventral ramus T_{12} is subcostal	In intercostal spaces between second and third layer of muscles	Muscles of, and skin over, intercostal spaces; lower intercostals supply muscles and skin of anterolateral abdominal wall; dermatome T_{10} flanks umbilicus
Inferior (recurrent) laryngeal	Vagus nerve	On right, loops around subclavian, on left, around arch of aorta to ascend in tracheoesophageal groove	Intrinsic muscles of larynx (except cricothyroid); sensory below level of vocal folds
Cardiac plexus	Cervical and cardiac branches of vagus and sympathetic trunk	From concavity of arch of aorta and posterior surface of heart, fibers extend along coronary arteries and toward sinoatrial node	Impulses pass to S–A node; parasympathetics—slow rate, reduce force of heart beat, constrict coronary arteries; sympathetics, opposite action
Pulmonary plexus	Vagus and sympathetic trunk	Plexus forms on root of lung and extends along bronchial subdivisions	Parasympathetics constrict bronchioles; sympathetics opposite effect
Esophageal plexus	Vagus	Distal to tracheal bifurcation vagi become plexiform around esophagus in abdomen reform as anterior and posterior vagal (gastric) trunks	Digestive tract—parasympathetic innervation (in general) increases peristalsis and secretions, and relaxes sphincters
Splanchnic nerves	Sympathetic trunk—greater (T_{5-9}), lesser (T_{10-11}), least (T_{12})	Penetrate thoracic diaphragm to terminate in ganglia adjacent to major branches of abdominal aorta	Digestive tract—sympathetic innervation (in general) decreases peristalsis, closes sphincters, and causes vasoconstriction

the esophageal hiatus of the diaphragm, where they are reconstituted from their plexiform arrangement. The left vagus becomes the **anterior,** and the right the **posterior** trunks as they extend into the abdominal cavity (Table 3-2).

Esophagus

The **esophagus** originates at the level of the cricoid cartilage or sixth cervical vertebra as a continuation of the pharynx. It descends through the thorax, posterior to

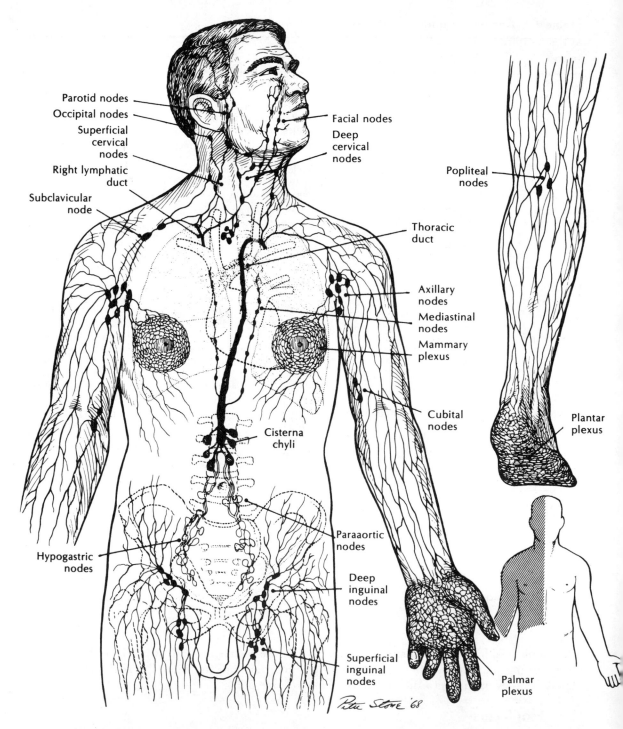

Parotid nodes
Occipital nodes
Superficial cervical nodes
Right lymphatic duct
Subclavicular node
Facial nodes
Deep cervical nodes
Thoracic duct
Axillary nodes
Mediastinal nodes
Mammary plexus
Cisterna chyli
Cubital nodes
Hypogastric nodes
Paraaortic nodes
Deep inguinal nodes
Superficial inguinal nodes
Palmar plexus
Popliteal nodes
Plantar plexus

Peter Stone '68

Fig. 3-17. Thoracic duct.

the trachea and anterior to the vertebral bodies, to pass through the **esophageal hiatus** of the diaphragm at the level of the tenth or eleventh thoracic vertebra.

ESOPHAGEAL CONSTRICTIONS

The esophagus has three distinct areas of normal narrowing. They are located 1) in the neck, at its junction with the pharynx, 2) in the mediastinum where the left main bronchus crosses the anterior surface of the esophagus, and 3) in the diaphragm at the esophageal hiatus. These three areas have clinical importance because disorders of the esophagus, like 1) swallowed foreign bodies, 2) strictures following drinking of caustic substances, 3) sites of perforations, and 4) cancers often occur at these normal constrictions.

Thoracic Duct

The **thoracic duct** begins in the abdomen as a dilatation, the **cisterna chyli,** which receives lymph drainage from the abdomen, pelvis, and inferior extremities (Fig. 3-17). Traversing the **aortic hiatus** of the diaphragm, it ascends in the thorax between the aorta and the azygos vein. In the neck it arches above the subclavian artery to empty into the junction of the subclavian and internal jugular veins. In addition to draining the lower half of the body, the thoracic duct receives the lymphatic drainage from the left side of the thorax, head, and left upper extremity. The **right lymphatic duct,** a short, 1 to 2 cm long vessel, is formed by the junction of the right jugular, subclavian, and mediastinal lymphatic trunks and empties at the origin of the right brachiocephalic vein. It drains the right side of the head and thorax, and the right upper extremity.

Thymus

The **thymus gland** is a prominent organ in the infant and reaches its greatest relative size at 2 years and greatest absolute size at puberty. It occupies the anterior mediastinum and the anterior part of the superior mediastinum. At puberty it begins to involute and is largely replaced by adipose tissue. It is a lymphoid organ whose principal function is lymphocyte and antibody production.

4

abdomen

Abdominal Wall

Surface Anatomy

The **linea alba** is a linear depression in the median plane extending from the xiphoid process to the pubic symphysis. It is formed by the fused insertions of the aponeuroses of the anterolateral muscles of the abdominal wall. The **umbilicus** (navel) is interposed in the linea alba and results from the closure of the umbilical cord shortly after birth, leaving a puckered, yet depressed scar. The **linea semilunaris** indicates the lateral extent of the rectus abdominis muscle and its sheath. The point where this line meets the right ninth costal cartilage indicates the position of the gallbladder. Transverse bands of connective tissue, the **tendinous inscriptions,** are interposed segmentally within the rectus abdominis and are visible as transverse lines in a muscular individual. The **xiphoid process** can be palpated in the midline at the thoracic outlet. The **crests of the ilia,** the **superior** and **inferior anterior iliac spines,** and the **pubic tubercle** are all palpable at the inferior extent of the abdominal wall. **Pubic hair** distribution differs in the two sexes. In the male it is dispersed in a diamond-shaped area and extends from the pubis to the umbilicus, while in the female it is triangular, with the base of the triangle above the mons pubis.

Muscles and Rectus Sheath

The anterolateral abdominal wall between the thoracic outlet and the innominate bone is composed of skin, superficial fascia, the three abdominal muscles (**external abdominal oblique, internal abdominal oblique,** and **transversus abdominis**) and their fascial envelopes. Anteriorly the **rectus abdominis** muscle and its sheath contribute to the abdominal wall, while posteriorly the muscular components of the wall are the **quadratus lumborum** and the **psoas major** and **minor** muscles (Table 4-1).

The muscles of the anterolateral wall are disposed in three layers. The fibers of the outermost external oblique pass inferomedially, those of the intermediate internal oblique, inferolaterally, and those of the innermost transversus abdominis, transversely. Their insertions by broad aponeuroses into the midline envelop the rectus abdominis and form the rectus sheath (Fig. 4-1). Nerves and vessels to the anterolateral wall course between the internal oblique and transversus abdominis muscles. The posterior abdominal wall muscles, namely, the quadratus lumborum and the psoas major and minor, are the deepest lying structures in the small of the back and extend from the twelfth rib to the iliac crest, with the psoas muscles continuing inferiorly into the iliac fossa.

The **rectus sheath** extends from the xiphoid process and adjacent costal cartilages to the pubic bone (Fig. 4-2). It holds the rectus abdominis muscle in place but does not restrict its movement owing to the presence of anterior and posterior fascial clefts. The **anterior wall** of the rectus sheath covers the muscle from end to end, but the **posterior wall** is incomplete superiorly and inferiorly. In the uppermost part of the abdomen the anterior wall of the sheath is formed by the aponeurosis of the external oblique; the

Table 4-1. Muscles of the Abdominal Wall

Muscle	Origin	Insertion	Action	Nerve
External abdominal oblique	External surface of lower eight ribs	Anterior half of iliac crest and linea alba	Compresses and supports abdominal viscera; rotates and flexes vertebral column	Lower five intercostals and subcostal
Internal abdominal oblique	Lateral two-thirds of inguinal ligament, iliac crest, and lumbodorsal fascia	Cartilages of lower three or four ribs, linea alba, and by conjoined tendon into pubis	Compresses and supports abdominal viscera; flexes and rotates vertebral column	Lower five intercostals, subcostal, and first lumbar
Transversus abdominis	Lateral one-third of inguinal ligament, iliac crest, lumbodorsal fascia, and cartilages of lower six ribs	Linea alba and by conjoined tendon into pubis	Compresses and supports abdominal viscera	Lower five intercostals, subcostal, and first lumbar
Rectus abdominis	Xiphoid process and fifth to seventh costal cartilages	Crest and symphysis of pubis	Tenses abdominal wall and flexes vertebral column	Lower five intercostals and subcostal
Pyramidalis	Front of pubis and anterior pubic ligament	Linea alba midway between umbilicus and pubis	Tenses linea alba	Subcostal
Quadratus lumborum	Lumbar vertebrae, lumbodorsal fascia, and iliac crest	Lower border of twelfth rib and transverse processes of upper four lumbar vertebrae	Draws last rib toward pelvis and flexes vertebral column laterally	Subcostal and first three or four lumbar
Psoas major	Transverse processes, intervertebral discs and bodies of all lumbar vertebrae	Lesser trochanter of femur	Flexes and medially rotates thigh; flexes vertebral column	Second and third lumbar
Psoas minor	Bodies and intervening discs of twelfth thoracic and first lumbar vertebrae	Pectineal line and iliopectineal eminence of pelvis	Flexes vertebral column	First lumbar

posterior wall in this area is absent. From the lower margin of the thoracic outlet to the midpoint between the umbilicus and pubis, the aponeurosis of the internal oblique splits into two laminae. In this area the anterior wall of the sheath is formed by the aponeurosis of the external oblique and the superficial lamina of the aponeurosis of the internal oblique, while the posterior wall is composed of the deep lamina of the aponeurosis of the internal oblique and the aponeurosis of the transversus abdominis. Inferiorly the anterior wall is formed by the aponeuroses of all three abdominal muscles, whereas posteriorly the rectus abdominis rests directly on the **endoabdominal fascia,** which lines the entire abdominal cavity. The **arcuate line** (semicircular line of Douglas) is the lower edge of the aponeurotic components of the posterior wall of the rectus sheath. **Contents of the sheath** include the rectus abdominis and pyramidalis muscles,

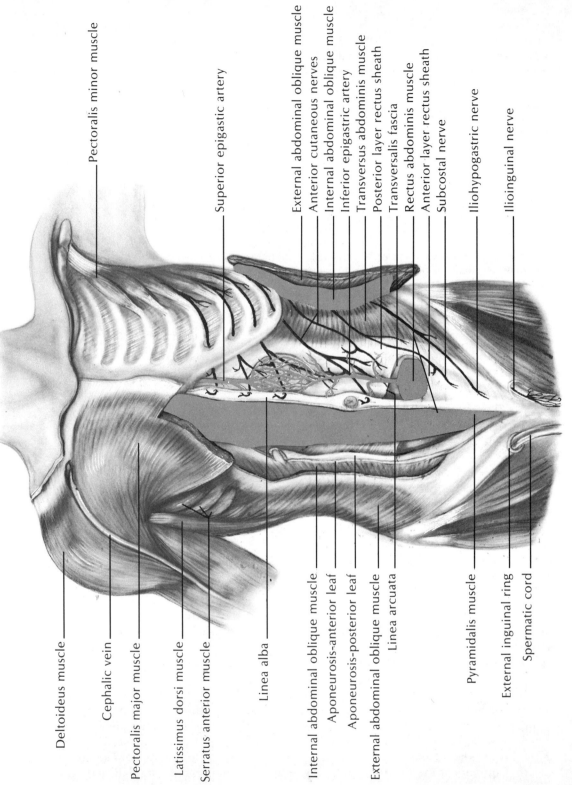

Pectoralis minor muscle

Superior epigastric artery

External abdominal oblique muscle
Anterior cutaneous nerves
Internal abdominal oblique muscle
Inferior epigastric artery
Transversus abdominis muscle
Posterior layer rectus sheath
Transversalis fascia
Rectus abdominis muscle
Anterior layer rectus sheath
Subcostal nerve
Iliohypogastric nerve
Ilioinguinal nerve

Deltoideus muscle

Cephalic vein

Pectoralis major muscle

Latissimus dorsi muscle

Serratus anterior muscle

Linea alba

Internal abdominal oblique muscle
Aponeurosis-anterior leaf
Aponeurosis-posterior leaf
External abdominal oblique muscle
Linea arcuata

Pyramidalis muscle

External inguinal ring

Spermatic cord

Fig. 4-1. Abdominal wall musculature and rectus sheath. (After Healey, JE Jr., Seybold

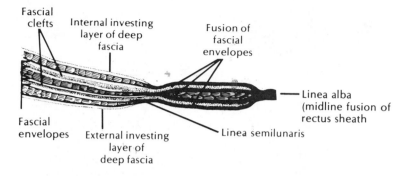

Fig. 4-2. The rectus sheath.

terminations of the lower five intercostal and subcostal nerves, and the inferior and superior epigastric arteries and their venae comitantes.

Fasciae

The superficial fascia of the abdominal wall is divided into a **superficial (Camper's) layer,** with abundant adipose tissue, and a **deep membranous (Scarpa's) layer.** In the male, Scarpa's fascia extends over the pubic symphysis and thickens to form the **fundiform ligament** of the penis, which extends inferiorly to attach to the dorsum and sides of this structure. Camper's fascia extends over the inguinal ligament as the **superficial fascia of the thigh.** Over the scrotum the superficial layer loses its fat and fuses with the membranous layer to become the **dartos tunic** of the scrotum. This same combined layer, devoid of fat, elongates to ensheathe the penis. Inferiorly Scarpa's fascia passes over the inguinal ligament and is continuous with the **fascia cribrosa** in the thigh, while over the perineum it becomes the **superficial perineal (Colles')** **fascia.**

Inguinal Region

Inferiorly the abdominal muscles contribute to the formation of the inguinal ligament and inguinal canal (Fig. 4-3 and Fig. 4-4). The **inguinal (Poupart's) ligament** is formed by the inferior free border of the aponeurosis of the external abdominal oblique muscle. The inguinal ligament attaches laterally to the **anterior superior iliac spine** and medially to the **pubic tubercle** and adjacent area. As it attaches medially, it splits to form a triangular gap with a **superior** and an **inferior crus.** Intercrural fibers transform this triangular gap into the **superficial inguinal ring.** Extensions from the deep surface of the inguinal ligament pass to the superior pubic ramus to gain additional attachment and separate structures passing from the abdomen into the thigh.

The **lacunar (Gimbernat's) ligament** is a portion of the medial end of the inferior crus that rolls under the spermatic cord to attach to the **pectineal line** on the superior pubic ramus lateral to the pubic tubercle. The lacunar ligament is triangular in form and has a lateral crescentic base and an apex directed medially toward the pubic tubercle. The **pectineal ligament (of Cooper)** is a lateral prolongation from the lacunar ligament. It forms a strong narrow band that attaches along the **iliopectineal line.** A

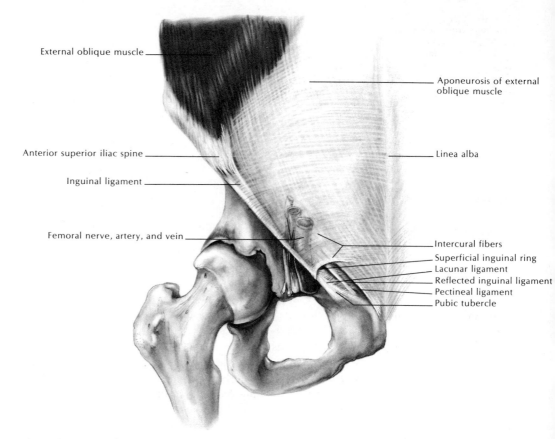

External oblique muscle

Aponeurosis of external oblique muscle

Anterior superior iliac spine

Linea alba

Inguinal ligament

Femoral nerve, artery, and vein

Intercural fibers
Superficial inguinal ring
Lacunar ligament
Reflected inguinal ligament
Pectineal ligament
Pubic tubercle

Fig. 4-3. Inguinal region.

tendinous medial expansion, continuous with the lacunar ligament, sweeps beneath the superficial inguinal ring to the linea alba as the **reflected** (reflex) **inguinal ligament.**

The **superficial inguinal ring,** easily palpable in the male living subject, allows passage of the spermatic cord in the male (round ligament of the uterus in the female) from the inguinal canal to the scrotum (labium majus). The **deep inguinal ring** is an opening in the endoabdominal fascia. It lies 1 to 2 cm above the midinguinal point and is immediately lateral to the inferior epigastric vessels. The **inguinal canal** is a narrow channel about 3 to 4 cm in length, passing between the deep and superficial rings. It is formed when the **processus vaginalis** evaginates the abdominal wall into the scrotum prior to the descent of the testis. The inguinal canal, directed inferiorly, medially, and anteriorly, has a floor, roof, and anterior and posterior walls. In the male it transmits the spermatic cord, cremasteric vessels, genital branch of the genitofemoral nerve, and cutaneous branches of the ilioinguinal nerve. In the female the round ligament of the uterus, with its vessels and nerves, transverse the canal.

The grooved surface of the inguinal ligament, reinforced laterally by the pectineal ligament and medially by the lacunar ligament, forms the **floor** of the inguinal canal. The **anterior wall** is formed by the aponeurosis of the external abdominal

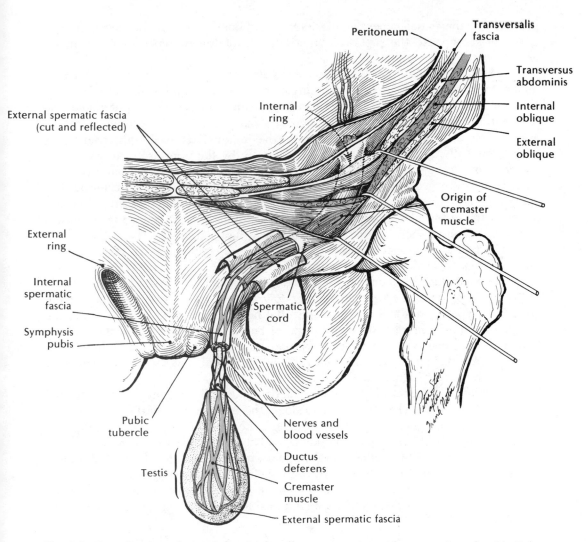

Fig. 4-4. Contributions from abdominal wall to spermatic cord covers. (Langley LL, Telford IR, Christensen JB: Dynamic Anatomy and Physiology, 5th ed. New York, McGraw-Hill, 1980)

oblique muscle. In its lateral half it is supplemented by the muscle fibers of the internal abdominal oblique muscle. The canal is **roofed** by the arching fibers of the internal abdominal oblique and its medial continuation, the cremaster muscle. The endoabdominal fascia covers the entire **posterior wall.** It is reinforced medially by the fusion of the aponeuroses of the internal oblique with the transversus abdominis muscle to form the **conjoint tendon** (falx inguinalis).

The wall of the inguinal canal and the superficial and deep rings are potential areas of weakness in the abdominal wall through which inguinal herniae pass. The anterior wall of the canal is strongest opposite the deep inguinal ring because of the presence of the muscle fibers of the internal oblique; similarly, the posterior wall is strongest opposite the superficial ring because of the presence of the conjoint tendon.

A **hernia** is an abnormal protrusion of any structure beyond its normal site. In the inguinal region herniae are classified as direct or indirect, depending upon their course through the abdominal wall. A **direct inguinal hernia** bypasses the deep inguinal ring to penetrate the posterior wall of the inguinal canal medial to the inferior epigastric artery. This area of direct inguinal hernia, the **inguinal (Hesselbach's) triangle,** is formed medially by the rectus abdominis, laterally by the inferior epigastric artery, and inferiorly by the inguinal ligament. The direct inguinal hernia occurs more frequently in elderly people and is usually limited to the region of the superficial ring. The wall of the hernial sac of a direct hernia is composed, from within outward, of peritoneum, extraperitoneal connective tissue, endoabdominal fascia, cremasteric muscle and fascia, external spermatic fascia, subcutaneous fascia, and skin.

The **indirect inguinal hernia** is three times more common in the male than the direct hernia. It traverses the deep inguinal ring (lateral to the inferior epigastric artery) and passes through the inguinal canal to exit through the superficial ring into the scrotum. Thus, the indirect inguinal hernia follows the same course as the spermatic cord. This type is also known as the **congenital hernia,** since failure of closure of the processus vaginalis predisposes to this type of hernia. The coverings of the indirect hernia are the same as the coverings of the spermatic cord. The position of the inferior epigastric artery, immediately medial to the deep inguinal ring, aids in distinguishing direct from indirect inguinal herniae.

The internal aspect of the anterior abdominal wall presents five longitudinal ridges of peritoneum. The single median ridge, the **median umbilical fold,** encloses a slender fibrous cord (median umbilical ligament), the remnant of the **urachus.** It passes from the apex of the bladder to the umbilicus between the peritoneum and the endoabdominal fascia. Two **medial umbilical folds,** representing the **obliterated umbilical arteries** (lateral umbilical ligaments) continue from the superior vesicular branch of the internal iliac artery to pass to the umbilicus. The **inferior epigastric arteries** and **veins** raise ridges of peritoneum (**lateral umbilical folds**) as they pass toward the rectus abdominis.

Arteries and Nerves

The **inferior epigastric artery,** a branch of the external iliac, pierces the endoabdominal fascia to ascend behind the rectus abdominis, but within the rectus sheath, to anastomose with the **superior epigastric artery** from the internal thoracic. Branches of the superior and inferior epigastric arteries include muscular branches to the rectus muscle and cutaneous branches to overlying skin. Additional branches of the inferior epigastric artery include the **cremasteric** supplying the cremaster muscle, which passes through the deep inguinal ring into the inguinal canal, and **pubic branches** ramifying on the pubic bone. The **deep iliac circumflex artery,** the second branch of the external iliac, passes laterally to ramify on and supply the iliacus muscle in the iliac fossa.

Innervation of muscles and skin of the anterolateral abdominal wall is derived from the ventral rami of spinal nerves from the seventh thoracic to the first lumbar, as thoracoabdominal, subcostal, iliohypogastric, and ilioinguinal nerves. The **thoracoabdominal** (lower intercostals) and **subcostal nerves** pass deep to costal cartilages and descend medially between the transversus abdominis and the internal abdominal oblique muscles. They supply abdominal muscles and terminate as lateral and anterior cutaneous branches in line with the more superior intercostal nerves.

The **iliohypogastric nerve** (T_{12} and L_1) emerges at the upper lateral border of the psoas major, crosses the quadratus lumborum to the crest of the ilium, where it pierces the transversus abdominis muscle. It gives **lateral cutaneous branches,** which supply the skin immediately above the iliac crest and the gluteal region, and **anterior cutaneous branches,** which pierce the aponeurosis of the external abdominal oblique 2 to 3 cm above the subcutaneous inguinal ring to innervate the skin of the hypogastric region. The **ilioinguinal nerve** (L_1) follows a course parallel but inferior to the iliohypogastric nerve, piercing the internal abdominal oblique and accompanying the spermatic cord through the superficial inguinal ring. It supplies skin over the upper medial part of the thigh, mons pubis or root of the penis, and scrotum or labium majus.

Lumbar Plexus

The **lumbar plexus** is formed by the ventral rami of the first four lumbar spinal nerves with a contribution from the twelfth thoracic nerve (Fig. 4-5). Part of the fourth lumbar nerve joins with the fifth lumbar nerve to form the lumbosacral trunk, which contributes to the formation of the sacral plexus. The lumbar plexus differs from the brachial plexus in that no intricate interlacing of fibers occurs. Leaving intervertebral foramina, the nerves pass obliquely outward behind the psoas major and anterior to the quadratus lumborum muscles to form their several branches.

The **iliohypogastric** and **ilioinguinal nerves** described above are the first two branches of the lumbar plexus. **Muscular branches** arising segmentally as independent twigs pass from all four lumbar spinal nerves to the quadratus lumborum, and from the second and third lumbar nerves to the psoas major and minor muscles. The **genitofemoral nerve** (L_1 and L_2) passes through the substance of the psoas major, emerges from its medial border close to the lower lumbar vertebrae to descend on the anterior surface of this muscle. It divides into two branches, a **genital** and a **femoral.** The former pierces the endoabdominal fascia to pass through the inguinal canal behind the spermatic cord to supply the cremaster muscle and the skin of the scrotum. The femoral branch descends on the external iliac artery to enter the femoral sheath. The **lateral femoral cutaneous nerve** (L_2 and L_3) emerges at the middle of the lateral border of the psoas major to cross the iliacus muscle obliquely toward the anterior superior iliac spine. It courses deep to the inguinal ligament to be distributed to the skin of the lateral aspect of the thigh. The **obturator nerve** (L_2, L_3, and L_4) descends through the substance of the psoas major to emerge from the medial border of this muscle near the brim of the pelvis. It passes behind the common iliac vessels, along the lateral wall of

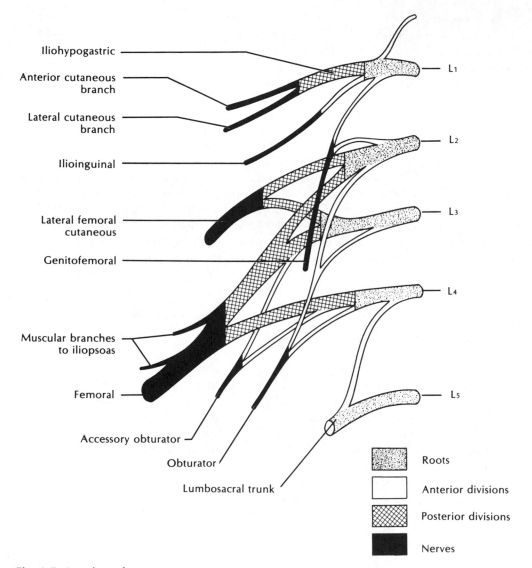

Iliohypogastric

Anterior cutaneous
branch

Lateral cutaneous
branch

Ilioinguinal

Lateral femoral
cutaneous

Genitofemoral

Muscular branches
to iliopsoas

Femoral

Accessory obturator

Obturator

Lumbosacral trunk

L_1

L_2

L_3

L_4

L_5

Roots

Anterior divisions

Posterior divisions

Nerves

Fig. 4-5. Lumbar plexus.

the pelvis superoanterior to the obturator vessels, to leave the pelvis through the obturator canal. Its distribution will be considered with the inferior extremity.

The **femoral nerve** (L_2, L_3, and L_4), the largest branch of the lumbar plexus, descends through the fibers of the psoas major to emerge from its lateral border. It gives a few filaments to the iliacus and psoas major and minor, then continues inferiorly passing deep to the inguinal ligament to enter the thigh. The ventral ramus of the fourth lumbar nerve divides into an upper and lower division. The upper division contributes to the lumbar plexus, the lower division joins the fifth lumbar to form the **lumbosacral trunk** passing to the sacral plexus.

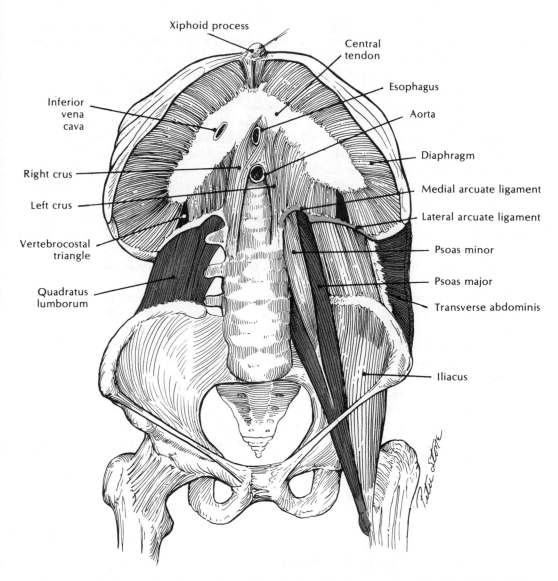

Xiphoid process

Central tendon

Esophagus

Aorta

Diaphragm

Medial arcuate ligament

Lateral arcuate ligament

Psoas minor

Psoas major

Transverse abdominis

Iliacus

Inferior vena cava

Right crus

Left crus

Vertebrocostal triangle

Quadratus lumborum

Fig. 4-6. Diaphragm and posterior abdominal wall muscles. (Langley LL, Telford IR, Christensen JB: Dynamic Anatomy and Physiology, 5th ed. New York, McGraw-Hill, 1980)

Diaphragm

The **diaphragm** is a movable musculotendinous partition between the thoracic and the abdominal cavities. It forms the concave roof of the abdominal cavity and the convex floor of the thoracic cavity (Fig. 4-6). The diaphragm rises higher on the right side than on the left due to the larger right lobe of the liver. The central portion is apo-neurotic and forms the strong **central tendon,** which is indistinctly divided into three leaflets. The muscle fibers of the diaphragm originate at the margins of the thoracic outlet and insert into the central tendon. The short and narrow **sternal portion** arises as small slips from the back of the xiphoid process; the extensive **costal portion,** from the inner surface of the lower six costal cartilages, and the **vertebral portion,** from the ar-cuate ligaments and the upper lumbar vertebrae as a pair of muscular crura. Each crus is a thick, fleshy bundle that tapers inferiorly and becomes tendinous. The **left crus** attaches to the upper two lumbar vertebrae and the intervening vertebral disc, the **right crus** to the upper three lumbar vertebrae and intervening discs. Fibers of each crus spread out and ascend to attach to the central tendon, with the right crus encircling the esophagus. The lateral, medial, and median arcuate ligaments give par-tial origin to the diaphragm. The unpaired **median arcuate ligament,** opposite the twelfth thoracic vertebra, arches between the right and left crura and crosses the aorta as it enters the abdomen. The paired **medial arcuate ligaments** are highly arched and pass laterally from the tendinous part of each crus to curve across the psoas major muscle and attach to the tip of the transverse process of the first or sec-ond lumbar vertebra. The paired **lateral arcuate ligaments** stretch across the quad-ratus lumborum from the tip of the transverse process of the first or second lumbar vertebra to the tip and lower margin of the twelfth rib. That portion of the dia-phragm attaching to the lateral arcuate ligament is thin and sometimes devoid of muscular fibers, with only connective tissue separating the pleura from the renal fat. Through this weakened area, the **vertebrocostal triangle,** abdominal contents may herniate into the thorax.

The continuity of the diaphragm is interrupted by three large and several small apertures. Between the crura, at the level of the twelfth thoracic vertebra, the **aortic hiatus** is bridged anteriorly by the median arcuate ligament. The aortic hiatus trans-mits the aorta, azygos vein, and thoracic duct. At the level of the tenth thoracic verte-bra the oval **esophageal hiatus** is situated obliquely behind the central tendon, 2 to 3 cm to the left of the midline, and is surrounded by the right crus of the diaphragm. It transmits the esophagus, vagal trunks, and esophageal branches of the left gastric ves-sels.

ESOPHAGEAL (HIATAL) HERNIA

The most common site of diaphragmatic herniae is through an en-larged esophageal hiatus. An hiatal hernia is usually acquired and occurs most often in middle-aged individuals whose principal com-plaints are "heartburn," upper abdominal fullness, and often accompa-nied by regurgitating of gastric contents into the mouth. These symp-toms are usually relieved by sitting up or elevating the head and shoulders. The acquired hiatal hernia is caused by the cardia of the stomach protruding into the mediastinum through the weakened, en-larged esophagel hiatus in the diaphragm. Since the stomach may move back and forth through the hiatus, it is also called a sliding her-nia.

At the level of the eighth thoracic vertebra the wide **caval opening** is located within the central tendon of the diaphragm about 3 cm to the right of the median plane. It transmits the inferior vena cava, terminal branches of the right phrenic nerve, and lymph vessels. Additional structures passing between the thorax and the abdomen include the superior epigastric vessels, musculophrenic vessels, the lower five thoracoabdominal nerves, and the subcostal vessels and nerves. The sympathetic trunk lies behind the medial arcuate ligament, and the three splanchnic nerves pierce the crus on their side of origin to enter the abdomen.

Innervation to the diaphragm is derived from the **phrenic nerve** and from twigs of the **lower intercostal nerves.** The **inferior phrenic artery,** a branch of the abdominal aorta, is the chief arterial supply to the diaphragm. Additional blood is supplied from the **pericardiophrenic branch** of the internal thoracic artery, from irregular twigs of the thoracic aorta, and peripherally, through branches from the **musculophrenic artery.**

Abdominal Cavity

The **abdominal cavity,** the largest cavity in the body, is bounded anteriorly by the rectus abdominis, laterally by the external, internal, and transverse abdominal, and more inferiorly by the iliacus muscles. The abdominal cavity is bonded posteriorly by the vertebral column and the psoas major and minor, and quadratus lumborum muscles. Superficially, it is subdivided, for descriptive purposes, into nine regions by two horizontal and three vertical arbitrary lines (or planes) (Fig. 4-7 A and B). The horizontal lines are the **transpyloric,** at the level of the pylorus of the stomach, and the **intertubercular,** at the level of the iliac tubercles. The vertical lines are the midline and two lateral lines that bisect the clavicles. The resulting **subdivisions of the abdomen** are from superior to inferior, **right** and **left hypochondriac** and **middle epigastric, right** and **left lumbar** and **middle umbilical,** and **right** and **left inguinal** and **middle pubic.**

Peritoneal Cavity

The **peritoneum** is a serous membrane that lines the walls of the abdominal cavity (Fig. 4-8). Developmentally, abdominal and pelvic viscera invaginate into the abdominal cavity carrying the peritoneum before them. This results in a covering over organs of **visceral peritoneum,** which is continuous with the **parietal peritoneum** lining the abdominal walls. The layers of apposing peritoneum between viscera and body wall or between two organs form the **mesenteries,** or visceral **ligaments,** of the abdominal cavity. The disappearance, fusion, shifting, shortening, or redundant growth of these peritoneal folds during development divides the peritoneal cavity into two distinct parts, the greater and lesser sac. The **lesser sac** (omental bursa) is situated posterior to the lesser omentum, stomach, and gastrocolic ligament. It is limited inferiorly by the transverse colon with its mesocolon and bounded on the left by the gastrolienal and lienorenal ligaments. To its right, the omental bursa communicates through the **epiploic foramen (of Winslow)** with the **greater sac** of the peritoneal cavity. In the male, the peritoneum forms a closed cavity, while in the female, it communicates with

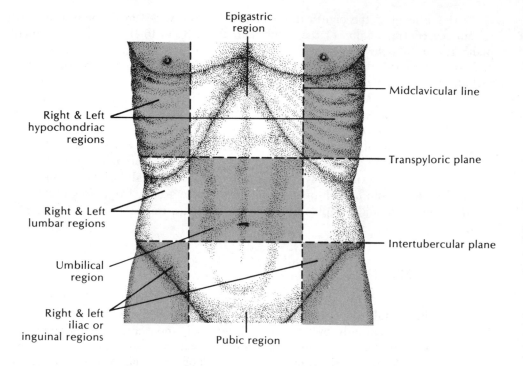

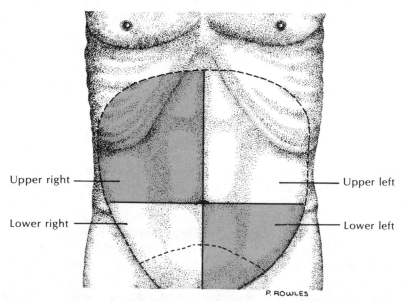

Fig. 4-7. A. Regions of the abdomen. **B.** Quadrants of the abdomen.

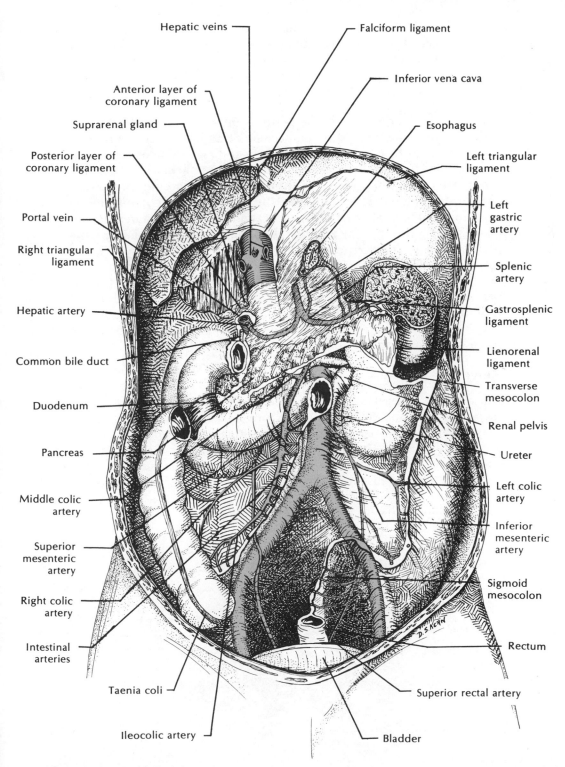

Hepatic veins

Falciform ligament

Anterior layer of
coronary ligament

Inferior vena cava

Suprarenal gland

Esophagus

Posterior layer of
coronary ligament

Left triangular
ligament

Portal vein

Left
gastric
artery

Right triangular
ligament

Splenic
artery

Hepatic artery

Gastrosplenic
ligament

Common bile duct

Lienorenal
ligament

Duodenum

Transverse
mesocolon

Renal pelvis

Pancreas

Ureter

Middle colic
artery

Left colic
artery

Superior
mesenteric
artery

Inferior
mesenteric
artery

Right colic
artery

Sigmoid
mesocolon

Intestinal
arteries

Rectum

Taenia coli

Superior rectal artery

Ileocolic artery

Bladder

Fig. 4-8. Peritoneal cavity.

the exterior through the openings of the uterine tubes. Structures within the abdominal cavity not suspended from the body wall by a mensentery or from other viscera by a visceral ligament are **retroperitoneal** in position (Fig. 4-9).

PERITONITIS

Peritonitis is inflammation of the peritoneum. It may be general or localized. It is characterized by an accumulation of a large amount of peritoneal fluid (ascites) containing fibrin and many leucocytes (pus). In the supine patient, the infected fluid tends to collect at two sites: 1) the pelvic cavity, and 2) in the right posterior subphrenic space. Tapping or draining of excess fluid from the abdomen is accomplished by inserting, under local anesthesia, a trocar and cannula or a needle and plastic tubing through the anterior abdominal wall, and aspirating the peritoneal fluid (paracentesis). When a patient with peritonitis is examined, stretching of the infected parietal peritoneum is very painful. The pain is especially severe when the digital pressure over the inflamed area is suddenly released, because this causes the abdominal wall to rebound suddenly. Clinically, this is called rebound tenderness, which localizes the site of inflammation of the parietal peritoneum and often occurs over the infected organ.

Nerves

Parasympathetic innervation of the abdominal viscera is derived from vagal and sacral nerves. The **left** and **right vagus nerves** form a plexus around the esophagus and, upon passing through the diaphragm, are reconstituted as the anterior and posterior vagal trunks, respectively. The **anterior vagal trunk** innervates the liver, gallbladder, bile ducts, pylorus of the stomach, duodenum, and pancreas. The **posterior vagal trunk** supplies the rest of the stomach and then joins the superior mesenteric plexus. Here it

Fig. 4-9. Epiploic foramen and related structures.

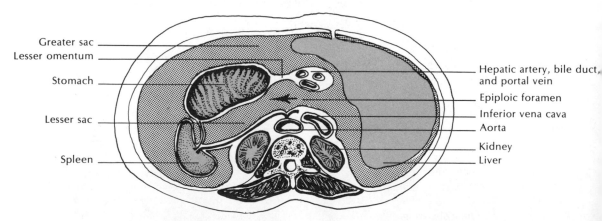

intermingles with sympathetic filaments and is distributed to the kidney, small intestine, and ascending and transverse colon. The distal portion of the large intestine beyond the splenic flexure is innervated by **sacral parasympathetic nerves.** Postganglionic parasympathetic cell bodies are usually located within the substance of the organ they supply.

The principal sources of sympathetic innervation to abdominal viscera are the **greater** (T_5 to T_9), **lesser** (T_{10} and T_{11}) and **least** (T_{12}), **splanchnic nerves.** Additional sympathetic fibers, the **lumbar splanchnic** and **sacral splanchnic nerves,** arise directly from the lumbar or sacral segments of the vertebral chain. The sympathetic nerves form plexuses around the main branches of the abdominal aorta. The large **celiac ganglia** and **plexus** lie adjacent to the celiac artery; smaller plexuses surround the phrenic, hepatic, gastric, splenic, superior renal, and inferior mesenteric, gastric, suprarenal, and gonadal arteries. Ganglia associated with these plexuses contain postganglionic sympathetic cell bodies, but fibers distributed by the plexus consist of both sympathetic and parasympathetic components.

Gastrointestinal Tract

STOMACH

The **stomach** is the first abdominal subdivision of the alimentary canal. Its size, position, and configuration are determined by its physiologic state, the impingement of other abdominal viscera, and general body build. Classically, it is described as pear-shaped, with its blunt upper end related to the left dome of the diaphragm (Fig. 4-10). The upper and lower ends of the stomach are relatively fixed. Its midportion moves as the position of the other viscera or its contents may require.

GASTRIC ULCER

In a gastric ulcer, one type of peptic ulcer, the mucosa is eroded or lost and a "craterlike" depression penetrates to various depths of the gastric wall. Although the cause of gastric ulcers is unknown, one prominent theory is the loss of the mucous protection of the mucosa and back diffusion of HCL acid. Common complications of gastric ulcers are bleeding, recurrent pain, and gastric outlet obstruction. Empirically, surgeons have discovered that removal of 50% of the stomach (hemigastrectomy) will cure most patients with gastric ulcer.

It is subdivided into the cardiac portion, fundus, body, and pyloric portion, with anterior and posterior surfaces, and greater and lesser curvatures. The esophagus joins the stomach at the **cardiac orifice,** with the limited **cardiac portion** located adjacent to this opening. The **fundus** is the full-rounded uppermost part above the level of the esophageal junction, while the main part of the stomach, the **body,** lies below the fundus. The **pyloric region** includes the dilated antrum, the pyloric canal, and the py-

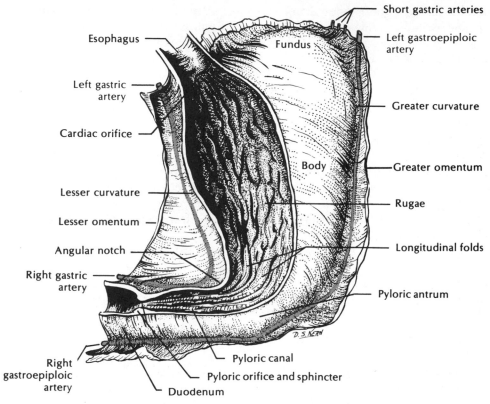

Esophagus

Left gastric artery

Cardiac orifice

Lesser curvature

Lesser omentum

Angular notch

Right gastric artery

Right gastroepiploic artery

Fundus

Short gastric arteries

Left gastroepiploic artery

Greater curvature

Body

Greater omentum

Rugae

Longitudinal folds

Pyloric antrum

D.S.KERN

Pyloric canal

Pyloric orifice and sphincter

Duodenum

Fig. 4-10. Stomach.

loric orifice, which is surrounded by a thickened muscular coat, the **pyloric sphincter.** The anterior surface of the stomach is closely related to the diaphragm, the left lobe of the liver, and the left rectus abdominis muscle. The posterior surface is related to many structures, which collectively make up the "bed of the stomach." These include the body of the pancreas, splenic artery, medial border of the left kidney, left suprarenal gland, spleen, diaphragm, transverse colon, and the transverse mesocolon.

PYLORIC STENOSIS

Pyloric stenosis is due either to true hypertrophy, or a continuous spasm of the pyloric sphincter of the stomach. It appears in infants usually between the second and twelfth weeks of life and is much more frequent in the male than female. Clinical signs include projectile vomiting of food, no substantial bowel movements, and a loss of weight. Such a condition is life-threatening and requires surgical relief by longitudinal division of the pyloric sphincter, which allows the stomach to empty food easily into the duodenum.

The **lesser curvature** of the stomach is concave and affords attachment for the **lesser omentum** (gastrohepatic and hepatoduodenal ligaments), a double layer of peritoneum extending between the stomach and the liver. At the lesser curvature the peritoneum separates to cover the anterior and posterior surfaces of the stomach. From the convex **greater curvature** of the stomach the serosal coverings fuse to form the **greater omentum,** which extends inferiorly in an apronlike fashion to cover most of the abdominal contents. Inferiorly it reflects back on itself and ascends to pass over the transverse colon and attach to the posterior body wall. That portion of the greater omentum between the transverse colon and the greater curvature of the stomach is the **gastrocolic ligament.** Between the transverse colon and the body wall, the greater omentum fuses with the transverse mesocolon. Below the transverse colon its layers (two descending and two ascending) fuse to form a four-layered segment of the greater omentum.

Blood Vessels and Nerves

The stomach receives a rich blood supply from all branches of the **celiac artery** (Fig. 4-11). This vessel originates from the anterior aspect of the uppermost part of the abdominal aorta as a short trunk and gives rise to splenic, common hepatic, and left

Fig. 4-11. Celiac trunk.

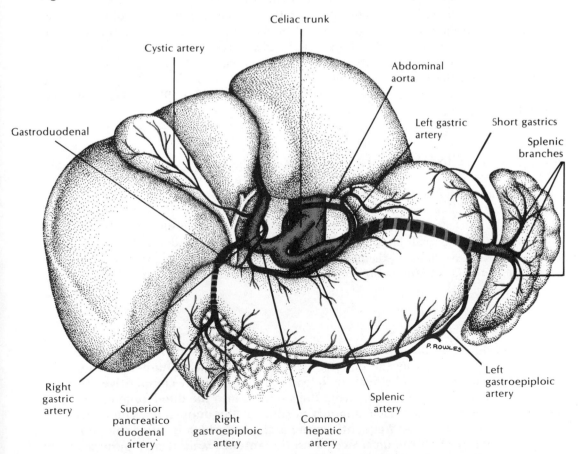

gastric arteries. The **left gastric,** the smallest of the three branches, ascends retroperitoneally behind the lesser sac toward the esophagus to pass forward and then descend between the layers of the lesser omentum. It parallels and supplies the lesser curvature of the stomach, gives branches to the lower esophagus, and terminates by anastomosing with the **right gastric artery.** The latter vessel arises as a terminal branch of the hepatic artery. It enters the lesser omentum from its retroperitoneal position and passes to the left, between the layers of the lesser omentum along the lesser curvature, to supply the stomach. The splenic artery sends **short gastric branches** and the **left gastroepiploic branch** to the stomach. The former pass through the gastrolienal ligament to the upper part of the greater curvature of the stomach. The **left gastroepiploic,** a long branch passing first through the gastrolienal ligament and continuing between the layers of the greater omentum, supplies the greater curvature of the stomach and the greater omentum before anastomosing with the **right gastroepiploic artery.** The latter vessel is a branch of the gastroduodenal from the hepatic artery. It passes between the layers of the greater omentum to supply the left portion of the greater curvature of the stomach. Veins draining the stomach correspond in position to the arteries and are named similarly, except for the **pyloric vein,** which is the companion of the right gastric artery, and the **coronary vein,** the companion of the left gastric artery.

As the esophagus enters the abdomen the plexiform arrangement of the left and right vagi around the esophagus is reconstituted as the anterior and posterior vagal trunks. They are distributed with the celiac and superior mesenteric plexuses and provide parasympathetic innervation to the abdominal viscera. Parasympathetic innervation through the vagal components extends distally to the level of the descending colon. Sympathetic innervation is also distributed through the celiac and superior mesenteric plexuses.

SMALL INTESTINE

The **small intestine,** a convoluted tube extending from the pylorus to the ileocecal valve, is subdivided into the duodenum, the jejunum, and the ileum. It is situated centrally in the abdominal cavity and is flanked laterally and superiorly by the large intestine. As the small intestine proceeds distally it gradually diminishes in diameter. Anteriorly it is related to the greater omentum and the anterior abdominal wall, and posteriorly to the posterior abdominal wall, the pancreas, the kidney, and occasionally the rectum.

DUODENAL ULCER

A duodenal ulcer, another type of peptic ulcer, results from excessive HCL secretion by the parietal cells of the stomach. Duodenal ulcers often occur in individuals who lead a stressful, highly competitive life. The acid chyme pouring from the stomach into the duodenum causes the ulcer in the first (superior) portion of the duodenum. Duodenal ulcers occur most frequently in the first 2 cm, which is called the duodenal bulb. If the ulcer occurs on the anterior wall, it may perforate into the peritoneal cavity. If it occurs on the posterior wall, it may erode the head of the pancreas or the gastroduodenal artery, and cause massive hemorrhage. Patients with duodenal ulcers who have such serious complications are usually treated surgically. The bleeding vessel is ligated; the perforation is closed; and the duodenal ulcer is most commonly treated with a vagotomy and pyloroplasty.

The **duodenum,** the first segment (about 25 cm), has the widest lumen and the thickest wall of any region of the small intestine (Fig. 4-12). It follows a C-shaped course and is divided into four portions. The first (superior) portion, arising from the pylorus of the stomach, passes posteriorly and superiorly, and at the neck of the gallbladder makes a sharp inferior bend to become the second (descending) part. It is connected by the **hepatoduodenal ligament** to the porta hepatis. The bile and the pancreatic ducts penetrate the posteromedial surface of the second part to form the **hepatopancreatic ampulla (of Vater).**

The second segment passes inferiorly and opposite the third and fourth lumbar vertebrae bends to the left to become the third (horizontal) part.

Fig. 4-12. Duodenum, pancreas, and gallbladder.

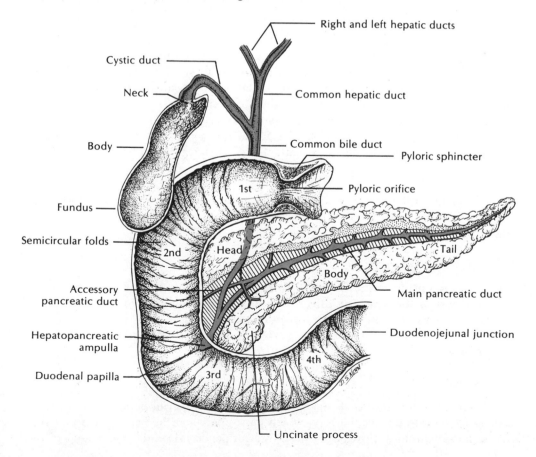

The third portion passes horizontally across the vertebral column, crura of the diaphragm, inferior vena cava, and aorta to ascend slightly and become the fourth (ascending) part. The fourth and terminal part of the duodenum passes upward to the left side of the aorta and head of the pancreas to bend sharply at the duodenojejunal flexure, where it becomes the jejunum. Except for about 2 to 3 cm at its origin and termination, the duodenum is entirely retroperitoneal. A fibromuscular band, the **ligament of Trietz,** attaches the small intestine to the posterior abdominal wall at the duodenojejunal flexture. The internal structure of the duodenum presents the **duodenal papilla** at the entrances of the common bile duct and the pancreatic duct.

MECKEL'S DIVERTICULUM

Meckel's diverticulum is the most common anomaly of the small intestine. It is the result of failure of the yolk sac duct to atrophy. If present, it consists of a blind pouch of varying lengths and extends from the ileum, about a meter proximal to the ileocecal valve. It may end freely or be attached to the anterior abdominal wall. When inflamed, the diverticulum may produce symptoms similar to appendicitis. Since it is prone to infections, it is generally removed incidentally if discovered during an abdominal operation. It is more common in men than in women.

The junctional area between the jejunum and the ileum has no gross morphologic points of distinction. The gross division is arbitrary, however, histologic characteristics change so that sections taken from the two areas can be easily distinguished. The **jejunum** is wider and has thicker walls than the ileum and prominent circular folds (plicac circulares). The **ileum** has fewer circular folds, has aggregates of lymph nodules (Peyer's patches), and terminates by joining the medial aspect of the cecum at the **ileocecal valve.** The jejunum and ileum are attached to the posterior abdominal wall by an extensive fold of peritoneum, the fan-shaped **mesentery,** which projects 20 to 25 cm into the abdominal cavity. The **root of the mesentery,** 15 to 18 cm long, is attached obliquely along the posterior abdominal wall and crosses successively the horizontal part of the duodenum, the aorta, the inferior vena cava, the right ureter, and the right psoas major and minor muscles.

INTUSSUSCEPTION

Intussusception is a term that denotes the invagination or telescoping of a part of the intestine into the lumen of another part of the intestine. It is a common cause of bowel obstruction in infants. It most frequently involves the prolapse of the ileum into the lumen of the cecum.

Arteries

The duodenum receives its blood supply from the **superior pancreaticoduodenal branch** of the gastroduodenal artery, which passes between the head of the pancreas and the duodenum to its distribution. The **inferior pancreaticoduodenal branch** of the superior mesenteric artery follows a similar course to give twigs to the duodenum. Twelve to fifteen **intestinal** branches arise from the superior mesenteric artery. They lie parallel to each other between the layers of the mesentery. Distally they form a series of two to five loops or arcades from which small branches arise that encircle and supply the intestine. The latter are referred to as the **arteria recti** (straight arteries) of the intestine.

VOLVULUS

Volvulus is a twisting of the intestine either around its mesenteric axis, an adjacent intestinal coil, or a postoperative adhesive band. The sigmoid colon and the cecum are predisposed to this condition, especially if their mesenteries are unusually long, which affords these gut segments greater mobility. The twisting may return to normal or reduce spontaneously, but if it persists, the blood supply to the bowel is impaired and gangrene may set in. If volvulus blocks the superior mesenteric artery virtually all of the supply to the small intestine and half of the large intestine is lost.

LARGE INTESTINE

The **large intestine** extends from the ileum to the anus (Fig. 4-13). It is about 1.5 meters in length and diminishes in diameter from its origin toward its termination. It differs from the small intestine in its greater luminal size, more fixed position, the presence of **sacculations,** or haustra, the presence of fat tabs on its external coat, the **appendices epiploicae,** and the disposition of its external longitudinal muscular coat into three longitudinal bands, the **taenia coli.**

The **cecum** is a large blind pouch 5 to 8 cm long at the beginning of the large intestine. The ileum opens into it medially through a longitudinal slit, the **ileocecal orifice,** guarded by the **ileocecal valve.** Below the orifice, the vermiform appendix opens into the cecum, which usually lies in the right iliac fossa immediately above the left half of the inguinal ligament.

The **vermiform appendix,** a blind tube about 5 mm thick and 10 cm long, is suspended by a mesentery, the mesoappendix, which is continuous with the mesentery of the small intestine. The appendix has no fixed position, but commonly moves with the cecum and is most usually inferior and posterior to it.

APPENDICITIS

The vermiform appendix is quite mobile and its location variable. It most frequently lies either lateral or anterior to the cecum. Other positions include: medially towards the ileum, inferiorly towards the middle of the inguinal ligament, or posteriorly behind the cecum (retrocecal). The surgeon must know these various positions as he searches for the appendix through a small muscle splitting incision. Since an inflamed appendix may perforate, and spew pathogenic organisms into

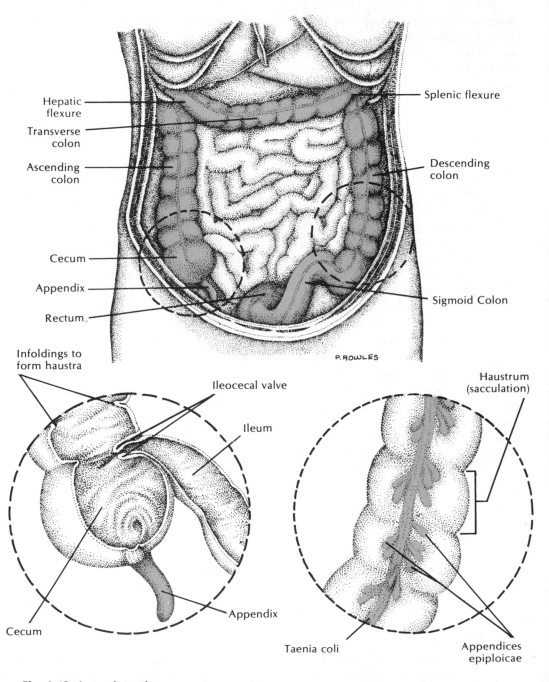

Hepatic flexure

Splenic flexure

Transverse colon

Ascending colon

Descending colon

Cecum

Appendix

Sigmoid Colon

Rectum

P. ROWLES

Infoldings to form haustra

Ileocecal valve

Haustrum (sacculation)

Ileum

Cecum

Appendix

Taenia coli

Appendices epiploicae

Fig. 4-13. Large intestine.

the peritoneal cavity that may cause generalized peritonitis and death, it must be located and removed to prevent the progression of these infectious complications.

The **ascending colon** passes superiorly on the posterior abdominal wall from the right iliac fossa to become the transverse colon as it bends to the left at the **hepatic flexure.** It is normally retroperitoneal, but it may possess a limited mesentery. The **transverse colon** extends to the left from the hepatic flexure. It passes transversely across the superior region of the abdominal cavity to the spleen, where it makes a sharp inferior bend called the **splenic flexure** to become the descending colon. It is always suspended from the posterior abdominal wall by the **transverse mesocolon,** which varies in length and thereby permits the location of the transverse colon to vary in position from just inferior to the liver, to the level of the iliac fossae. From the splenic flexure the **descending colon** passes inferiorly along the lateral border of the left kidney in the angle between the psoas major muscle and the quadratus lumborum, to become the **sigmoid colon** at the level of the iliac crest. This latter portion, usually situated within the pelvis, may be displaced into the abdominal cavity due to its long mesentery. Being fixed in position only at its junctions with the descending colon and the rectum, the sigmoid colon curves on itself, resulting in an S-shaped configuration from which it receives its name. From the sigmoid colon, the **rectum** continues inferiorly to follow the sacral curvature to the pelvic diaphragm, where it makes a ninety-degree bend to become the **anal canal.** Externally the rectum has no sacculations. The taenia coli spread out to form a uniform outer muscular coat, and the peritoneum covers only its anterior and lateral aspects. From the rectum, the peritoneum reflects onto the bladder in the male to form the **rectovesical pouch,** and onto the uterus in the female as the **rectouterine pouch.**

MEGACOLON OR HIRSCHSPRUNG'S DISEASE

Megacolon, or Hirschsprung's disease, is a congenital condition. It is usually diagnosed in infants and young children. In this condition the large intestine, usually the sigmoid colon, is greatly enlarged, hypertrophied, and often elongated, causing tremendous abdominal distension. Megacolon is caused by a failure of ganglia to develop in Auerbach's and Meissner's plexuses. Absence of these ganglia results in loss of muscular tone and peristalsis in the involved gut wall and causes pronounced constipation and distension. Successful surgical treatment of the disease consists of resecting the affected colon and bringing well-innervated colon to the distal rectum for an anastomosis. If successful, normal bowel function is restored.

The **anal canal** is surrounded by the levator ani and external anal sphincter muscles as it passes through the pelvic diaphragm posterior to the perineal body (Fig. 4-14). The **external anal sphincter** is disposed into a **subcutaneous portion** just deep to the skin and a **superficial part** attaching to the tip of the coccyx and perineal body. A

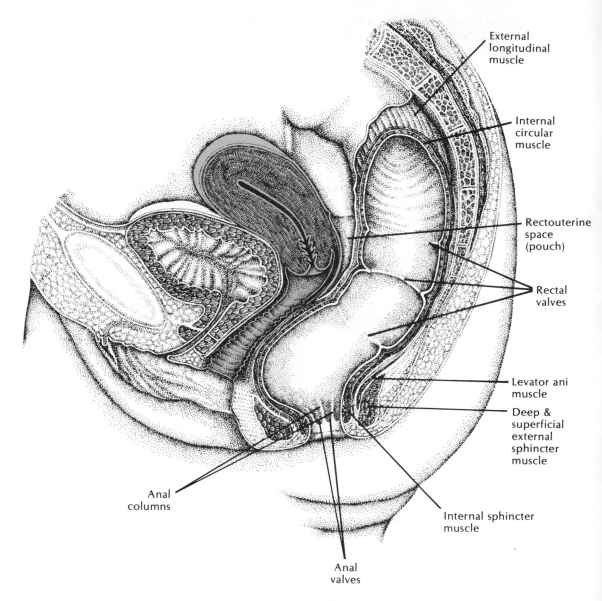

External
longitudinal
muscle

Internal
circular
muscle

Rectouterine
space
(pouch)

Rectal
valves

Levator ani
muscle

Deep &
superficial
external
sphincter
muscle

Internal sphincter
muscle

Anal
columns

Anal
valves

Fig. 4-14. Anal canal, uterus, and urinary bladder.

deep portion associated with the puborectalis muscle (a portion of the levator ani) also attaches anteriorly to the perineal body. Internally the mucosa of the upper half of the anal canal forms five to ten vertical folds designated as the **anal** or **rectal columns.** Each fold contains small veins that result in internal hemorrhoids if they become over-distended. The lower ends of these columns are joined by small cresentic folds of mucosa, the **anal valves** (anal sinuses), with the scalloped **pectinate line** present at their lower limit. In this region the anal mucosa merges with the skin of the anus as the **mu-**

cocutaneous line. Venous varicosities in the area below the pectinate line result in external hemorrhoids.

Arteries

The blood supply to the large intestine is from branches of the **superior** and **inferior mesenteric** and internal iliac arteries. The **middle colic branch** of the superior mesenteric artery passes between the layers of the transverse mesocolon to divide into right and left branches supplying the transverse colon. Arising at about the middle of the superior mesenteric artery, the **right colic branch** courses to the right behind the peritoneum toward the middle of the ascending colon where it divides into ascending and descending branches. They supply the ascending colon and then anastomose, respectively, with the middle colic and ileocolic arteries. The **ileocolic artery,** the terminal branch of the superior mesenteric, sends a colic branch to supply the cecum and the appendix, and small ileal branches to the distal portion of the ileum. The **inferior mesenteric artery** arises from the anterior aspect of the aorta, 3 to 5 cm above its bifurcation. Coursing inferiorly in a retroperitoneal position, the inferior mesenteric artery gives origin to the **left colic artery.** This vessel passes to the left and divides into ascending and descending branches to supply the left half of the transverse colon and the descending colon, respectively. Three or four **sigmoid branches** of the inferior mesenteric artery pass within the sigmoid mesocolon to supply the lower part of the descending and sigmoid colon. The inferior mesenteric then terminates as the **superior rectal artery,** which continues inferiorly between the layers of the sigmoid mesocolon to ramify on the upper part of the rectum. Additional blood to the rectum is derived from the **middle rectal branch** of the internal iliac and the **inferior rectal branch** of the internal pudendal arteries.

Liver

The **liver,** the largest gland in the body, is roughly wedge-shaped, with the base of the wedge directed to the right (Fig. 4-15). The **superior (convex) surface** continues undemarcated into the posteroinferior surface. The **anterior surface** of the liver is triangular, slightly convex, and related to the anterior body wall, diaphragm, and ribs. The **posterior surface** is roughly triangular and markedly concave from left to right as it passes in front of the vertebral column. The oblong **inferior (visceral) surface** is very uneven and, when hardened *in situ,* bears the impressions of structures in contact with it. Anteriorly the inferior surface is separated from the anterior surface by a sharp border.

The **porta hepatis** is a relatively deep and wide area, approximately 5 cm long, through which the portal vein, hepatic artery, bile ducts, nerves, and lymphatics pass. The porta hepatis separates the **quadrate lobe** inferiorly from the **caudate lobe** superiorly, and its boundaries serve as the site of attachment of the **hepatoduodenal ligament.** The **fissure for the ligamentum venosum** extends superiorly from the left end of the porta hepatis. It holds the ligamentum venosum, which is the remnant of the fetal

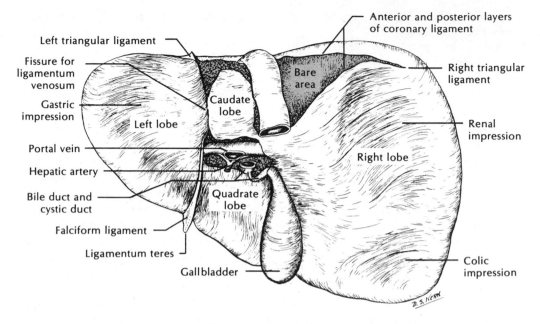

Fig. 4-15. Visceral surface of the liver.

ductus venosus. It affords attachment for the **gastrohepatic ligament,** which, with the hepatoduodenal ligament, forms the **lesser omentum.** The liver is partially divided by the fissures for the ligamentum venosum and ligamentum teres into a large **right** and a small **left lobe.** Two circumscribed areas on the medial aspect of the right lobe are further demarcated by the **fossae for the gallbladder** and the **inferior vena cava.** Between the above fissures and fossae the porta hepatis divides a superior segment, the **caudate lobe,** from an inferior segment, the **quadrate lobe.** The latter two subdivisions are part of the right lobe of the liver. The larger right lobe rises higher than the left, pushing the right dome of the diaphragm upward and thus partially displacing the right lung.

Structures on the posterior surface of the liver include the fissure for the ligamentum venosum, the fossa for the inferior vena cava, and the caudate lobe, which projects into the lesser peritoneal sac. Between the layers of the coronary ligament, the posterior surface is not covered by peritoneum and is in direct contact with the endoabdominal fascia of the diaphragm as the "bare area" of the liver. On the inferior surface the left lobe presents the gastric impression for the stomach and the **tuber omentale,** a bulging prominence above the lesser curvature of the stomach. The fossa for the gallbladder, impressions of the duodenum, right colic flexure, and right kidney are associated with the inferior surface of the right lobe.

Ligaments

The **falciform ligament** is a thin, sickle-shaped anteroposterior fold consisting of two apposing layers of peritoneum. One of its three borders is attached to, and reflected over, the anterior surface of the liver. Another border is attached to, and reflected over, the diaphragm and the anterior abdominal wall to the level of the umbilicus, while the third inferior border is free and encloses the **ligamentum teres.** The latter represents the obliterated umbilical vein. At the upper extent of the border of the falciform ligament attaching to the anterior surface of the liver, the peritoneal layers diverge laterally and reflect onto the diaphragm. The right reflection forms the **anterior layer of the coronary ligament,** which passes laterally to bend sharply at the **right triangular ligament,** where it becomes the **posterior layer of the coronary ligament.** The peritoneum forming the coronary ligament reflects from the liver onto the diaphragm to enclose an area devoid of peritoneum. Here the liver is in direct contact with the diaphragm and is designated as the "bare area." The left divergence of the falciform ligament, the **left triangular ligament,** reflects onto the left lobe corresponding to, and continuous posteriorly with, the posterior layer of the coronary ligament. The two folds of peritoneum composing this left divergence are not widely separated.

Gallbladder and Biliary Tree

The **gallbladder** is a small piriform sac that serves as the reservoir for bile (see Fig. 4-12). It holds 60 to 150 ml of fluid. It lies in a small fossa on the visceral (inferior) surface of the liver, and is divided into the fundus, body, and neck. The inferior extremity of the sac, the **fundus,** is wide, usually protrudes beyond the inferior margin of the liver, and is covered by peritoneum. It is in contact with the transverse colon, the anterior body wall, and the ninth costal cartilage at the point the latter is crossed by the linea semilunaris. The anterior surface of the **body** of the gallbladder is in direct contact with the liver, while the posterior surface and sides of the body are covered by peritoneum. Posteriorly the body is related to the transverse colon and the second part of the duodenum. The narrow **neck** of the gallbladder is continuous with the cystic duct. Both structures are closely applied to the liver, and related inferiorly to the first part of the duodenum. The 2 to 3 cm long **cystic duct** enters the hepatoduodenal ligament at the right end of the porta hepatis and runs a short distance with, and then joins, the common hepatic duct to form the common bile duct.

Bile, secreted by the liver cells, is carried away from the liver lobules by way of **bile canaliculi** to **intralobular ductules.** These unite to become **interlobar ducts,** which in turn join to form **right** and **left hepatic ducts,** whose junction forms the **common hepatic duct** at the porta hepatis. This latter duct, joined by the cystic duct, forms the **common bile duct.** It descends in the free margin of the lesser omentum anterior to the portal vein and lateral to the hepatic artery. The common bile duct continues inferiorly between the duodenum and the head of the pancreas, and terminates by uniting with the pancreatic duct before penetrating the second part of the duodenum. The musculature in the wall of the common bile duct thickens to form a **sphincter** at its junction with the pancreatic duct. Within the wall of the duodenum the duct expands slightly as the **hepatopancreatic ampulla (of Vater),** which bulges the mucous membrane of the gut inward to form the **duodenal papilla** with the duct opening at its summit (Fig. 4-16).

GALLSTONES

Gallstones result from precipitations, especially cholesterol, of chemical compounds of the bile. These precipitations become lodged in the

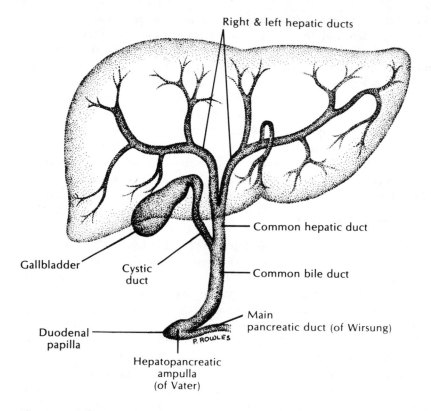

Fig. 4-16. Biliary tree.

gallbladder or the biliary ducts. Their development is usually accompanied by repeated inflammations of the gallbladder. Stones lodging along the excretory duct system may cause an obstruction or stasis of bile flow. Acute, severe pain or biliary colic is caused by this obstruction with stretching of the duct system and smooth muscle spasm. Biliary obstruction distends the ducts and results in a cessation of flow of the bile. The individual becomes jaundiced (yellow) due to the impaired clearance of bilirubin from the blood.

Arteries

Blood is supplied to the liver by the hepatic artery, a branch of the celiac trunk. The **hepatic artery** runs inferiorly and to the right. It courses retroperitoneally behind the lesser sac to reach the first part of the duodenum. Here it continues below the epiploic foramen to gain a position between the layers of the hepatoduodenal ligament and passes to the porta hepatis. Within the hepatoduodenal ligament it divides into the **left** and **right hepatic branches,** which supply their respective hepatic lobes. It gives a third branch, the **cystic artery,** to supply the cystic duct and the gallbladder. The cystic artery may arise from the right hepatic artery.

Portal Circulation

The **portal vein** originates behind the neck of the pancreas by the union of the **splenic** and **superior mesenteric veins.** It ascends behind the first part of the duodenum to pass in the free margin of the lesser omentum and divides into right and left branches to enter their respective lobes of the liver. These branches eventually subdivide to gain a position at the margin of the liver lobule, where they course with the branches of the hepatic artery and the bile duct. The portal vein drains the spleen, pancreas, gallbladder, and all of the alimentary canal distal to the stomach, except for the lower rectum and anal canal. Tributaries of the portal system, in addition to the splenic and superior mesenteric veins, include the **short gastric, right** and **left gastroepiploic,** and **coronary veins** draining the stomach, small **pancreatic veins** from the pancreas, and the **inferior mesenteric vein** draining blood from the upper rectum and the distal half of the colon. The **hepatic veins** draining the liver begin as **central veins** of the liver lobules. These small vessels drain into **sublobular veins** as tributaries of the hepatic veins, which empty into the inferior vena cava before it passes through the diaphragm.

PORTAL-SYSTEMIC VENOUS ANASTOMOSIS

Tributaries of the portal vein communicates with systemic veins in several locations, particularly the lower ends of the esophagus and rectum, and the umbilicus. Ordinarily these channels are collapsed with little blood flow because blood drains from the abdominal viscera through the portal vein to the liver. If venous flow through the liver is impeded by a blood clot or scarring in the liver (cirrhosis), the portal vein pressure rises markedly (Fig. 4-17). Portal hypertension dilates the veins of the portal system, including some or all of these anastomotic channels. This can cause varicosities of the umbilicus (caput medusae), esophagus (esophageal varices), and rectum (hemorrhoids).

Pancreas

The **pancreas** is an elongated endocrine and exocrine gland that lies obliquely on the upper part of the posterior abdominal wall. It extends from the concavity of the duodenum to the spleen (see Fig. 4-12). It is soft and pliable, contains a minimum of connective tissue, and is subdivided into a head, neck, body, and tail. The flattened, expanded **head** nestles in the concavity of the duodenum. Posteriorly the head is related to the aorta and inferior vena cava, with the common bile duct enbedded in its lateral margin. It is related anteriorly to the superior mesenteric vessels and is in contact with the transverse colon. The **uncinate process** is an extension of the head, and the latter is continuous with the **neck** (isthmus) of the pancreas. Laterally the anterior surface of the **body** forms part of the stomach bed. It bulges slightly in the median plane as the **tuber omentale** immediately inferior to the celiac artery. The splenic ar-

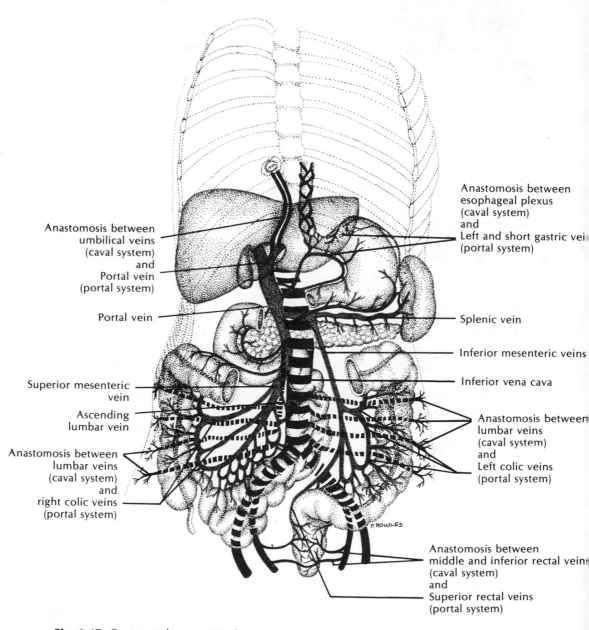

Fig. 4-17. Portacaval anastomosis.

tery courses along the superior border of the pancreas, and the transverse mesocolon reflects from the posterior abdominal wall along its anterior surface. Posteriorly the body of the pancreas rests on the aorta, the superior mesenteric artery, the left crus of the diaphragm, the left psoas muscles, and the left kidney. The **tail** is thick and blunt and related to the hilus of the spleen.

DIABETES MELLITUS

Diabetes mellitus is a deficiency of insulin production from the islet cells of the pancreas. Insulin is essential in carbohydrate metabolism. It facilitates cellular uptake of glucose and the conversion of glucose to glycogen. Loss of these functions result in increased blood sugar levels (hyperglycemia), increased sugar in the urine (glycosuria), increased urine output (polyuria), thirst, hunger, emaciation, and weakness. Diabetes mellitus is controlled by strict dieting, by properly spaced injections of insulin, or by the oral administration of antidiabetic drugs.

Two ducts drain the pancreas. The **main pancreatic duct** courses along the entire length of the gland, emerging to join the common bile duct and pierce the duodenal wall 7 to 10 cm beyond the pylorus, as the duodenal papilla. The much smaller **accessory pancreatic duct** commmonly appears in the neck as an offshoot from the main pancreatic duct and usually opens independently into the duodenum 2 to 3 cm above the duodenal papilla.

Arteries and Nerves

The pancreas receives blood from small twigs of the **splenic artery** as it courses along the superior border of the pancreas, from the **superior pancreaticoduodenal branch** of the gastroduodenal artery, and from the **inferior pancreaticoduodenal branch** of the superior mesenteric artery. Innervation to the gland is derived from an extension of the **celiac plexus** surrounding its arterial supply.

Spleen

The **spleen** (lien) is an oblong, flattened, highly vascular organ. It is located in the left hypochondriac region of the abdomen, behind the stomach and inferior to the diaphragm. It has diaphragmatic and visceral surfaces, superior and inferior extremities, and anterior, posterior, and inferior borders. The convex **diaphragmatic surface** is molded to fit the diaphragm. The **visceral surface** is divided by a ridge into an anterior (gastric) portion and an inferior (renal) portion. A fissure for the passage of vessels and nerves, the **hilus**, separates the gastric from the renal portions. The gastric portion is concave and in contact with the posterior wall of the stomach, while the somewhat flattened renal portion is related to the left kidney and left suprarenal gland. The **superior extremity** is directed toward the vertebral column at the level of the eleventh thoracic vertebra. The flat, triangular **inferior extremity** (colic surface) rests on the left colic flexure in contact with the tail of the pancreas. The notched **anterior border** is free and sharp and separates diaphragmatic and gastric surfaces. The rounded blunt **posterior border** demarcates diaphragmatic from renal surfaces, while the **inferior border** divides diaphragmatic from colic surfaces.

The spleen is almost entirely surrounded by peritoneum and held in position by two peritoneal ligaments. The short **lienorenal ligament** extends from the upper half of the left kidney to the hilus and contains the splenic artery and vein, sympathetic nerves, and lymphatics. The **gastrolienal ligament** transmits the gastro-epiploic and the short gastric artery and vein, nerve filaments, and lymphatics to the stomach as it passes forward from the hilus to become continuous with the greater omentum.

Arteries

The **splenic artery,** the largest branch of the celiac trunk, follows a tortuous course along the superior border of the pancreas behind the posterior wall of the lesser sac. It terminates in five or six branches that pass through the lienorenal ligament to supply the spleen. The artery gives **pancreatic branches** in its retroperitoneal position and sends the **left gastroepiploic** and **short gastric** branches through the gastrolienal ligament to supply the stomach. Figure 4-18 demonstrates artery distribution of the abdomen.

REMOVAL OF THE SPLEEN

Although the spleen is well protected from traumatic injuries, it is the most frequently damaged organ from blunt abdominal trauma; particularly from severe blows over the lower left chest or upper abdomen, that fracture protecting ribs. Such a crushing injury may rupture the spleen, which causes severe intraperitoneal hemorrhage and shock. This requires a prompt splenectomy to keep the patient from bleeding to death.

Kidney

The **kidney,** a retroperitoneal structure embedded in fascia and fat, is ovoid in outline, with the medial border markedly concave. It lies obliquely in the upper part of the posterior abdominal wall. The left kidney is opposite the twelfth thoracic and upper three lumbar vertebrae, and the right is slightly lower. The middle of the hilus is approximately 5 cm from the median plane, with the upper pole of the kidney closer to the midline than the lower. The kidney presents anterior and posterior surfaces, medial and lateral borders, and upper and lower poles.

Relations of the **anterior surface** vary on each side. On the right the anterior surface is related to the liver, duodenum, right colic flexure, and small intestine. On the left side the surface is related to the spleen, splenic vessels, left colic flexure, and small intestine. The portions of the kidney in contact with the suprarenal gland, pancreas, colon, and duodenum are devoid of peritoneum, while the remainder of the anterior surface is covered by peritoneum. The **posterior surface** of each kidney has similar relations, being embedded in areolar and fatty tissue and is entirely devoid of peritoneum. The kidney rests on the lower part of the diaphragm, arcuate ligaments, psoas

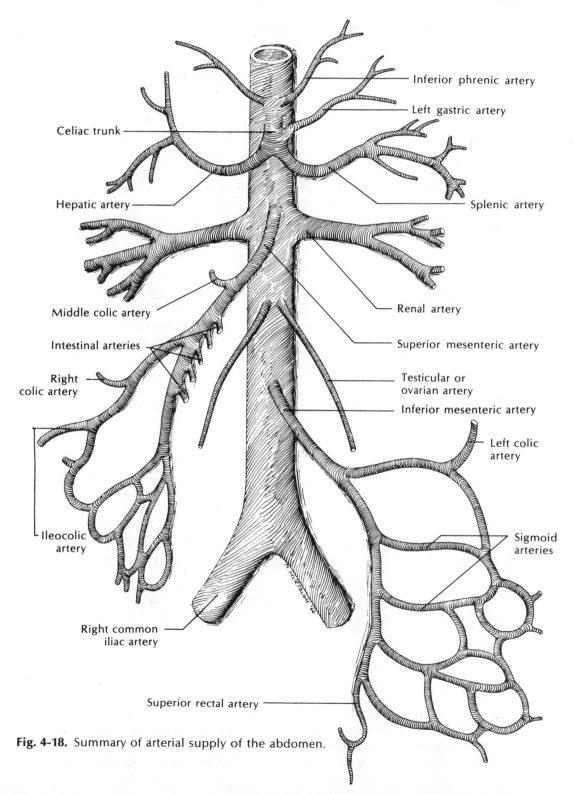

Inferior phrenic artery

Left gastric artery

Celiac trunk

Splenic artery

Hepatic artery

Middle colic artery

Renal artery

Intestinal arteries

Superior mesenteric artery

Right colic artery

Testicular or ovarian artery

Inferior mesenteric artery

Left colic artery

Ileocolic artery

Sigmoid arteries

Right common iliac artery

Superior rectal artery

Fig. 4-18. Summary of arterial supply of the abdomen.

major, quadratus lumborum, and transversus abdominis muscles, one or two lumbar arteries, and subcostal, ilioinguinal, and iliohypogastric nerves.

The adipose tissue that completely surrounds the kidney is separated by a thin membranous sheet into **perirenal fat** adjacent to the kidney and **pararenal fat** external to this membrane. From the lateral border of the kidney this membrane splits to pass anterior and posterior to the kidney as the **renal fascia.** Traced medially, the anterior layer of renal fascia passes anterior to the renal vessels, over the aorta, and becomes continuous with the same layer of the opposite side. Superiorly it passes over the suprarenal gland to become continuous with the posterior layer, which is the thicker layer of the two, and passes medially behind the aorta and the vena cava to unite with connective tissue over the vertebral column. Inferiorly the renal fascia fuses with the extraperitoneal connective tissue and loses its identity. Thus, in emaciation a loss of the fat surrounding the kidney may be accompanied by a **downward displacement** (ptosis) of the kidney because of the nonclosure of this fascia inferiorly.

NEPHROPTOSIS

The fat surrounding the kidney is a major factor in stabilization of this organ. In emaciation loss of this fat may permit the kidney to displace inferiorly. If this occurs it may cause a "kink" in the ureter because the ureter is firmly bound by parietal peritoneum to the posterior abdominal wall. Urine will then accumulate in the renal pelvis and result in an hydronephosis. A similar mobility of the kidney may occur from a blow to the lumbar region of the back ("kidney punch" in boxing), or jarring of the body in riding trail bikes. The latter is the rationale for racing motorcyclists to wear wide supportive belts.

Internal Anatomy

On sagittal section the internal anatomy of the kidney presents, adjacent to the cortex, a fibrous tunic (**capsule**) and an inner medullary region (Fig. 4-19). The **cortex** consists of granular appearing tissue containing glomeruli and elements of the nephron. At intervals extensions of cortical tissue project centrally between the pyramids of the medulla as the **renal columns.** The **medulla** is composed of a series of eight to sixteen conical masses, the renal or medullary **pyramids,** containing the collecting tubules. The bases of the pyramids are directed toward the cortex and their apices converge at the **renal sinus,** forming within the sinus the prominent **renal papillae.** Four to thirteen cup-shaped, sleevelike projections surround one or more renal papillae as the **minor calyces.** They join to form two or three **major calyces,** which empty into, and are continuous with, the funnel-shaped **renal pelvis,** the proximal dilatation of the ureter (Fig. 4-20).

HYDRONEPHROSIS

Hydronephrosis is the distention of the pelvis and calyces of the kidney. It is caused by an obstruction of the lower urinary tract, and a backup of urine into these structures. If the condition persists, permanent damage to the parenchyma of the kidney results. One common

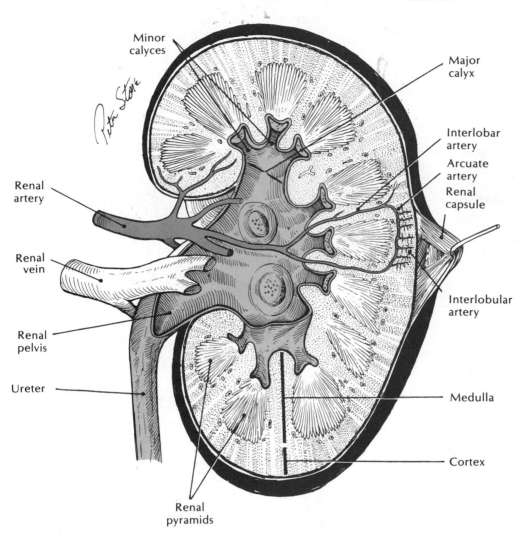

Fig. 4-19. Sagittal section of kidney. (Langley LL, Telford IR, Christensen JB: Dynamic Anatomy and Physiology, 5th ed. New York, McGraw-Hill, 1980)

cause of ureteral obstruction is a kidney stone. Hydronephrosis may also be caused by an abnormal position of the kidney (nephroptosis) that allows the ureter to kink. Hydronephrosis is also an infrequent complication of pregnancy. The developing fetus exerts pressure on the ureter as the latter crosses the brim of the pelvis. This may obstruct urine flow.

Blood Vessels and Nerves

Arising from the aorta about 2 to 3 cm below the origin of the superior mesenteric artery, the **renal artery** passes transversely to reach the hilus of the kidney. The

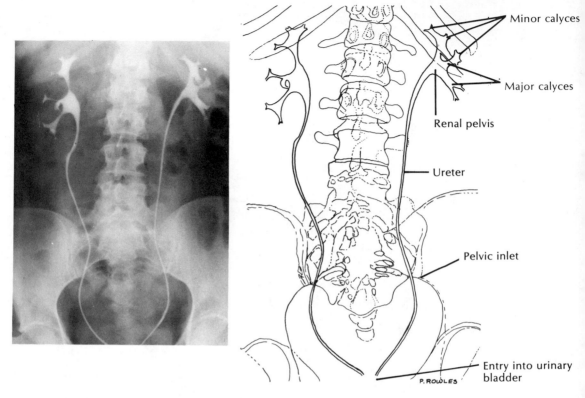

Fig. 4-20. Anteroposterior pyelogram.

right renal artery, longer and lower than the left, passes behind the inferior vena cava, the head of the pancreas, and the second part of the duodenum to reach the kidney. The left vessel passes behind the renal vein, the pancreas, and the splenic vein to enter the kidney. Each gives an **inferior suprarenal branch** as it enters the hilus, then divides into interlobar branches that traverse the renal sinus. The **interlobar branches** course in the renal columns to reach the cortex of the kidney where they bifurcate at right angles to give rise to **arcuate arteries.** The latter course parallel to the cortex, to anastomose and form arcades. From the arcades **interlobular arteries** pass peripherally into the cortex, giving a series of **afferent arterioles** to the glomeruli.

The large renal veins are anterior to the renal arteries. The right vein empties directly into the inferior vena cava. After receiving suprarenal, renal, left gonadal, and left phrenic tributaries, the left renal vein also joins the inferior vena cava. Innervation to the kidneys is from the renal autonomic plexus, whose fibers accompany the renal artery into the kidney (Table 4-2).

RENAL HYPERTENSION

In many kidney diseases blood flow through the organ is compromised. The ischemic kidney reacts by secreting the enzyme, renin,

Table 4-2. Nerve Distribution to Abdomen

Nerve	Origin	Course	Distribution
Thoracoabdominal nerves	Continuations of lower intercostal nerves	Pass between second and third layers of muscle	Anterolateral abdominal muscles and overlying skin (T_{10} dermatome flanks umbilicus); periphery of diaphragm
Iliohypogastric	Posterior divisions of lumbar plexus	Crosses posterior abdominal wall to pierce transversus abdominis	Skin over iliac crest and adjacent buttock and hypogastric regions
Ilioinguinal	Anterior divisions of lumbar plexus	Passes between second and third layers of muscle to reach and traverse inguinal canal	Skin of scrotum (labium majus), adjacent thigh and mons pubis
Segmental branches	Ventral rami of all lumbar nerves	Short twigs to successive segments of muscles of posterior abdominal wall	Quadratus lumborum and psoas muscles (iliacus by twigs from femoral)
Autonomic plexus	Formed around branches of abdominal aorta and their subsequent branches; parasympathetic (preganglionic) from vagi and pelvic splanchnics; sympathetic (postganglionic) by splanchnics from lower levels of sympathetic trunk	Follow arteriolar branchings to walls of digestive organs; myenteric and submucosal plexuses are locations of postganglionic parasympathetic nerve cell bodies	Parasympathetics increase peristalsis and secretion, relax sphincters; sympathetics opposite action

which activates a plasma protein fraction, angiotension I. The latter is converted by another enzyme into angiotension II, which is a powerful vasoconstrictor. This causes a narrowing of arterioles over the entire body. The resulting high blood pressure (renal hypertension) ensures a continuance of a sufficiently high blood pressure for producing an increased volume of urine. In this way, the needs of the kidney are met but at the price of chronic high blood pressure.

Ureter

The **ureter** carries urine from the renal pelvis to the urinary bladder. It is approximately 25 cm long, has thick muscular walls, and a narrow lumen. It is situated half in the abdominal and half in the pelvic cavities. Descending retroperitoneally on the psoas major muscle, it passes anterior to the bifurcation of the iliac artery and posterior to the gonadal and right or left colic arteries. The ureter then turns anteriorly at the level of the ischial spine to reach the posterior aspect of the urinary bladder.

NORMAL CONSTRICTIONS

Normal constrictions of the ureter are located: 1) at the lower end of the pelvis of the kidney, or the ureteropelvic junction, 2) at the pelvic brim where it crosses the iliac vessels, and 3) at its oblique entry through the bladder wall. Kidney stones tend to become lodged at these narrow points in the ureter.

In the male, the ureter is crossed on its medial side by the ductus deferens and, as it approaches the bladder, lies anterior to the upper end of the seminal vesicle. In the female, it is related to the posterior border of the ovary. Along the lateral pelvic wall it turns anteromedially to pass close to the vagina and cervix in its course toward the bladder. The ureter receives its **blood** and **nerve supply** regionally from the renal, gonadal, and vesicular arteries and nerves.

SURGICAL TRAUMA TO URETERS

The pelvic ureters are the most vulnerable pelvic organs in surgery performed on other pelvic structures, such as the uterus. During an abdominal hysterectomy, the ureter may be accidentally ligated with the ovarian vessels. It may be crushed by the clamp applied to the serous layers on the broad ligament, or ligated as a tie is placed on the uterine artery and the transverse cervical ligament. While the possibility of ureteral injury is greatest in gynecologic procedures, the risk of injury is ever present in any operation that involves the pelvis or lower abdomen.

Suprarenal Gland

The **suprarenal (adrenal) gland** is an endocrine organ adjacent to the upper pole of the kidney. On the right side the gland is related to the diaphragm posteriorly, the inferior vena cava medially, and the liver laterally. On the left side it rests on the left crus of the diaphragm anteriorly, the stomach superiorly, and the pancreas inferiorly. The two glands are separated from each other by the celiac axis and plexus and are enclosed in the renal fascia. The suprarenal gland is supplied by the six to eight small **superior suprarenal** twigs from the phrenic artery; the **middle suprarenal,** a branch of the aorta, and the **inferior suprarenal,** a branch of the renal artery. Veins draining the gland correspond to the arteries; however, the principal drainage is by way of a prominent single **central vein** that emerges from the hilus to empty on the left side into the renal vein, and on the right side into the inferior vena cava.

5

perineum
and pelvis

Perineum

SURFACE ANATOMY

The **perineum,** conforming to the deeply lying outlet of the pelvis, is a diamond-shaped area at the lower end of the trunk between the thighs and buttocks (Fig. 5-1). It is bounded anteriorly by the pubic symphysis, laterally by the ischial tuberosities, and posteriorly by the coccyx. A line passing transversely between the ischial tuberosities through the central point of the pereineum, divides it into an anterior **urogenital triangle** and a posterior **anal triangle.** The **central point of the perineum,** located between the anus and the urethral bulb in the male, and between the anus and the vestibule in the female, overlies the deeply placed **perineal body.** In the male, a slight median ridge, the **perineal raphe,** passes forward from the anus to become continuous with the **median raphe** of the scrotum and the **ventral raphe** of the penis.

FASCIAE AND SUPERFICIAL PERINEAL COMPARTMENT

The **superficial fascia** of the perineum, consisting of a superficial fatty layer and a deeper membranous layer, is continuous with the superficial fascia of the abdominal wall, thigh, and buttock (Fig. 5-2). In the male the fat is lost as the superficial fascia is extended over the scrotum, where both superficial and membranous layers fuse and gain smooth muscle fibers to form the **dartos tunic** of the scrotum. In the female the fat in the superficial layer increases as it passes over the labia majora and the mons pubis. The **membranous layer** (superficial perineal or Colles' fascia) is attached posteriorly to the posterior margin of the urogenital diaphragm and laterally to the ischiopubic rami. Anteriorly in the male it is continuous with the dartos tunic and, at the superior extent of the scrotum, with the membranous layer of the superficial fascia of the abdomen. In urethral rupture fascial attachments of the membranous layer prevent urine from passing beyond the posterior margin of the urogenital diaphragm or into the thigh, but may allow it to spread a considerable distance into the subcutaneous tissue over the abdomen.

The **superficial perineal compartment,** between the external perineal fascia and the superficial layer of the urogenital diaphragm, contains the crura of the penis or clitoris, the bulb of the penis or vestibule, the superficial transverse perineal muscles, and perineal vessels and nerves. The **external perineal (Gallaudet's) fascia** covers the ischiocavernosus, bulbocavernosus, and superficial transverse perineal muscles and is firmly adherent to the inferior layer of the urogenital diaphragm. Within the superficial perineal compartment, the root of the penis, consisting of two crura and the bulb, is composed mostly of erectile tissue. The crura of the penis are attached to the inferior aspect of the ischiopubic rami and the inferior layer of the urogenital diaphragm and are covered by the ischiocavernosus muscles. Anteriorly they converge toward each other to form the two corpora cavernosa of the penis. The bulb of the penis (urethral bulb) is attached to the inferior layer of the urogenital diaphragm and surrounds the urethra. It is covered by the bulbocavernosus muscle and passes forward to lie inferior to the corpora cavernosa of the penis as the corpus spongiosum penis.

The **perineal body** (tendinous center of the perineum) is a fibromuscular mass located in the median plane between the anal canal and the urogenital diaphragm, with which it is fused. Several muscles attach, at least in part, to it, including the superficial and deep transverse perinei, the bulbocavernosus, the central portion of the levator ani, and the external anal sphincter. The perineal body is of special importance in the female, as it may be torn or damaged during parturition.

Fig. 5-1 Subdivisions of the male perineum.

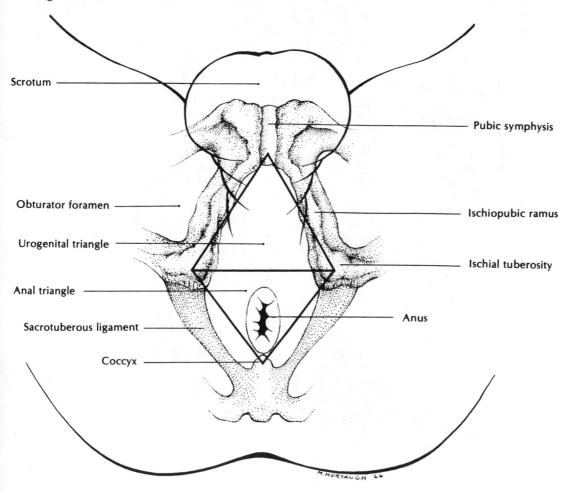

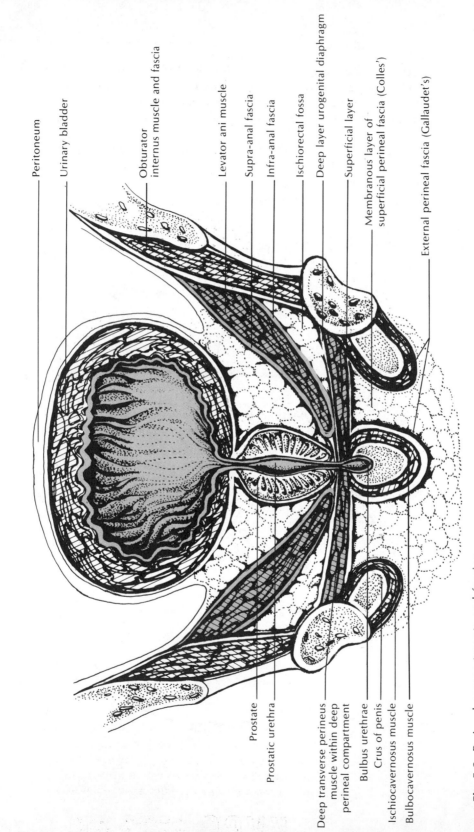

Peritoneum

Urinary bladder

Obturator
internus muscle and fascia

Levator ani muscle

Supra-anal fascia

Infra-anal fascia

Ischiorectal fossa

Deep layer urogenital diaphragm

Superficial layer

Membranous layer of
superficial perineal fascia (Colles')

External perineal fascia (Gallaudet's)

Prostate

Prostatic urethra

Deep transverse perineus
muscle within deep
perineal compartment

Bulbus urethrae

Crus of penis

Ischiocavernosus muscle

Bulbocavernosus muscle

Fig. 5-2. Perineal compartments and fasciae.

UROGENITAL DIAPHRAGM AND DEEP PERINEAL COMPARTMENT

The **urogenital diaphragm** is situated within the infrapubic angle between the ischiopubic rami (see Fig. 5-2). It is formed by connective tissue membranes spanning this area as the superficial and deep layers of the urogenital diaphragm and by the structures located between them. The **superficial layer of the urogenital diaphragm** (inferior perineal membrane) is composed of strong bands of fibrous connective tissue that pass between ischiopubic rami and separate the deep perineal compartment from the external genitalia. A similar structure, the **deep layer of the urogenital diaphragm** (superior perineal membrane), also passes between the ischiopubic rami and separates the deep perineal compartment from the anterior recess of the ischiorectal fossa. Anteriorly these membranes fuse to form the **transverse ligament of the pelvis,** and posteriorly they fuse at the posterior extent of the deep transverse perineus muscle.

The **deep perineal compartment** is the area between the superficial and deep layers of the urogenital diaphragm. It contains the deep transverse perineus and external urethral sphincter muscles, the internal pudendal vessels and pudendal nerve, and in the male the bulbourethral (Cowper's) glands.

MUSCLES (TABLE 5-1)

Muscles of the perineum include the external anal sphincter, the structures within the deep perineal compartment, the superficial transverse perineal muscle, and muscles covering the crura and bulb of the penis or clitoris and vestibule.

The **external anal sphincter** is a ring of voluntary muscle dispersed around the distal part of the anal canal. It is arranged into subcutaneous, superficial, and deep portions. The **subcutaneous part** surrounds the anal orifice, attaching to the skin of the anus. The **superficial segment** extends from the tip of the coccyx to the perineal body. Encircling the anus, it holds the anus in the median plane. The larger **deep portion** surrounds and attaches to the anal canal like a heavy collar. Its muscle fibers are closely associated superiorly with the levator ani muscle (puborectalis), and decussate anteriorly to interlace with the superficial transverse perineal muscle, while other fibers insert into the perineal body.

Within the superficial perineal compartment, thin sheetlike muscles, the median bulbocavernosus and two ischiocavernosi, cover, respectively, the bulb and the two crura of the penis. At the posterior extent of the urogenital diaphragm, superficial transverse perinei muscles arise from the ischial tuberosities and insert into the central tendon of the diaphragm.

The **bulbocavernosus** originates at the central tendon of the perineum and the median raphe on the superficial (ventral) aspect of the bulb. Its fibers spread over the bulb and the corpus spongiosum to insert into the superficial perineal membrane. The muscle fibers interdigitate with those of the opposite side, and the most distal fibers spread over and attach to the corpus cavernosum penis. In the female the counterparts of the bulbocavernosus covering the bulb of the penis in the male are separate muscles

Table 5-1. Muscles of the Perineum

Muscle	Origin	Insertion	Action	Nerve
External anal sphincter	Skin and fascia surrounding and tip of coccyx	Central tendon of perineum	Closes anus	Inferior rectal
Bulbocavernosus	*Male*—median raphe, ventral surface of bulb, and central tendon of perineum *Female*—central tendon of perineum	*Male*—corpus cavernosum, subpubic triangle, and root of penis *Female*—dorsum of clitoris and superficial layer of urogenital diaphragm between crura of clitoris	*Male*—compresses bulb and bulbous portion of urethra; anterior fibers believed to act in erection *Female*—compresses bulb and vaginal orifice	Perineal branch of pudendal
Ischiocavernosus	Pelvic surface of inferior ramus of ischium surrounding crus	Crus near pubic symphysis	Maintains erection of penis or clitoris by compression of crus	Perineal branch of pudendal
Superficial transverse perineus	Ramus of ischium near tuberosity	Central body of perineum	Supports central body of perineum	Perineal branch of pudendal
Deep transverse perineus	Internal aspect of inferior ramus of ischium	Median raphe, central tendon of perineum, and external anal sphincter	Fixes central body of perineum	Perineal branch of pudendal
External urethral sphincter	Inferior ramus of ischium	Fibers interdigitate around urethra	Closes and compresses urethra	Perineal branch of pudendal

In the female, an additional muscle of the deep perineal space is specialized as the constrictor vaginae and acts to compress the vagina and greater vestibular glands.

that cover the bulb of the vestibule on either side of the pudendal cleft. They may act as a weak sphincter of the vagina. Originating from the central tendon, their fibers pass anteriorly to insert into the body of the clitoris.

The **ischiocavernosus** originates from the tuberosity of the ischium and the ischiopubic ramus. It covers and inserts into the crus of the penis.

The **superficial transverse perineus,** a narrow muscular slip, passes from its origin on the inner surface of the ischial tuberosity to insert medially into the central tendon of the perineum.

Two muscles are located within the deep perineal space. Centrally the **external urethral sphincter** surrounds the membranous portion of the urethra. Peripherally, in the same plane, the **deep transverse perineus** passes from the ischiopubic rami to interdigitate with its opposite member and insert into a fibrous raphe.

ISCHIORECTAL FOSSA

Each of the paired lateral **ischiorectal fossae** (the area between the perineum and the pelvis) is bounded anterolaterally by the fascia of the obturator internus, the ischial tuberosity, and the ischiopubic ramus; medially by the inferior layer of fascia covering the levator ani, and posterolaterally by the fascia of the gluteus maximus. Each continues anteriorly toward the body of the pubis as an **anterior recess,** situated between the ischiopubic rami,the levator ani, and the deep layer of the urogenital diaphragm. Anterosuperiorly the anterior recess is limited at the point of origin of the levator ani. The **posterior recess** of each fossa extends posteriorly between the coccygeus muscle medially, and the gluteus maximus muscle and sacrotuberous ligament posterolaterally.

Each fossa is filled by an **ischiorectal fat pad,** composed of adipose tissue traversed by irregular connective tissue septa, which are continuous with the subcutaneous fatty layer of the perineum. The **pudendal (Alcock's) canal,** a fascial tunnel located on the medial aspect of the ischial tuberosity internal to the obturator internus muscle, is formed by a splitting of the obturator fascia. The pudendal canal transmits the pudendal nerves and internal pudendal vessels that supply structures in the urogenital triangle. The **inferior rectal vessels** and **nerves,** arising from nerves and vessels noted above, pass from the canal across the ischiorectal fossa to supply the lower rectum and anal canal.

EXTERNAL MALE GENITALIA

Penis

The **penis,** formed by two corpora cavernosa and the corpus spongiosum, is covered by skin and fascia (Fig. 5-3). The thin elastic skin is loosely connected to the deeper parts and extends over the distal end of the penis for a variable distance, as the **prepuce.** A narrow median fold extends from the inferior aspect of the glans to the prepuce, as the **frenulum.**

BALANITIS

Balanitis is an inflammation of the glans penis. This is significant (especially in the very young) in individuals that have not been circumcised, and have a narrow preputial opening. If the foreskin is retracted over a glans that is edematous and swollen from inflammation, it may act as a tourniquet, compress the veins, decreasing the venous return and thereby cause further enlargement to the glans. Surgical intervention may be necessary to release the constriction by the prepuce.

The **superficial fascia** is directly continuous with dartos tunic of the scrotum. The **deep (Buck's) fascia** forms a tubular investment of the shaft of the penis to the corona, thus surrounding the corpora with a strong capsule. The **suspensory ligament,** a strong,

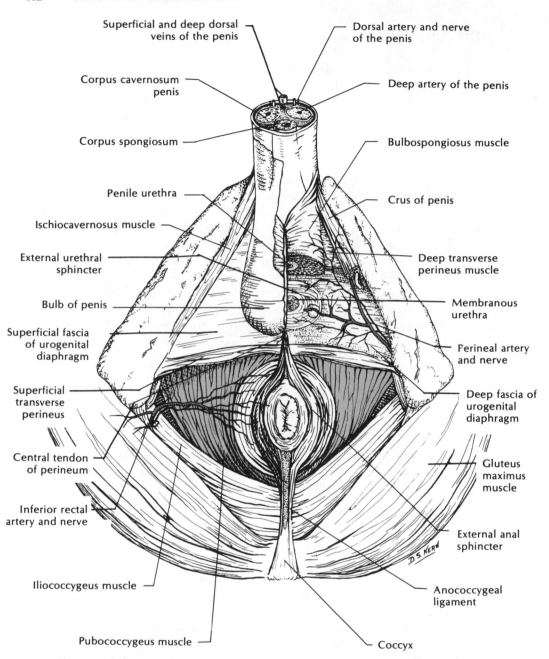

Fig. 5-3. Root of the penis.

fibroelastic triangular band of deep fascia, extends from the anterior border of the pubic symphysis to the penis, where it divides into right and left lamellae. The latter fuse to the deep fascia at the sides of the shaft of the penis. Vessels and nerves to the dorsum of the penis pass between the two lamellae.

The **corpora cavernosa penis** are a pair of elongated bodies extending from the perineum to the corona of the penis. They consist of **erectile tissue** (dilated vascular spaces) filled with blood, and each is surrounded by a dense white fibrous capsule, the **tunica albuginea.** The apposing sides of this capsule are imperfectly fused to form the **septum.** The **deep arteries of the penis** are terminal branches of the internal pudendal arteries and are located in the center of each corpus cavernosum penis. The **corpus spongiosum penis,** a structure similar to the corpus cavernosum penis, has a thinner, less dense capsule, lies in a groove on the under surface of the corpora cavernosa penis, and contains centrally the penile portion of the **urethra.** The anterior extremity of the corpus spongiosum is expanded to form the **glans.** The junctional area between the glans and the shaft of the penis is the **corona.**

The **superficial dorsal vein** of the penis lies in the superficial fascia in the median plane and drains into either the right or the left external pudendal vein. The **deep dorsal vein,** also in the median plane but deep to the deep fascia, passes between the lamellae of the suspensory ligament below the infrapubic ligament to drain into the pudendal and prostatic plexuses. Two **dorsal arteries,** terminal branches of the internal pudendal vessels, course lateral to the deep dorsal vein and supply the glans and skin. The dorsal arteries are accompanied by two **dorsal nerves,** the terminal branches of the pudendal nerve, which course lateral to the dorsal arteries and supply twigs to the skin and glans.

ERECTION

Enlargement of the penis or clitoris is under the influence of the parasympathetics (nervi erigentes) and results from stimulation of the genitalia or erotic thoughts. The physiological mechanism involves a relaxation of the smooth muscle in the walls of arteries supplying the cavernous spaces, thereby enlarging their caliber and increasing the volume of blood flowing into the organ. The inelasticity of the ensheathing tunica albuginea results in increased pressure on the thin walled veins draining the cavernous tissue and tumescence ensues. Following orgasm the sympathetics take over, the arteries become constricted, arterial flow is reduced and the penis gradually returns to its flaccid condition.

Scrotum

The **scrotum,** a pendulous purselike sac of skin and fascia, contains the testis, the epididymis, and the spermatic cord. The skin is rugose, contains smooth muscle fibers and sebaceous glands, and is covered with sparse hair. **Dartos tunic** forming part of the wall of the scrotum, is a continuation of the two layers of the superficial fascia of the abdomen (Camper's and Scarpa's layers), which fuse into a single layer. It is devoid of fat, highly vascular, and interspersed with smooth muscle fibers. The **septum** (a midline extension of deep fascia) is incomplete superiorly. It passes into the interior of the scrotum to divide it into two chambers.

HYDROCELE

A hydrocele is a collection of fluid in the serous cavity of the scrotum. It may be congenital due to the persistence of a communication between the processus vaginalis and the peritoneal cavity. By gravity, peritoneal fluid drains slowly into the processus with no pain or distress. The hydrocele may attain huge size, greatly distending the scrotum. An acquired hydrocele often arises from inflammation of the serous covering of the testis.

Spermatic Cord

The **spermatic cord** is formed at the deep inguinal ring and passes through the inguinal canal to exit at the superficial ring and enter the scrotum, where it is attached to the testis (Fig. 5-4). It contains the vas deferens, a hard cordlike tube that transmits spermatozoa from the epididymis to the urethra, the testicular artery from the aorta, and the artery of the vas deferens. Other structures within the spermatic cord include the pampiniform plexus of veins which drain on the right side into the inferior vena cava and on the left side to the renal vein (through internal spermatic veins), lymph vessels draining to nodes at the aortic bifurcation, sympathetic nerves from the renal and aortic plexuses, and the remnant of the processus vaginalis.

VASECTOMY

Vasectomy is usually performed by sectioning or ligating the ductus (vas) deferens. A bilateral procedure blocks the passage of sperm from the testis to the urethra. Although the vasectomized male is sterile, he is able to engage in normal sexual activities and does not show signs of male hormone deficiency since the hormonal activity of the testis is not impaired in this procedure.

The **coverings of the spermatic cord** are continuous with the fasciae of the abdominal wall. The **external spermatic fascia,** attached to the crura of the inguinal ligament, is continuous above with the fascia of the external oblique muscle. The **cremasteric fascia,** with interspersed skeletal muscle fibers, is an extension of the internal oblique and its fascia. The **internal spermatic fascia** is continuous with the endoabdominal (transversalis) fascia of the abdominal cavity, and the **subserous fascia** is continuous with the extraperitoneal connective tissue.

CREMASTERIC REFLEX

The cremasteric reflex is elicited by stroking the skin over the medial side of the thigh, which causes contraction of the cremaster muscle. In a positive cremaster reflex, the testis is drawn upwards towards the superficial inguinal ring. This muscle forms the middle component of the coverings of the spermatic cord, and is an extension from the internal

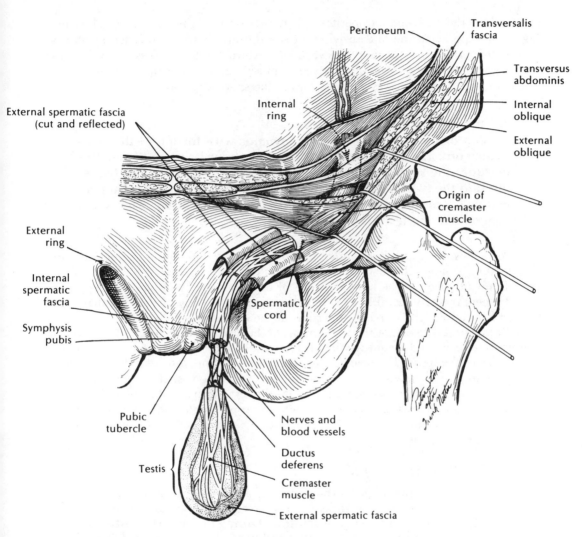

Fig. 5-4. Contributions from abdominal wall to spermatic cord coverings. (Langley LL, Telford IR, Christensen JB: Dynamic Anatomy and Physiology, 5th ed. New York, McGraw-Hill, 1980)

abdominal oblique. As the processus vaginalis evaginates the abdominal wall, some of the muscle fibers are drawn into the scrotum by the testis, where they become incorporated into the scrotal wall as a looplike muscular investment, the cremasteric muscle.

Testis

The **testes** are two, flattened, oval glands suspended in the scrotum by the spermatic cords that produce sperm and certain male sex hormones. During early development they are attached to the upper posterior body wall behind the peritoneum. In late fetal life they descend retroperitoneally to pass through the inguinal canal, emerging at the superficial ring to descend into the scrotum. They are preceded by an evagination of peritoneum, the **processus vaginalis.** External to the processus vaginalis, the wall of the developing scrotum is derived from several layers of the abdonimal wall. These are described above as the coverings of the spermatic cord.

TORSION OF THE TESTIS

Torsion of the testis is a term synonymous with torsion of the spermatic cord. The testis undergoes rotation about its vertical axis. This occludes the spermatic veins but usually not the arteries. The tissues distal to the twist becomes edematous, and develop an infiltrating hematoma from the distended veins. If arterial circulation is impaired, gangrene of the testis may occur.

As the testis enters the scrotum it invaginates into the processus vaginalis to form a parietal layer lining the inner surface of the internal spermatic fascia, and a visceral layer firmly attached to the front and sides of the testis and epididymis. The posterolateral aspect of the visceral layer is tucked between the body of the epididymis and the testis to form a slitlike recess, the **sinus of the epididymis.**

In longitudinal section the testis presents the **tunica albuginea,** a dense, tough, fibrous, outer coat, and the **mediastinum,** a longitudinal thickened ridge along the posterior edge of the testis. The mediastinum is traversed by nerves, veins, lymph vessels, and a network of channels, the **rete testis.** Radiating fibrous septa, passing from the mediastinum to the tunica albuginea, separate the testis into about 250 **lobules,** which contain the **seminiferous tubules.** These unite to form **straight tubules,** passing to the rete testis within the mediastinum. The rete testis continues as the **efferent ducts** to the head of the epididymis.

DESCENT OF THE TESTIS

The testis develops high on the posterior abdominal wall. In late fetal life it descends retroperitoneally, traverses the inguinal canal, and shortly before birth enters the scrotum. During descent the testis carries with it the nerves, blood vessels, and lymphatics it acquired during development. Hence, pain from testicular disease may be referred to the renal region, and conversely, kidney disease may cause scrotal pain. Furthermore, testicular cancer will spread initially through lymphatics to the upper lumbar and para-aortic lymph nodes, and only much later to the inguinal nodes.

Anomalies arise from imperfect descent of the testis. It may become arrested in descent anywhere along its path, for example, within the abdomen, along the inguinal canal, especially at the superficial ring, or high in the scrotum. Failure of the testis to descend fully into the scrotum (cryptorchidism) will result in an inhibition of spermatogenesis. Normal sperm development occurs only in the scrotum where the temperature is about 5° F lower than in the abdomen. Indirect inguinal hernias are frequently associated with an undescended testis.

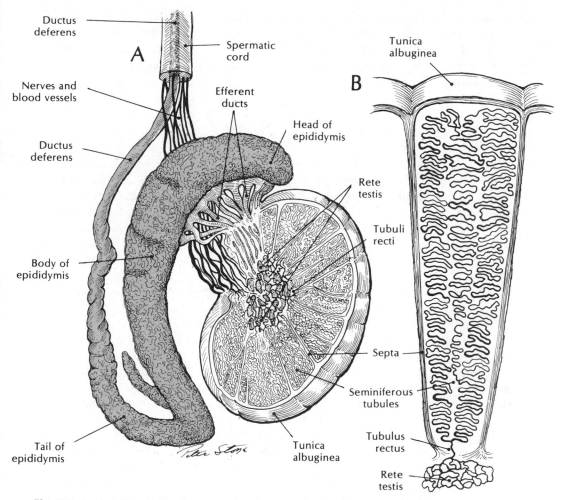

Fig. 5-5. A. Epididymis and testis. Testis is shown in sagittal section. **B.** Lobule of testis. (Langley LL, Telford IR, Christensen JB: Dynamic Anatomy and Physiology, 5th ed. New York, McGraw-Hill, 1980)

Epididymis

The comma-shaped **epididymis** is composed of a **head,** which receives fifteen to twenty efferent ducts from the rete testis; a **body,** separated from the posterior aspect of the testis by the sinus of the epididymis; and a **tail,** which is continuous with the ductus deferens (Fig. 5-5). The spirally coiled efferent ducts form a series of small masses, the **lobules** of the epididymis.

EPIDIDYMITIS

Epididymitis (inflammation of the epididymis) is usually caused by tuberculosis, gonorrhea, or high fever. If the inflammatory process is bi-

lateral, sterility may result. Since the seminiferous tubules of the testis are terminal extensions of the duct of the epididymis (through the efferent ducts and rete testis), inflammation involving one of these structures often spreads to the other, particularly from epididymitis to orchitis.

EXTERNAL FEMALE GENITALIA (FIG. 5-6)

Subcutaneous fat anterior to the pubic symphysis forms a rounded median eminence, the **mons pubis** (mons veneris). The **pudendal cleft,** a midline fissure in the urogenital triangle, is flanked by the **labia majora,** two elongated swellings that are the lateral boundaries of the vulva. The labia majora converge anteriorly at the mons pubis to unite at the lower border of the symphysis pubis as the **anterior commissure.** Posteriorly the labia majora do not unite; however, the forward projection of the perineal body gives the appearance of a posterior commissure, which lies between the vagina and the anus.

The **labia minora** are two thin folds of skin devoid of hair and subcutaneous fat but richly supplied with blood vessels and nerve endings. They flank the vaginal orifice and diverge posteriorly to blend with the labia majora. A transverse fold of skin, the **fourchette,** passes between the posterior terminations of the labia minora. Anteriorly each labium minus divides into two small folds that extend above and below the distal extremity of the clitoris. These folds unite with similar folds of the opposite side to form dorsally the **prepuce,** and ventrally the **frenulum of the clitoris.**

The cleft between the labia minora, the **vestibule,** receives the openings of the vagina, the urethra, and the ducts of the greater vestibular glands. The **external urethral orifice,** a median slitlike aperture, opens posterior to the glans clitoris. The urethral margins are slightly everted. The minute **paraurethral ducts (of Skene)** open into the pudendal cleft.

The vaginal opening is located posterior to the urethral orifice and is narrowed in the virgin by a crescent-shaped fibrovascular membrane, the **hymen.** The **fossa navicularis** is that portion of the floor of the vestibule between the vaginal orifice and the fourchette. **Greater vestibular glands (of Bartholin)** are located bilaterally between the labia and the posterior part of the vestibule at the minora and the vaginal opening and during coitus are compressed to release a mucuslike secretion to lubricate the lower end of the vagina.

BARTHOLIN'S GLAND CYST

Bartholin's gland cyst develops when the duct of the gland becomes blocked, creating a cyst or an abscess. Cysts are best treated by complete excision. An abscess is usually caused by gonorrhea. It often gets started during sexual excitement when the duct is patent and the gonoccoceal organism is present. The cyst may form a relatively large mass that protrudes into the vestibule because the gland is within the superficial perineal compartment, and the superficial fascia limiting the compartment is distensible.

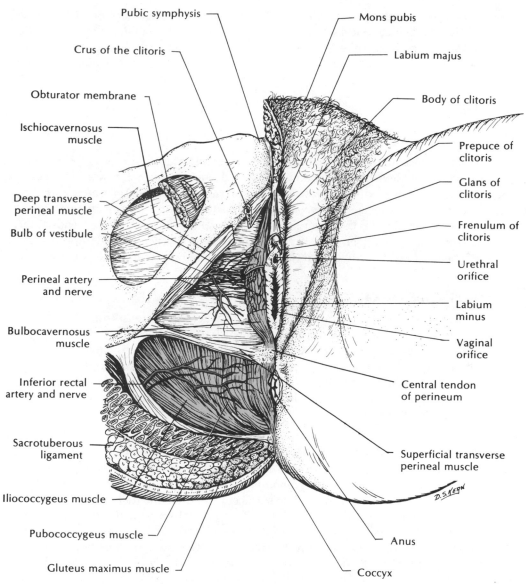

Pubic symphysis

Crus of the clitoris

Obturator membrane

Ischiocavernosus muscle

Deep transverse perineal muscle

Bulb of vestibule

Perineal artery and nerve

Bulbocavernosus muscle

Inferior rectal artery and nerve

Sacrotuberous ligament

Iliococcygeus muscle

Pubococcygeus muscle

Gluteus maximus muscle

Mons pubis

Labium majus

Body of clitoris

Prepuce of clitoris

Glans of clitoris

Frenulum of clitoris

Urethral orifice

Labium minus

Vaginal orifice

Central tendon of perineum

Superficial transverse perineal muscle

Anus

Coccyx

D.S.KERN

Fig. 5-6. Vulva.

Paired elongated masses of erectile tissue forming the **bulbs of the vestibule** are located at the sides of the vaginal orifice and are attached to the superficial layer of the urogenital diaphragm. They are covered by the bulbocavernosus muscle.

The **clitoris,** an erectile organ corresponding anatomically to the male penis, is composed of a body, two crura, and a glans. It differs from the penis in that it is smaller and is not traversed by the urethra. The **body,** formed by the union of the crura, is entirely embedded in the tissues of the vulva and suspended from the pubic symphysis by the **suspensory ligament.** The **crura** of the clitoris are attached to the perineal surface of the ischiopubic rami and to the inferior layer of the urogenital diaphragm. They are covered by the ischiocavernosus muscles. The glans is a small, rounded elevation at the free end of the body. It, like the crura, is composed of erectile tissue and contains abundant sensory nerve endings.

Pelvis

The funnel-shaped **pelvis,** that portion of the trunk inferoposterior to the abdominal cavity, is bounded by the innominate, sacral, and coccygeal bones. The **pelvis minor,** or true pelvic cavity, is below the brim of the pelvis, while the **pelvis major,** or false pelvic cavity, is located between the iliac fossae and is a part of the abdominal cavity. The pelvis contains the lower part of the alimentary canal, the distal ends of the ureters, the urinary bladder, and in the female, most of the internal reproductive organs. The peritoneal cavity extends from the abdominal cavity into the pelvic cavity.

PELVIC MENSURATION

Pelvic mensuration is a procedure that measures the size of the inlet or outlet of the pelvis. The dimensions and shape of the bony pelvis can be most accurately determined by x-ray studies. Because of the danger of radiation to the fetus such procedures may be contraindicated even though such information would be helpful to the obstetrician in determining if the baby can be delivered vaginally or whether a cesarean section is indicated.

For descriptive purposes, the pelvic boundaries can be divided into the lateral and posterior walls and the floor (Fig. 5-7). The **bony framework** of the true pelvis is formed by the portion of the paired innominate bones below the arcuate line, the sacrum, and the coccyx. Most of the pelvic surface of the innominate bone is covered by the obturator internus muscle and its fascia. The gap between each innominate bone and the sacrum is partially filled by the **sacrotuberous** and **sacrospinous ligaments.** The former attaches superiorly to the posterior iliac spine, the lower sacrum, and the upper coccyx, while inferiorly its fibers converge to attach to the ischial tuberosity. An extension of the sacrotuberous ligament, the **falciform process,** passes to the lower margin of the ramus of the ischium. The triangular **sacrospinous ligament** lies internal to the sacrotuberous ligament, with its base attached to the lower sacrum and the upper coccyx, and its apex attached to the ischial spine. The sacrotuberous and sacrospinous ligaments convert the lesser and greater sciatic notches into foramina, with the sacrospinous ligament demarcating the **lesser** from the **greater sciatic foramen.** The latter transmits the piriformis muscle, the superior and inferior gluteal vessels and nerves, the internal pudendal vessels, and the pudendal, sciatic, and posterior femoral cutaneous nerves. The lesser sciatic foramen is traversed by the tendon of the obturator internus, the internal pudendal vessels and pudendal nerve, and nerves to the obturator internus.

The curved **posterior wall** of the pelvis, which faces anteroinferiorly, is composed of the sacrum and coccyx, and is covered internally by the piriformis and coccygeus

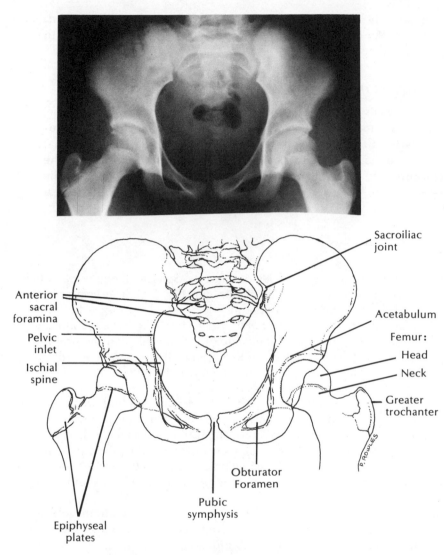

Fig. 5-7. X-ray and schematic of an adolescent pelvis.

muscles and their fasciae. The **floor** of the pelvis includes all structures giving support to pelvic viscera, that is, the peritoneum, the pelvic and urogenital diaphragms, and structures associated with them. Posteriorly the rectum passes through the pelvic floor. The urethra in the male, and the urethra and vagina in the female, penetrate the floor anteriorly. The **peritoneum** covers pelvic viscera to a variable extent, reflecting from the rectum onto the urinary bladder to form the **rectovesical pouch** in the male, and onto the uterus in the female to form the **rectouterine pouch.** Anterior to the recto-uterine pouch, the peritoneum reflects from the anterior surface of the uterus onto the bladder and forms the **uterovesical pouch.** Subserous connective tissue between the peritoneum and the pelvic diaphragm varies in thickness and contains blood vessels and nerve plexuses to pelvic viscera, the lower part of the ureter, and the proximal part of the ductus deferens. Localized connective tissue thickenings form ligaments that aid in the support of various organs.

PARACENTESIS IN THE RECTOUTERINE SPACE

The rectouterine space (pouch of Douglas) is the inferiormost portion of the peritoneal cavity. Therefore, blood and pus in the peritoneal cavity tend to settle in this region, causing pelvic abscesses that must be drained. Drainage may be accomplished by opening the abscess through the wall of the vagina (paracentesis). The needle is passed through the posterior vaginal fornix into the peritoneal cavity.

MUSCLES

The levator ani and coccygei muscles, and fasciae covering the upper and lower surfaces of these muscles form the **pelvic diaphragm** (Fig. 5-8). The **levator ani** is a wide, thin, curved sheet of muscle, variable in thickness, which forms the muscular floor of the true pelvis and separates the pelvic cavity from the ischiorectal fossae. A narrow midline gap permits passage of the vagina in the female and the urethra and rectum in both sexes. The levator ani is subdivided into three parts, the pubococcygeus, puborectalis, and iliococcygeus muscles, which may be differentiated by the position, direction, and attachment of their fibers.

The **pubococcygeus** forms the main part of the levator ani and originates from the posterior aspect of the body of the pubis and tendinous arch of the levator ani. Its more lateral fibers insert into the perineal body, the wall of the anal canal, and the anococcygeal body. In the male, its most medial fibers insert into the prostate as the **levator prostatae muscle,** while in the female these fibers insert into the urethra and vagina as the **pubovaginalis muscle.** Fasciculi of the latter encircle the urethra and vagina to form the **sphincter vaginea.**

The most conspicuous portion of the levator ani, the **puborectalis muscle,** passes posteriorly from the pubis and unites with fibers from the opposite side to form a muscular sling behind the rectum near its anorectal junction.

DEFECATION

In order to pass a solid stool the axis of the rectum and anal canal must assume a more or less straight line. Therefore, defecation necessitates a relaxation of the puborectalis (rectal sling), which allows the anal canal to become aligned with the rectum.

The **iliococcygeus,** although extending over a relatively large area, is often the most poorly developed portion of the levator ani. It originates from the tendinous arch and the ischial spine and passes obliquely inferiorly to insert into the sides of the coccyx and the anococcygeal raphe.

Located posterior to the levator ani, the **coccygeus** (ischiococcygeus) may be

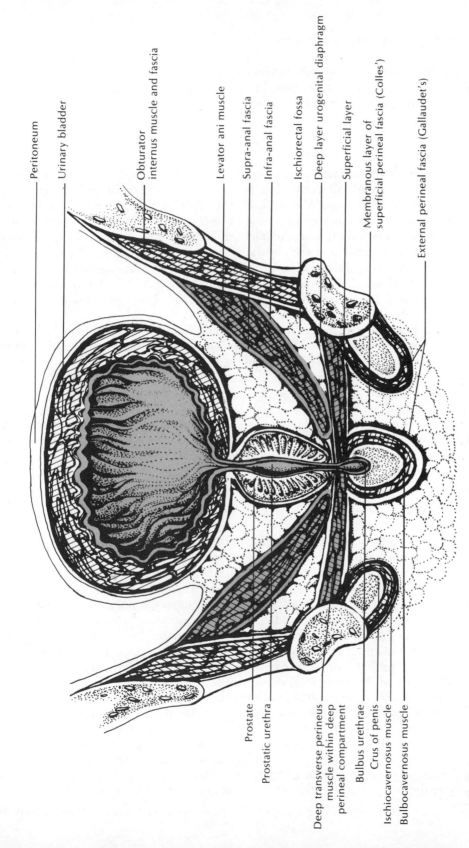

Peritoneum

Urinary bladder

Obturator internus muscle and fascia

Levator ani muscle

Supra-anal fascia

Infra-anal fascia

Ischiorectal fossa

Deep layer urogenital diaphragm

Superficial layer

Membranous layer of superficial perineal fascia (Colles')

External perineal fascia (Gallaudet's)

Prostate

Prostatic urethra

Deep transverse perineus muscle within deep perineal compartment

Bulbus urethrae

Crus of penis

Ischiocavernosus muscle

Bulbocavernosus muscle

Fig. 5-8. Perineal compartments and fasciae.

present only as tendinous strands. It arises from the pelvic aspect of the ischial spine, and its fibers spread out to insert into the lateral and lower margins of the sacrum and the upper part of the coccyx. It lies on the internal aspect of the sacrospinous ligament.

FASCIAE

The **endopelvic fascia** is a continuation of the endoabdominal fascia, which passes onto the lateral pelvic wall as the iliac fascia to become the **obturator fascia** at the brim of the pelvis. The latter, the most definite layer of fascia in the pelvis, lines the internal surface of the obturator internus and, at the margins of this muscle, fuses with the periosteum. Inferiorly it joins the falciform process of the sacrotuberous ligament and, at the anterior margin of the obturator foramen, fuses with the obturator membrane to form the floor of the obturator canal. The obturator fascia gives origin to most of the levator ani and, below the origin of this muscle, forms the lateral walls of the ischiorectal fossae. Internal to the ischial tuberosity, the obturator fascia splits to form the pudendal canal.

The **supra-anal fascia**, attaching at the tendinous arch of the levator ani, covers the pelvic surface of the levator ani and coccygeus muscle. **Infra-anal fascia**, thinner than the above, covers the lower surface of the levator ani and coccygeus muscles. Anteriorly a midline thickening of these fasciae forms the **puboprostatic (pubovesical** in the female) **ligament;** posteriorly the fasciae thin out to cover the coccygeus and fuse with the sacrospinous ligament.

The thin **obturator membrane** closes the obturator foramen except anterosuperiorly, where a gap, the **obturator canal,** transmits the obturator nerve and vessels to the adductor region of the thigh. The obturator membrane also gives partial origin to the obturator internus and externus muscles.

NERVES

Innervation to pelvic structures is from the **sacral** and **coccygeal spinal nerves** and the **sacral portion of the autonomic nervous system.** Each of the five sacral and the coccygeal nerves divides into dorsal and ventral rami within the sacral canal. Dorsal rami of the first through fourth sacral nerves pass through the posterior sacral foramina and divide into a medial muscle branch to the erector spinae muscles and a lateral cutaneous branch that forms a series of loops to give perforating branches (clunial nerves) to the skin over the buttock. Dorsal rami of the fifth sacral and coccygeal nerves pass through the sacral hiatus to supply the skin over the coccyx and around the anus. The ventral rami of the first four sacral nerves pass through sacral foramina into the pelvis, while the ramus of the fifth sacral nerve passes between the coccyx and the sacrum. The ventral rami of the first and second sacral nerves are the largest components of the sacral plexus, and thereafter the ventral rami decrease in size. The ventral ramus of the coccygeal nerve passes below the rudimentary transverse process of the coccyx to form, with the fourth and fifth sacal nerves, the small coccygeal plexus. The second, third, and fourth sacral nerves give off **pelvic splanchnic nerves,** (nervi erigentes), which transmit parasympathetic preganglionic fibers to the pelvic autonomic plexuses supplying the large intestine below the level of the splenic flexure, as well as the pelvic viscera.

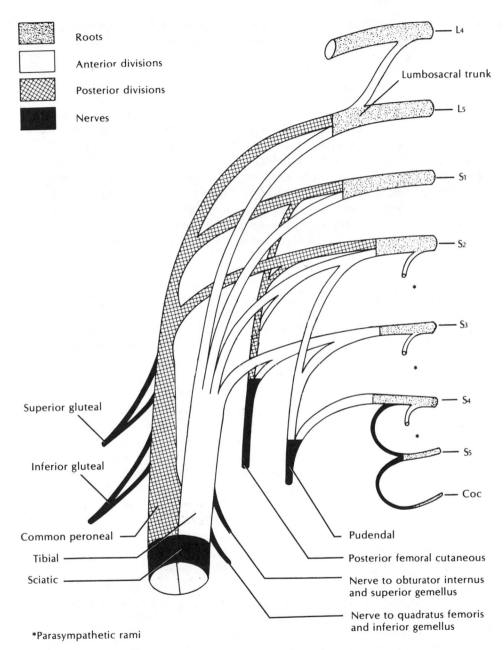

Roots

Anterior divisions

Posterior divisions

Nerves

L4

Lumbosacral trunk

L5

S1

S2

*

S3

*

S4

*

S5

Coc

Superior gluteal

Inferior gluteal

Common peroneal

Tibial

Sciatic

Pudendal

Posterior femoral cutaneous

Nerve to obturator internus
and superior gemellus

Nerve to quadratus femoris
and inferior gemellus

*Parasympathetic rami

Fig. 5-9. Sacral plexus.

Sacral Plexus

The **sacral plexus,** situated largely anterior to the piriformis muscle, is formed by the ventral rami of the fourth and fifth lumbar nerves, and the first four sacral nerves (Fig. 5-9). The ventral ramus of the fourth lumbar nerve divides into an upper and lower segment. The latter joins with the ventral ramus of the fifth lumbar to form the

lumbosacral trunk, which descends to join the sacral plexus. Twelve named nerves are described as arising from the sacral plexus. Seven are directed to the buttock and lower limb, and five supply pelvic structures.

GLUTEAL INJECTIONS

Since the gluteal region is a common site for needle injections, the sciatic nerve may be injured by a poorly placed intramuscular injection. Injury is usually avoided if the injection is made in the upper outer quadrant of the buttock, which is far removed from the sciatic nerve and large blood vessels. Most nerve damage from injections affect the peroneal division of the sciatic nerve, which causes loss of power to dorsiflex the ankle or to extend the toes, and is called foot drop. Other complications of parenteral injections include embolism, hematoma, abscess, intravascular injection of drugs, and sloughing of skin.

The **superior gluteal nerve** (L_4, L_5, and S_1) passes posteriorly through the greater sciatic foramen above the level of the piriformis muscle. It is accompanied in its course by the superior gluteal artery and vein. It innervates the gluteus medius and minimus muscles and terminates in the tensor fascia lata muscle. The **inferior gluteal nerve** (L_5, S_1, and S_2) passes posteriorly through the greater sciatic foramen below the level of the piriformis muscle to supply the gluteus maximus.

The **nerve** to the **quadratus femoris** and **inferior gemellus** muscles (L_4, L_5, and S_1) leaves the pelvis through the greater sciatic foramen inferior to the piriformis muscle. The **nerve** to the **obturator internus** (L_5, S_1, and S_3) exits by way of the same foramen and, after giving a branch to the superior gemellus, curves around the ischial spine to pass through the lesser sciatic foramen to supply the obturator internus.

The **posterior femoral cutaneous nerve** (S_1, S_2, and S_3) also leaves the pelvis through the greater sciatic foramen, passing inferior to the piriformis muscle. It accompanies the inferior gluteal vessels to the inferior border of the gluteus maximus. It courses down the thigh, superficial to the biceps femoris but deep to the fascia lata. It innervates skin over the posterior aspect of the thigh and leg.

Inferior clunial (perforating cutaneous) **nerves** (S_2 and S_3) pierce the gluteus maximus and deep fascia midway between the coccyx and the ischial tuberosity to supply skin of the lower gluteal region.

The largest nerve in the body, the **sciatic** (L_4, L_5, S_1, S_2 and S_3), enters the gluteal region through the lower part of the greater sciatic foramen. It traverses the gluteal region in the interval between the greater trochanter of the femur and the ischial tuberosity and descends under cover of the gluteus maximus. It is related anteriorly, in sequence from superior to inferior, to the nerve to the quadratus femoris, the tendon of the obturator internus, the two gemelli, and the quadratus femoris muscles. As it enters the posterior compartment of the thigh, it descends on the posterior aspect of the adductor magnus muscle to divide into the **tibial (medial popliteal)** and **common peroneal (lateral popliteal) nerves.** Its distribution will be considered with the inferior extremity.

The **nerve to the piriformis** (S_1 and S_2) enters the anterior aspect of this muscle directly, while the **nerve to the levator ani and coccygeus** (S_3 and S_4) descends on the deep aspect of these muscles to innervate them. The **nerve to the external anal sphincter** (S_4) passes either through the coccygeus or between this muscle and the levator ani to continue forward in the ischiorectal fossa. It supplies the external anal sphincter and the skin surrounding the anus.

The **pudendal nerve** (S_2, S_3, and S_4) supplies most of the perineum. It emerges from the pelvis through the greater sciatic foramen inferior to the piriformis muscle, crosses the posterior aspect of the ischial spine medial to the internal pudendal artery, then passes through the lesser sciatic foramen to enter the pudendal canal. Its branches include several inferior rectal twigs, which pass across the ischiorectal fossa to supply the rectum and anal canal. It also gives rise to perineal nerves, which supply cutaneous innervation to the perineum, and muscular branches to muscles in the urogenital triangle. Its terminal branch is the dorsal nerve of the penis or clitoris. The **inferior rectal nerve** may arise anywhere along the course of the pudendal nerve but usually pierces the fascia forming the pudendal canal to cross the ischiorectal fossa and to supply the external anal sphincter and the overlying skin. It also innervates the lining of the anal canal below the pectinate line. The **perineal nerve** divides into a **superficial** and a **deep branch** within the pudendal canal. The former supplies **posterior scrotal** or **labial nerves,** which are distributed to the posterior aspect of the scrotum or the labium majus and to the skin of the perineum. The **deep branch** pierces the medial wall of the pudendal canal to supply twigs to the levator ani and the external anal sphincter. It also sends branches to structures in the superficial perineal space, namely, the bulbocavernosus, ischiocavernosus, superficial transverse perineus muscles, and the bulb of the penis. The deep perineal subsequently pierces the posterior margin of the urogenital diaphragm to supply the deep transverse perineus muscle, the sphincter urethra, and the corpus cavernosum penis. It then continues anteriorly to pass between the lamellae of the suspensory ligament of the penis. The terminal branch of the pudendal is the **dorsal nerve of penis,** which runs forward on the dorsum of the penis supplying the skin, prepuce, and glans of that organ (Table 5-2).

PUDENDAL NERVE BLOCK

Pudendal nerve block is perhaps the safest anesthesia for childbirth and is also effective for procedures on the female external genitalia. The primary innervation to the skin and muscles of the perineum is the pudendal nerve. Anesthesia is possible through two routes. In the perineal approach, a needle is inserted through the skin of the perineum towards the ischial tuberosity. The anesthetic is infiltrated just medial to and behind the ischial spine, where the pudendal nerve lies in the pudendal (Alcock's) canal. In the transvaginal approach, the needle is passed through the lateral vaginal wall to a point just medial to the ischial spine. In both procedures, it is necessary to block the pudendal nerves bilaterally to obtain adequate anesthesia.

Table 5-2. Nerve Distribution to Perineum and Pelvis

Nerve	Origin	Course	Distribution
Pudendal	Anterior divisions of sacral plexus	Passes through greater sciatic foramen below level of piriformis; into ischiorectal fossa via lesser sciatic foramen; in ischiorectal fossa traverses pudendal canal, to reach UG diaphragm	Supplies perineum and external genitalia
Inferior rectal	Pudendal in ischiorectal fossa	Passes transversly across ischiorectal fossa	External anal sphincter
Superficial perineal	Pudendal at posterior border of UG diaphragm	Traverses superficial perineal compartment to reach skin	Muscles of superficial compartment (ischiocavernosus, bulbocavernosus, superficial transverse perineus) and skin of perineum
Deep perineal	Continuation of pudendal at posterior border of UG diaphgram	Traverses deep perineal compartment	Muscles of deep compartment (deep transverse perineus, external urethral sphincter, sphincter vaginae)
Dorsal nerve of penis (clitoris)	Terminal branch of pudendal at anterior part of UG diaphragm	Dorsum of penis or clitoris	Skin of penis or clitoris
Autonomic plexuses	Form around branches of internal iliac artery; parasympathetic (preganglionic) from pelvic splanchnic ($S_{2,3,4}$); sympathetic (postganglionic) from sacral splanchnics	Follow arteriolar branchings to walls of organs; postganglionic parasympathetics are in walls of organs	Parasympathetics— nerve of erection (vasodilatation); contraction of detrusor muscle of bladder and muscular wall of rectum; relaxation of internal sphincters of rectum and bladder. Sympathetics have opposite action and aid in ejaculation.

Coccygeal Plexus

The **coccygeal plexus** is formed by the lower division of the ventral ramus of the fourth sacral nerve, the ventral ramus of the fifth sacral, and the coccygeal nerve. It supplies twigs to the sacrococcygeal joint and the skin over the coccyx.

Autonomic Nerves

The **pelvic splanchnic nerves (nervi erigentes)** are slender filaments passing from the second, third, and fourth sacral nerves. They transmit **preganglionic parasympathetic fibers** to the pelvic plexuses and have their cell bodies at the above-named levels of the spinal cord. They also supply parasympathetic innervation to the large intestine distal to the splenic flexure.

Pelvic autonomic plexuses receive their **sympathetic contribution** from either the inferior extension of the vertebral trunk, or the downward continuation of the preaortic plexuses. The sacral portion of the sympathetic trunk consists of three or four ganglia lying on the anterior aspect of the sacrum, just medial to the sacral foramina. It ends in a fusion of the two sympathetic trunks in the midline to form the **ganglion impar.** Gray rami communicantes pass to sacral and coccygeal spinal nerves and are distributed by way of those nerves to the inferior extremity, perineum, and pelvis.

The **superior hypogastric autonomic plexus** is located between the common iliac arteries. It receives its component fibers from the lower lumbar splanchnic nerves and descending fibers of the inferior mesenteric plexus. It continues inferiorly, divides, and passes to either side of the rectum as the **inferior hypogastric plexus.** Further subdivisions are designated according to the vessels they follow or the organs they supply, as for example, the middle rectal, prostatic, vesical, uterovaginal, and cavernous plexuses of the penis or clitoris. These plexuses contain postganglionic sympathetic fibers, postganglionic parasympathetic cell bodies and fibers, and visceral efferent (sensory) fibers.

ARTERIES

The **internal iliac (hypogastric),** the smaller of the two terminal branches of the common iliac artery, arises at the level of the lumbosacral articulation (Fig. 5-10). It usually divides into an **anterior** and **posterior trunk** before giving origin to its several named branches. Branches from the trunks are not constant, but usually all visceral branches (the vesicular, uterine, vaginal, and rectal) arise from the anterior trunk, as do the obturator, inferior gluteal, and internal pudendal arteries.

In the male, the **superior vesical (umbilical) artery** gives off a branch close to its origin, the **artery to the ductus deferens,** before it continues along the lateral pelvic wall to the apex of the bladder. This small branch supplies the lower end of the ureter, the ductus deferens, the seminal vesicle, and part of the bladder. The superior vesical artery gives one or more twigs to the bladder before becoming a solid cord, the obliterated umbilical artery, which is embedded in a fold of peritoneum on the internal aspect of the anterior abdominal wall as it extends toward the umbilicus.

The **obturator artery** passes along the lateral pelvic wall on the surface of the obturator internus to supply the muscle. With the obturator nerve and vein, it passes through the obturator canal to be distributed to the muscles in the adductor compartment of the thigh. The **uterine artery** passes inferiorly along the lateral pelvic wall and turns medially at the base of the broad ligament to pass to the cervicouterine junction. Here it gives rise to one or more **vaginal branches,** ascends on the uterus, which it supplies, and terminates as twigs to the uterine tube.

URETER/UTERINE ARTERY RELATIONSHIP

In gynecological surgery, especially hysterectomy, the relationship of the ureter to the uterine artery is critically important. The ureter is contiguous to the artery as it courses towards the bladder. Thus, the surgeon must dissect or visualize the ureter to avoid clamping, ligating, or cutting it if he needs to transect the uterine artery.

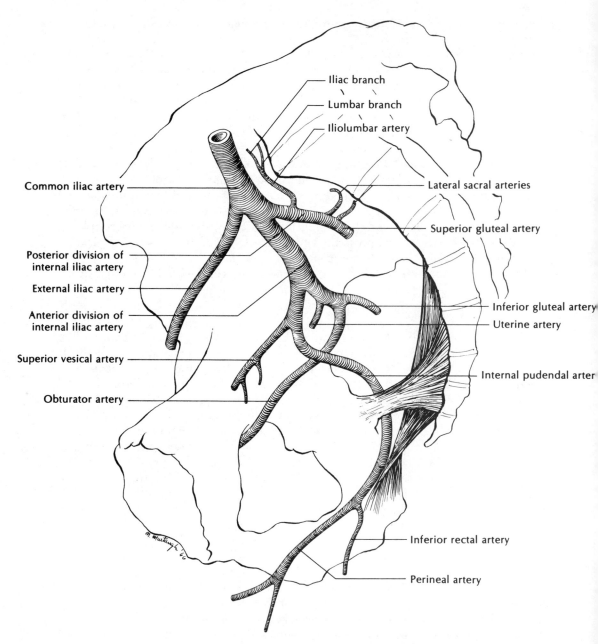

Fig. 5-10. Arterial supply of the pelvis.

The **middle rectal artery** is variable in both origin and size as it passes to the rectum, just superior to the pelvic diaphragm.

The **inferior gluteal artery** enters the gluteal region below the piriformis muscle at the lateral edge of the sacrotuberous ligament and supplies the gluteus maximus and other muscles in this area. It gives off the **artery to the sciatic nerve,** which passes inferiorly to enter the nerve. The **internal pudendal artery** emerges from the pelvic cavity, in company with the inferior gluteal artery and the pudendal nerve, then winds around the ischial spine to pass through the lesser sciatic notch. With the nerve and a corresponding vein, it courses in the pudendal canal to be distributed to the perineum.

The posterior trunk of the internal iliac artery gives rise to three branches. One of these, the **iliolumbar artery,** divides near its origin into an iliac and a lumbar branch. The **iliac branch** crosses the pelvic brim and passes deep to the common iliac or external iliac artery and the psoas major muscle to supply structures in the iliac fossa. The **lumbar branch** ascends, coursing parallel to the lumbosacral trunk, to supply the psoas muscles, and then gives twigs to the structures within the vertebral canal. Usually, two **lateral sacral arteries** pass into the vertebral canal to supply the cord, meninges, and spinal nerves.

The **superior gluteal artery** is the largest branch of the internal iliac. It leaves the pelvic cavity through the greater sciatic foramen above the level of the piriformis muscle to enter the gluteal region. It usually courses between the two elements of the lumbosacral trunk, or between the lumbosacral trunk and the first sacral nerve. Passing between the gluteus maximus and medius muscles, it divides into a superficial and a deep branch. The former supplies the gluteus maximus, and the latter further divides into superior and inferior branches to supply the gluteus medius and minimus.

URINARY PELVIC ORGANS

Urinary Bladder

The size, shape, and position of the **urinary bladder** varies with its contents (Fig. 5-11). The distended bladder is spherical, rises into the abdomen, and has an average capacity in the adult of about 500 ml. The empty bladder lies on the pubis and the adjacent pelvic floor at the level of the pelvic inlet. It is usually slightly lower in position in the female than in the male. The flattened posterior surface forms the **base (fundus)** of the bladder. The apical superior surface is covered with peritoneum and is convex when filled, but concave and resting on the other bladder surfaces when empty. The inferolateral surfaces rest on the pelvic diaphragm and are continuous with the superior surface at the **apex,** from which the **urachus** (median umbilical ligament) extends to the umbilicus. The **body** of the bladder lies between the apex and the fundus, while the neck surrounds the internal urethral orifice.

The **anterior surface** of the bladder has no peritoneal covering and faces anteroinferiorly toward the pubic symphysis, from which it is separated by the **prevesical space (of Retzius).** The **inferolateral surface** is separated from the levator ani and the obturator internus muscles by extraperitoneal tissue enclosing the vesicular vessels. In the male, it is related posteriorly to the ductus deferens, and its **posterior surface** is in direct contact with the anterior wall of the rectum, the ampulla of the ductus deferens, and the seminal vesicles. The neck is related inferiorly to the prostate and, in the female, adherent to the cervix of the uterus and the anterior wall of the vagina.

PROLAPSE OF THE BLADDER
Prolapse of the bladder is usually a result of repeated obstetric trauma. If the pelvic floor is damaged from laceration of the perineum and sep-

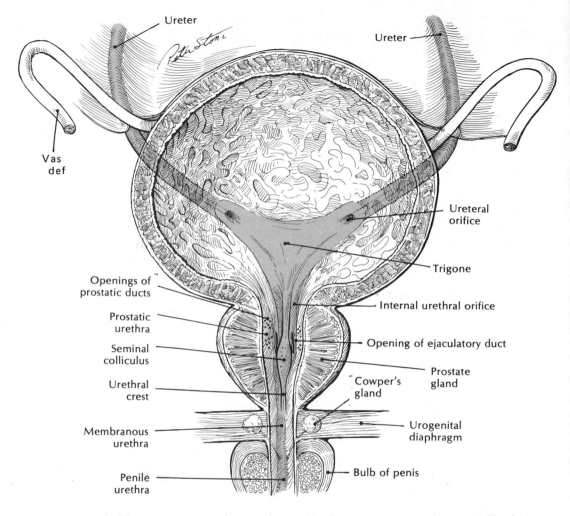

Fig. 5-11. Bladder, prostate, urethra, and associated structures. (Langley LL, Telford IR, Christensen JB: Dynamic Anatomy and Physiology, 5th ed. New York, McGraw-Hill, 1980)

aration of the levator ani muscle, the pelvic diaphragm (floor) sags downward and becomes funnel shaped. Thus, the most inferior portion of the bladder will be below the opening of the urethra. In such conditions, recurrent infections may occur due to residual retention of urine. Retrograde infection may involve the kidney, resulting in nephritis. Surgical intervention may be necessary to realign the bladder to its normal anatomic position.

The **neck** is the least movable portion of the bladder. It is firmly anchored to the pelvic diaphragm and is continuous, in the male, with the prostate, where an external groove demarcates the separation of the two organs. The **medial puboprostatic (pubo-**

vesicular) ligament passes from the body of the pubis to the anterior aspect of the prostate in the male, or the neck of the bladder in the female. The **lateral puboprostatic (pubovesicular) ligament** extends from the tendinous arch of the levator ani to the capsule of the prostate or the neck of the bladder. Lateral ligaments (condensations of subserous fascia) from the base of the bladder pass posterolaterally and posteriorly to continue as the **rectovesical folds** in the male, and the **rectouterine folds** in the female.

In the empty bladder the internal surface is thrown into folds and modified at the posterior aspect to form a smooth triangular area, the **trigone.** The angles of the trigone are marked by the orifices of the two ureters and the urethra. Between the ureteric orifices, a transverse ridge, the **interureteric fold** (plica interureterica), is formed by the underlying musculature. Later extensions of this fold, **plicae uretericae,** are formed by the passage of the ureters through the wall of the bladder. The **uvula,** a median longitudinal ridge above and behind the internal urethral orifice, is formed by the underlying median lobe of the prostate. The **internal urethral orifice,** situated at the lowest point of the bladder, is encircled by a thickening of the smooth muscle of the wall of the bladder, the **internal urethral sphincter.**

MALE PELVIC ORGANS

Prostate

The **prostate,** situated behind the pubic symphysis and below the urinary bladder, is formed by smooth muscle and collagenous fibers, in which is embedded secretory glandular tissue (Fig. 5-11). The **apex** is the lowermost part of the prostate, and the **base,** penetrated by the internal urethral orifice, lies horizontally and fuses with the wall of the more superiorly located bladder. Peripherally a narrow groove separates the two organs. The convex **inferolateral surfaces** are surrounded by the prostatic plexus of veins, while the narrow **anterior surface** is separated from the pubis by the retropubic fat pad. The flattened triangular **posterior surface,** which may have a more or less prominent median groove, can be palpated rectally. The prostate consists of two lateral and a median lobe. No superficial demarcation is present between the **lateral lobes** which are connected by the isthmus anterior to the urethra. The **median lobe,** responsible for the formation of the uvula, is variable in size and projects inwardly from the upper part of the posterior surface between the ejaculatory ducts and the urethra. The enlargement of the median lobe may block urinary flow.

BENIGN HYPERTROPHY OF THE PROSTATE

Benign hypertrophy of the prostate is present in 20% of men over 50 years of age. The incidence increases about 10% with each decade of life, reaching about 60% in men over 80 years old. The clinical significance of this condition is not so much the size of the gland but whether it interferes with the passage of urine. Hypertrophy of the median lobe is most likely to cause urinary obstruction. Surgery is indicated if urination becomes difficult or impossible, or if the bladder retains a large volume of urine after urination. Benign hypertrophy may be a disposing factor in cancer of the prostate, which is the second

most common malignancy in the male. It is detected by performing a careful rectal exam.

Ductus Deferens

The **ductus deferens** begins as a continuation of the ductus epididymis and passes as a component of the spermatic cord to the deep inguinal ring. From the deep ring it ascends anterior to the external iliac artery, then turns posteriorly to enter the pelvic cavity. It passes onto the lateral pelvic wall, medial to the umbilical artery and obturator vessels, crosses the ureter and continues medially to reach the posterior aspect of the bladder. Near the base of the prostate it enlarges and becomes tortuous, as the **ampulla,** and then is joined by the duct of the seminal vesicle to form the **ejaculatory duct.** The latter penetrates the base of the prostate and passes anteroinferiorly to enter the prostatic portion of the urethra just lateral to the colliculus seminalis.

Seminal Vesicle

The **seminal vesicles** are two large, sacculated pouches approximately 5 cm in length. They consist of blind coiled tubes with several diverticula which secrete the seminal fluid. When the bladder is empty, the seminal vesicles lie horizontally; when the bladder is distended, they are nearly vertical in position. The upper parts of the seminal vesicles are separated from the rectum by the rectovesical pouch. The terminal parts of the ureters and ampullae of the ductus deferens are located medially, and the prostatic and vesical venous plexuses laterally.

Urethra

The male **urethra** extends from the bladder to the external urethral orifice in the glans penis (Fig. 5-11). It serves as a passage of both urine and semen and consists of the prostatic, membranous, and penile portions. The **prostatic portion** passes through the substance of the prostate where the posterior wall presents a longitudinal fold, the **urethral crest.** The latter is elevated at its midpoint into an enlargement, the **colliculus seminalis.** A small depression at the center, the **prostatic utricle,** corresponds developmentally to the vagina of the female. The ejaculatory ducts open as longitudinal slits at either side of the colliculus. More distally, small orifices of the ducts of the prostate are present.

The **membranous portion** penetrates the urogenital diaphragm, where it is surrounded by the **external urethral sphincter muscle** in the deep perineal compartment. Ducts of the two small **bulbourethral glands** lie at either side of the urethra in the deep pouch and pierce the inferior layer of the urogenital diaphragm to enter the penile urethra.

EXTRAVASATION OF URINE

Extravasation of urine occurs most frequently following rupture of the cavernous portion of the urethra. Such damage is common in straddle injuries, for example, striking the frame of a bicycle, the rail of a fence,

or the horn of a saddle. **The bulb of the penis often tears and urine escapes through the superficial perineal pouch into the subcutaneous perineal cleft. It then passes anteriorly in the fascial cleft to reach the anterior abdominal wall deep to Scarpa's (deep membranous layer of superficial) fascia. Since there is no anatomic barrier, the urine may reach the thorax unless surgical treatment intervenes. Urine does not pass down the thighs because of the attachment of superficial fascia to the inguinal ligament and ischiopubic rami. It does not pass posteriorly because of a similar fusion of superficial fascia to the posterior border of the urogenital diaphragm.**

The **penile portion** of the urethra is surrounded by the corpus spongiosum penis and, at the glans, flattens laterally to form the **fossa navicularis.** Numerous minute urethral glands (**of Littré**) open into this portion, with the largest of their orifices forming the **urethral lacunae (of Morgagni).**

The **female urethra,** approximately 3 to 4 cm long, extends inferiorly and slightly forward from the neck of the bladder to the external urethral orifice. It passes through the pelvic and urogenital diaphragms and opens between the labia minora, anterosuperior to the vaginal orifice and posteroinferior to the glans clitoris. The urethra is closed except during the passage of urine and is marked internally by longitudinal folds, the most prominent of which is located on the posterior aspect as the **urethral crest.** The female urethra is fused with the anterior wall of the vagina and fixed to the pubis by the **pubovesical ligament.**

INTERNAL FEMALE REPRODUCTIVE ORGANS (FIG. 5-12)

Ovary

The **ovary** is located within a depression, the **ovarian fossa,** on the lateral pelvic wall at the level of the anterior superior iliac spine. It is about the size and shape of an almond and presents medial and lateral surfaces, anterior and posterior borders, and tubal and uterine poles. The **lateral surface** is in contact with parietal peritoneum; the **medial surface,** adjacent to the uterine tube, is in contact with the coils of the ileum.

A ligament, the **mesovarium,** extends between the anterior border at the hilus, and the broad ligament. The posterior free border is related anteriorly to the uterine tube and posteriorly to the ureter.

The suspensory (infundibulopelvic) ligament of the ovary is attached to the **upper (tubal) extremity** with the opening of the uterine tube in close proximity. This peritoneal fold provides the pathway for the major vessels to the ovary. The **lower (uterine) pole,** is directed toward the uterus and attached to the uterus by the ovarian ligament (false ligament of the ovary).

The **mesovarium** is a short, two-layered mesenteric reflection of the posterior layer of the broad ligament of the uterus, which encloses the ovary. Between the broad ligament and the anterior border of the ovary the layers of the mesovarium are in apposition to each other. The **suspensory (infundibulopelvic) ligament** is a connective tissue condensation from the lateral pelvic wall to the superior pole of the ovary and sur-

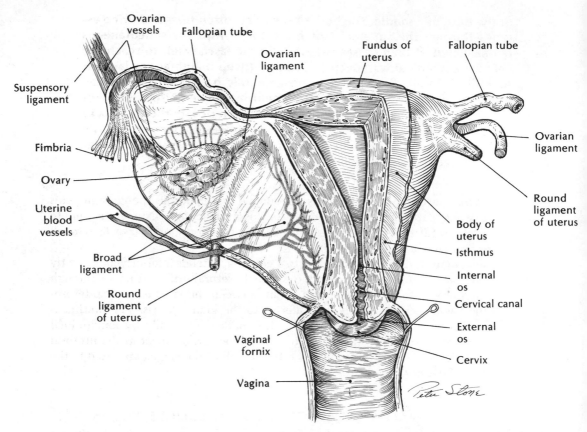

Fig. 5-12. Female reproductive organs with associated ligaments. (Langley LL, Telford IR, Christensen JB: Dynamic Anatomy and Physiology, 5th ed. New York, McGraw-Hill, 1980)

rounds the ovarian vessels and nerve. The (false) **ligament of the ovary** is a rounded cord containing some smooth muscle fibers that passes from the ovary to the uterus, where it attaches between the uterine tube and the round ligament of the uterus.

Uterine Tube

Uterine (fallopian) **tubes** convey ova from the ovaries to the uterine cavity and transmit spermatozoa in the opposite direction. Each tube is approximately 10 cm long and is located between the layers of the upper margin of the broad ligament. The uterine tube courses laterally from the uterus to the uterine end of the ovary to arch over and terminate close to the upper pole of the ovary.

UTERINE TUBE LIGATION

Uterine tube ligation is the tying off of the uterine tube. This prevents the sperm from ascending and the ovum from descending in the tube, thus fertilization is prevented.

The uterine tube is divided into four segments: the infundibulum, ampulla, isthmus, and uterine portion. The **infundibulum** is somewhat funnel shaped, with the abdominal or pelvic opening located at the outlet of the funnel, which has a number of irregular processes, the **fimbriae** projecting from its margins. The slightly tortuous **ampulla** is the longest portion and has relatively thin walls. The narrow, thick-walled **isthmus** is adjacent to the uterus. The **uterine (intramural) portion** is embedded in the wall of the uterus and opens into the uterine cavity.

TUBAL PREGNANCY

Tubal pregnancy is the most common type of ectopic gestation. Implantation of the fertilized ovum generally occurs in the ampulla but may occur in any region of the uterine tube. The trophoblastic cells of the conceptus invade the mucosa and musculature of the tube. As no decidua develops in the mucosa of the tube, as in the uterus, the very thin tubal wall usually ruptures from the pressure of the developing embryo. The resulting hemorrhage is life threatening and demands immediate surgical intervention.

Uterus

The **uterus** is a somewhat pear-shaped organ lying entirely within the pelvic cavity, with its narrow end directed inferoposteriorly. It is composed of the fundus, body, isthmus, and cervix. The **fundus** is the rounded upper portion above the level of the openings of the uterine tubes. The **body** is the main portion of the uterus, which constricts at the **isthmus** to become the **cervix.** The latter pierces the anterior wall of the vagina at the deepest aspect of the vagina. The entire organ forms an angle slightly greater than 90° with the vagina and usually inclines toward the right and is slightly twisted (Fig. 5-13).

The anterior surface of the uterus is separated from the urinary bladder by the **uterovesical pouch** and the posterior surface from the rectum by the **rectouterine pouch.** The right and left margins of the uterus are related to the **broad ligament** with its enclosed structures. The uterine cavity is widest at the entry of the uterine tubes and narrowest at the isthmus. The **cervical canal,** wider above than below, is an extension of the uterine cavity below the **internal ostium (os)** and opens into the vagina at the **external ostium (os).** Vertical folds are present on the anterior and posterior walls of the cervical canal, with **palmate folds** passing obliquely from the vertical folds.

The uterus derives much of its support from direct attachment to the surrounding organs, the vagina, rectum, and bladder, with additional support from peritoneal reflections. The **peritoneum** passes from the posterior surface of the bladder onto the isthmus of the uterus, to continue over its anterior surface, fundus, and posterior surface. It then passes onto the rectum to form the **uterovesical** and **rectouterine pouches** between these reflections. Laterally the peritoneum comes into apposition at the lateral border of the uterus and passes as the **broad ligament** to the pelvic wall, where it reflects to become parietal peritoneum. Superiorly the broad ligament encloses the uterine tubes and the ovaries.

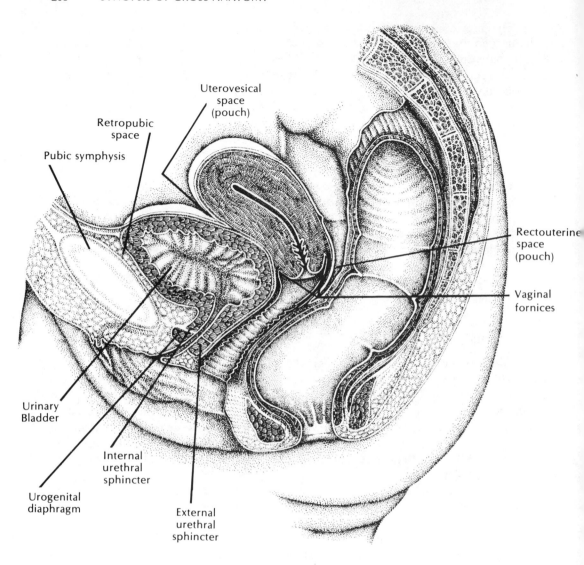

Fig. 5-13. Female pelvic organs.

That portion of the broad ligament between the level of the mesovarium and the uterine tubes is designated the **mesosalpinx.** The **mesovarium** is an extension of the posterior layer of the broad ligament, which encloses the ovary. The **mesometrium** is the remainder of the broad ligament below the level of the mesovarium.

The **round ligament** of the uterus is a narrow band of fibrous connective tissue between the layers of the broad ligament. It attaches to the uterus just inferior and anterior to the entrance of the uterine tube. It passes laterally toward the anterior abdominal wall, hooks around the inferior epigastric vessels, traverses the inguinal canal, and blends with the subcutaneous tissue of the labium majus. The **transverse cervical** (**cardinal or Mackenrodt's**) **ligament** is a thickened band of fascia and connective tissue together with smooth muscle fibers at the junction of the cervix and the vagina, which gives major support to the uterus.

PROLAPSE OF THE UTERUS

Prolapse of the uterus is usually a result of repeated obstetric trauma. If the pelvic floor is damaged from laceration of the perineum and separation of the levator ani muscle, the pelvic diaphragm (floor) sags downwards and becomes funnel shaped. The uterus becomes retroverted. This brings the longitudinal axis of the vagina and uterus in line. Thus, the uterus is able to descend along the vaginal canal, gradually inverting the vagina from above downward, causing prolapse of the vagina. If vaginal prolapse is not corrected, uterine prolapse may occur. In this condition the extruded uterus is often covered by the vagina.

Vagina

The **vagina** is a cylindrical musculomembranous tube, 7 to 10 cm in length. It extends from the vestibule to the uterus. It is constricted inferiorly, dilated at its middle, and narrowed superiorly where it surrounds and is attached to the periphery of the cervix of the uterus. Its posterior wall is longer than the anterior wall, and the space between the cervix and the vaginal wall is designated the **anterior** and **posterior fornices.** Anteriorly the vagina is related to the urethra and the fundus of the bladder; posteriorly it is separated superiorly from the rectum by the rectouterine pouch, and inferiorly from the anal canal by the perineal body; laterally it is attached to the levator ani and its fascia. The mucous membrane of the internal wall presents anterior and posterior longitudinal folds, the **vaginal columns,** and numerous transverse ridges, the **rugae vaginalis.** The entire structure forms an angle slightly greater than ninety degrees with the uterus.

PARACENTESIS

In peritonitis there is an accumulation of body fluid, blood, or pus in the peritoneal cavity. Drainage of the cavity is easily accessible in the female through the posterior fornix of the vagina. Only a few millimeters of tissue separate the vaginal vault from the rectouterine pouch of the peritoneal cavity. However, care must be taken to avoid penetrating the wall of the rectum with the needle.

6

inferior
extremity

Surface Anatomy

The inferior extremity is subdivided into the hip, thigh, knee, leg, ankle, and foot. It is limited superiorly by the iliac crest, inguinal ligament, symphysis pubis, ischiopubic ramus, sacrotuberous ligament, and coccyx. The **iliac crest** is easily palpable in its entirety and extends superiorly as high as the level of the fourth lumbar vertebra. Anteriorly the **anterior superior iliac spine** is palpable and affords the lateral attachment for the **inguinal ligament,** whose position is marked by the inguinal fold (groin). Posteriorly the **gluteal fold** delineates the inferior border of the gluteus maximus muscle. The **ischial tuberosities** are easily felt when the thighs are flexed, as are also the **ischiopubic rami** on the medial aspect of the thighs. Just lateral to the midline, the **pubic tubercles** are readily palpable, especially in the thin individual. Approximately a hand's breadth below the crest of the ilium, the **greater trochanter** of the femur can be palpated about 5 cm posterior to the anterior superior iliac spine.

The massive **quadriceps femoris muscle** tapers inferiorly over the front of the thigh to terminate in the **suprapatellar tendon,** which inserts into the margins of the subcutaneous **patella** and continues as the **patellar ligament** to insert into the **tibial tuberosity.** Laterally the tendons of the **biceps femoris** can be palpated and, with the medially situated tendons of the **semimembranosus** and **semitendinosus,** form prominent cords at the posterior aspect of the knee. With the thigh flexed, abducted, and laterally rotated, the outline of the **sartorius muscle** is visible anteriorly as it crosses the thigh obliquely.

The subcutaneous lateral portions of the **condyles of the femur** give width to the knee. Inferiorly the **head of the fibula** is easily felt at the lateral side of the knee. The **anterior border** and **medial surface of the tibia** are subcutaneous throughout the length of the bone and are continuous proximally with the medial condyle of the tibia. The **malleoli,** formed laterally by the fibula and medially by the tibia, are readily recognized at the ankle. Posteriorly the prominent **calcaneal tendon** serves as the insertion of the soleus and gastrocnemius muscles into the tuberosity of the calcaneus (heel bone).

Fasciae

The **superficial fascia** has special features only in the thigh and on the sole of the foot. In the thigh it contains considerable adipose tissue and varies in thickness, being relatively thick in the inguinal region. Here is it disposed into a **superficial fatty** and a **deeper membranous layer,** with superficial lymph nodes, the great saphenous vein, and smaller blood vessels lying between them. The deeper layer is rather prominent on the medial side, where it blends with the deep fascia and covers the saphenous opening as the **cribriform fascia.** The superficial fascia fuses with the femoral sheath and lacunar ligament superiorly and with the deep fascia laterally. In the sole of the foot the superficial fascia is greatly thickened by fatty tissue disposed into pockets extending inwardly from the skin. These **fibrous fat pads** are especially thick at weight-bearing sites, such as on the heel, ball of the foot, and pads of the toes, where they protect deeply lying structures.

The **deep fascia** of the thigh, the **fascia lata,** varies considerably in thickness and strength. It attaches superiorly to the inguinal ligament, the body of the pubis, the pubic arch, and the ischial tuberosity; laterally to the iliac crest; medially to the sacro-tuberous ligament, and posteriorly to the sacrum and the coccyx. The oval **saphenous opening** at the upper medial portion of the thigh presents a medial crescentic border. It is roofed by the cribriform fascia derived from the deep layer of the superficial fascia, which, in turn, is perforated by the great saphenous vein and other smaller vessels. At the medial side of the thigh the deep fascia is relatively thin, but it thickens laterally as the **iliotibial tract.** The latter is a wide, strong band that extends from the iliac crest to the lateral condyle of the tibia, the capsule of the knee joint, and the patellar ligament. The iliotibial tract affords insertion for the tensor fascia lata muscle and about three-fourths of the gluteus maximus muscle. With the body erect, the iliotibial tract serves as a powerful brace, which helps to steady the pelvis and keep the knee joint firmly extended. Internal extensions of the fasciae latae attach inwardly to the linea aspera of the femur as the **lateral, medial,** and **posterior intermuscular septa,** which separate the thigh into extensor, adductor, and flexor muscular compartments (Fig. 6-1).

Inferiorly the deep fascia attaches to the medial and lateral margins of the patella, the tibial tuberosity, the condyles of both the tibia and the femur, and the head of the fibula. At the posterior aspect of the knee joint the fascia forms a roof for the popliteal fossa. The deep fascia of the leg is firmly attached anteromedially to the subcutaneous shaft of the tibia and sends deep **intermuscular septa** which, with the interosseous membrane, divide the leg into extensor, peroneal, and flexor muscular compartments. The flexor (posterior) compartment is further subdivided by **transverse intermuscular septa** into deep, intermediate, and superficial portions. Inferiorly the deep fascia attaches at both malleoli and then continues onto the foot.

The deep fascia surrounding the ankle thickens to form bands, the extensor, flexor, and peroneal retinacula. These bind or hold the tendons of the leg muscles in place as they extend into the foot. The retinacula prevent bow-stringing of the tendons during contraction. The **extensor retinaculum** is subdivided into a superior and inferior portion. The **superior extensor retinaculum** (transverse crural ligament) is a strong band 2 to 3 cm wide situated immediately above the ankle joint. It is attached laterally to the fibula and medially to the tibia. The **inferior extensor retinaculum** (cruciate ligament) is Y- or V-shaped, with the Y lying on its side. The stem of the Y is attached laterally to the upper surface of the calcaneus and overlies the tendons of the extensor digitorum longus and peroneus tertius. The proximal band passes to the medial malleolus at the medial side of the ankle, while the distal band blends with the deep fascia over the dorsum of the foot. The **flexor retinaculum** (laciniate ligament) bridges the gap between the medial malleolus and the medial surface of the calcaneus and is firmly attached to both structures. Septa extend from the deep surface to the underlying bone and deltoid ligament to form osseofibrous tunnels transmitting structures from the posterior compartment of the leg. The **peroneal retinaculum** is subdivided into superior and inferior components and binds the peroneal tendons in place. Septa from the deep surface pass to the **peroneal trochlea,** forming two osseofibrous tunnels that separate the tendons of the peroneus longus muscle from the peroneus brevis. Both the superior and inferior peroneal retinacula extend from the lateral malleolus to the lateral surface of the calcaneus.

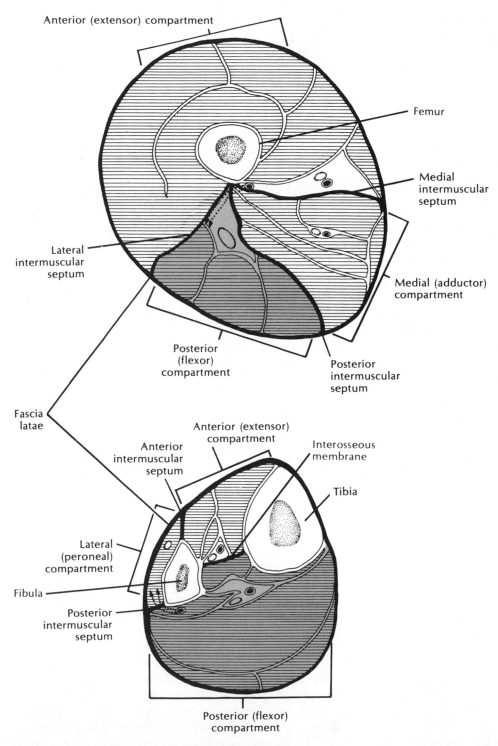

Fig. 6-1. Compartments of the thigh and leg.

Cutaneous Innervation

Skin over the gluteal region is supplied by **lateral branches of the iliohypogastric** and **subcostal nerves** derived from the lumbar plexus and twelfth thoracic nerve (Fig. 6-2). Branches from the dorsal rami, the **superior** and **middle clunial nerves,** supply the skin over the sacrum. The **inferior clunial nerves,** derived from the posterior femoral cutaneous nerve, pass superiorly over the lower border of the gluteus maximus to supply the skin of the inferior aspect of the buttocks (Fig. 6-3).

Cutaneous innervation of the anterior surface of the thigh is supplied by two branches of the femoral nerve, the **medial** and **intermediate femoral cutaneous nerves** (Fig. 6-4). The lateral aspect of the thigh is innervated by anterior and posterior branches of the **lateral femoral cutaneous nerve,** a direct branch of the lumbar plexus. The medial surface of the thigh is supplied by the medial femoral cutaneous and **cutaneous branches from the obturator** and **perineal nerves.** The **femoral branch of the genitofemoral** reaches a small area just below the inguinal ligament. The **genital branch of the genitofemoral,** together with the ilioinguinal nerve, distribute to the skin adjacent to the region of the superficial inguinal ring, the base of the penis and scrotum in the male, and the labia majora in the female. The posterior surface of the thigh is supplied by the **posterior femoral cutaneous nerve** from the sacral plexus.

PHANTOM PAIN

Phantom pain is often felt by patients who have had a limb amputated. They still experience pain or other sensations in the extremity as if the limb were still there. An explanation for this phenomenon is that sensory nerves, that previously received impulses from the now missing limb, are being stimulated by the trauma to the amputation stump. Stimuli from these nerves are interpreted by the brain as coming from the nonexistent (phantom) limb.

The femoral nerve continues into the leg as the **saphenous nerve.** It passes inferiorly to supply the anteromedial and medial surfaces of the leg, the medial side of the foot, and a small area on the medial aspect of the sole. Its **infrapatellar branch,** arising in the thigh, pierces the lower end of the sartorius to ramify on the anterior surface of the leg immediately below the knee. The lateral surface of the upper leg receives its cutaneous innervation from the lateral sural branch of the **common peroneal.** The distal third of the anterior surface of the leg and the dorsum of the foot are innervated by a continuation of the **superficial peroneal nerve,** except for the adjacent sides of the great and second toe, which are supplied by a cutaneous branch of the **deep peroneal.** The **sural nerve,** derived from both tibial and common peroneal nerves, distributes to the posterior surface of the leg, the lateral side of the foot, the little toe, and a small portion of the posterolateral surface of the sole.

The **calcaneal branch of the tibial nerve** supplies most of the plantar surface of the heel, while the anterior two-thirds of the sole is innervated by the **medial** and **lateral plantar nerves,** both cutaneous branches of the tibial. Their areas of distribution to

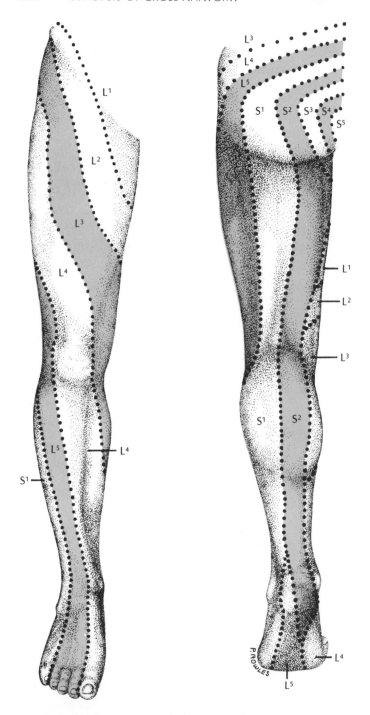

Fig. 6-2. Dermatomes of lower extremity.

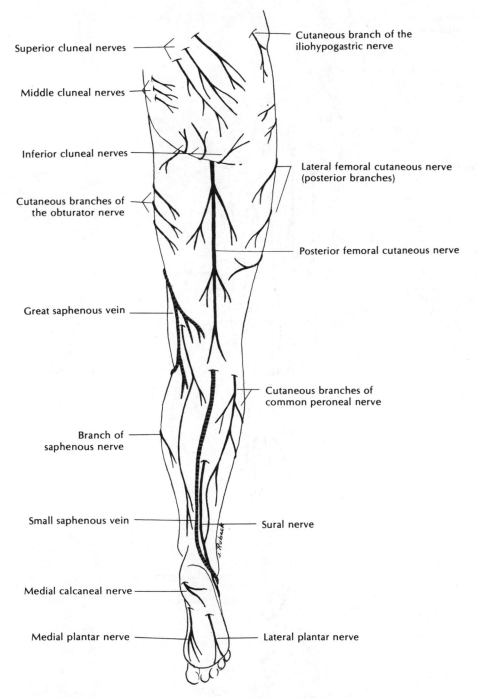

Fig. 6-3. Superficial venous drainage and cutaneous innervation of the posterior aspect of the inferior extremity.

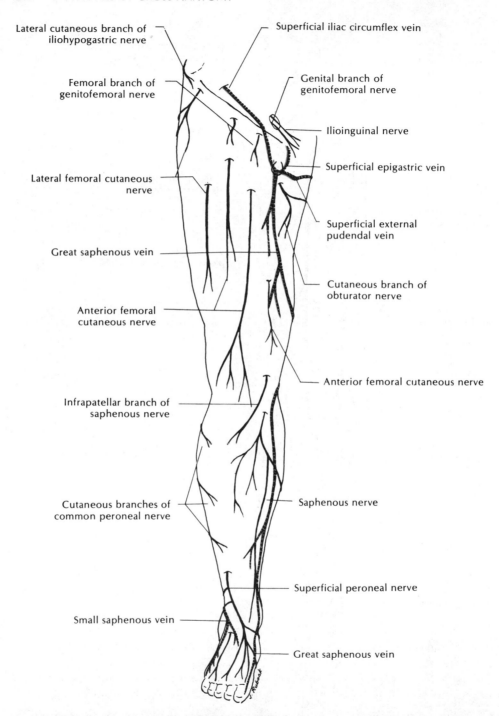

Lateral cutaneous branch of iliohypogastric nerve

Superficial iliac circumflex vein

Femoral branch of genitofemoral nerve

Genital branch of genitofemoral nerve

Ilioinguinal nerve

Lateral femoral cutaneous nerve

Superficial epigastric vein

Superficial external pudendal vein

Great saphenous vein

Cutaneous branch of obturator nerve

Anterior femoral cutaneous nerve

Anterior femoral cutaneous nerve

Infrapatellar branch of saphenous nerve

Cutaneous branches of common peroneal nerve

Saphenous nerve

Superficial peroneal nerve

Small saphenous vein

Great saphenous vein

Fig. 6-4. Superficial venous drainage and cutaneous innervation of the anterior aspect of the inferior extremity.

the sole are divided by an anteroposterior line passing through the midline of the fourth toe.

Venous Drainage

The **superficial venous drainage** of the inferior extremity is by way of two major vessels, the great and small saphenous veins (see Figs 6-3 and 6-4). The **great** (long) **saphenous vein** begins at the junction of the medial digital vein of the great toe and the dorsal venous arch of the foot. From its origin, it ascends anterior to the medial malleolus and along the medial border of the tibia to the knee, where it passes posterior to the medial epicondyle of the femur. It continues on the medial aspect of the thigh to the saphenous opening, where it pierces the cribriform fascia to empty into the femoral vein. In the inguinal region it receives the **superficial epigastric,** the **superficial iliac circumflex,** and the **superficial external pudendal veins.**

INTRAVENOUS THERAPY

The long saphenous is readily available for intravenous entry. It is usually used only in emergency situations because of the greater risk of developing phlebitis when veins of the lower extremity are cannulated.

The **small** (short) **saphenous vein** begins laterally at the dorsal venous arch of the foot and ascends posterior to the lateral malleolus. It continues along the posterior aspect of the leg to the popliteal fossa, where it penetrates the deep fascia to empty into the **popliteal vein.**

The **deep venous drainage** is by way of the **venae comitantes** of the corresponding arteries. Numerous anastomoses occur between the great and small saphenous veins and between the superificial and deep venous drainage.

VARICOSE VEINS OF THE LOWER EXTREMITY

One of the most common disorders of the vascular system is the dilation, elongation, and tortuosity of the superficial veins of the lower limb, called varicose veins. The principal cause of this affliction is an increased venous pressure with dilation of the veins, due largely to valvular incompetence and gravity from the upright position. The increased venous pressure and varicosities cause edema of the skin and subcutaneous tissue with decreased blood flow and poor healing. This venous stasis and edema predisposes the skin to develop ulcers following minor skin abrasions or injury. If the deep veins are patent and can handle the venous return from the foot and leg, then surgical removal or stripping of the superficial varicosed veins decreases the venous pressure in the skin and subcutaneous tissue, reduces the edema, and allows the stasis ulcers to heal.

Table 6-1. Muscles of the Gluteal Region

Muscle	Origin	Insertion	Action	Nerve
Gluteus maximus	Upper portion of ilium, posterior aspect of sacrum and coccyx, and sacrotuberous ligament	Gluteal tuberosity and iliotibial tract	Chief extensor and powerful lateral rotator of thigh	Inferior gluteal
Gluteus medius	Ilium between middle gluteal line and iliac crest	Greater trochanter and oblique ridge of femur	Abducts and rotates thigh medially	Superior gluteal
Gluteus minimus	Ilium between middle and inferior gluteal lines	Greater trochanter and capsule of hip joint	Abducts and rotates thigh medially	Superior gluteal
Tensor fascia lata	Iliac crest and anterior border of ilium	Iliotibial tract	Tenses fascia lata; assists in flexion, abduction, and medial rotation of thigh	Superior gluteal
Piriformis	Internal aspect of sacrum, greater sciatic notch, and sacrotuberous ligament	Upper part of greater trochanter	Rotates thigh laterally	Branches from and S_2
Gemelli superior and inferior	Superior, upper margin of lesser sciatic notch; inferior, lower margin of sciatic notch	Obliquely into either border of tendon of obturator internus and greater trochanter	Rotates thigh laterally	Branches from cral plexus
Quadratus femoris	Lateral border of ischial tuberosity	Posterior aspect of greater trochanter and adjoining shaft of femur	Rotates thigh laterally	L_4, L_5 and S_1

Gluteal Region

Muscles (Table 6-1)

The gluteal region extends from the iliac crest superiorly to the gluteal fold inferiorly. The mass of the buttocks is formed by the **gluteus maximus,** the largest muscle of the body, which overlies the more deeply placed **gluteus medius** and **gluteus minimus muscles.** The gluteus maximus is the chief extensor and the most powerful lateral rotator of the thigh. The medius and minimus act as abductors and medial rotators. The small lateral rotator muscles of the thigh, the **piriformis, superior** and **inferior gemelli,** together with the tendons of the **obturator internus** and **externus,** and **quadratus femoris,** underlie the gluteus maximus, are related to the back of the hip joint, and insert into the greater trochanter of the femur.

Arteries and Nerves

The **superior gluteal artery,** the largest branch of the internal iliac, courses posteriorly from its origin to pass between the lumbosacral trunk and the first sacral nerve. It leaves the pelvic cavity through the greater sciatic foramen above the level of the piriformis muscle (Fig. 6-5). It divides into a **superficial branch,** to supply the gluteus maximus from its deep surface, and a **deep branch** that passes between the gluteus medius and the gluteus minimus to supply both muscles, as well as the obturator internus, piriformis, levator ani, coccygeus muscles, and the hip joint.

The **inferior gluteal artery,** the larger of the two terminal branches of the internal iliac artery, passes posteriorly through the greater sciatic foramen below the level of the piriformis. It supplies the gluteus maximus and the lateral rotator muscles and gives a long slender branch that accompanies and supplies the sciatic nerve.

The **superior gluteal nerve** (L_4, L_5, and S_1) passes through the upper part of the greater sciatic foramen, courses anteriorly between the gluteus medius and the gluteus minimus to supply the gluteus medius, the gluteus minimus, and the tensor fascia lata. The **inferior gluteal nerve** (L_5, S_1, and S_2) leaves the pelvis by traversing the greater sciatic foramen, below the level of the piriformis, to supply the gluteus maximus (Table 6-2).

GLUTEAL INJECTIONS

Since the gluteal region is a common site for needle injections, the sciatic nerve may be injured by a poorly placed intramuscular injection. Injury is usually avoided if the injection is made in the upper outer quadrant of the buttocks, which is far removed from the sciatic nerve and large blood vessels. Most nerve damage from injections affect the peroneal division of the sciatic nerve, which causes loss of power to dorsiflex the ankle or to extend the toes and is called foot drop. Other complications of parenteral injections include embolism, hematoma, abscess, intravascular injection of drugs, and sloughing of skin.

Thigh

ANTERIOR COMPARTMENT OF THE THIGH

Femoral Sheath

The **femoral sheath** is a fascial funnel formed anteriorly by the endoabdominal fascia and posteriorly by the iliac fascia. It encloses the upper 3 to 4 cm of the femoral vessels and is situated deep to the inguinal ligament in a groove between the iliopsoas and the pectineus muscles. The interior of the sheath is divided into three compartments by two anteroposterior septa. The **medial compartment,** bounded medially by

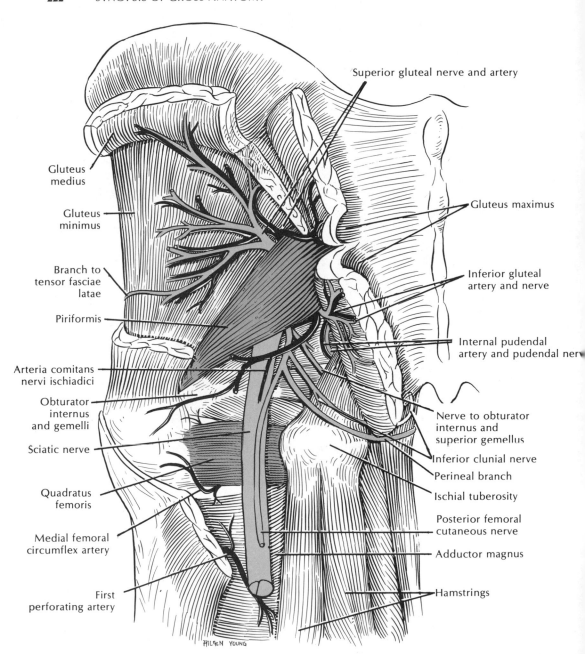

Superior gluteal nerve and artery

Gluteus medius

Gluteus minimus

Branch to tensor fasciae latae

Piriformis

Arteria comitans nervi ischiadici

Obturator internus and gemelli

Sciatic nerve

Quadratus femoris

Medial femoral circumflex artery

First perforating artery

Gluteus maximus

Inferior gluteal artery and nerve

Internal pudendal artery and pudendal nerve

Nerve to obturator internus and superior gemellus

Inferior clunial nerve

Perineal branch

Ischial tuberosity

Posterior femoral cutaneous nerve

Adductor magnus

Hamstrings

AILEEN YOUNG

Fig. 6-5. Gluteal region. (Hollinshead WH: Textbook of Anatomy, 3rd ed. New York, Harper & Row, 1974)

Table 6-2. Nerve Distribution to the Gluteal Region

Nerve	Origin	Course	Distribution
Clunial nerves (superior, middle, inferior)	Dorsal rami of sacral nerves and twigs from posterior femoral cutaneous	Pass through posterior sacral foramina; inferior clunials curve around gluteal fold	Skin over buttock
Posterior femoral cutaneous	Posterior and anterior divisions of sacral plexus	Passes through greater sciatic foramen, below level of piriformis; in thigh, lies on superficial aspect of biceps	Inferior clunial branches to skin over buttock; skin over posterior aspect of thigh
Superior gluteal	Posterior divisions of sacral plexus	Passes through greater sciatic foramen, above level of piriformis, to course between gluteus medius and minimus	Gluteus medius, minimus, and tensor fasciae latae
Inferior gluteal	Posterior divisions of sacral plexus	Passes through greater sciatic foramen, below level of piriformis, to course on deep aspect of gluteus maximus	Gluteus maximus

the crescentic base of the lacunar ligament, forms the **femoral canal** and contains only areolar connective tissue, a small lymph node, and lymphatic vessels. The **intermediate compartment** encloses the **femoral vein,** and the **lateral compartment,** the **femoral artery** and the **femoral branch of the genitofemoral nerve.** Structures passing deep to the inguinal ligament, lateral to the femoral sheath, include the iliopsoas muscle and the femoral and lateral femoral cutaneous nerves.

FEMORAL HERNIA

A femoral hernia results from the potential weakness of the lower abdominal wall at the femoral ring. This condition may allow entrance of a viscus, for example, a loop of small intestine, or a part of the omentum into the thigh. As the hernia sac passes through the femoral ring, which is medial to the common femoral vein and just below the inguinal ligament, the hernia protrudes subcutaneously into the femoral triangle. If it continues to enlarge it will often take a recurrent course upward over the inguinal ligament. The femoral hernia is always acquired and not congenital. It is found most frequently in the female, probably due to the wider female pelvis and larger femoral canal. Loss of muscle tone of the abdominal musculature and stretching from multiple pregnancies are also predisposing factors in the higher incidence of femoral herniae in females.

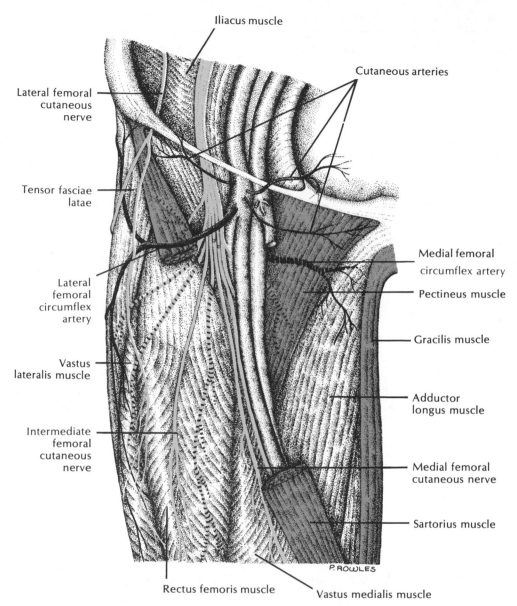

Fig. 6-6. Femoral triangle and adductor canal.

Femoral Triangle

The **femoral triangle** covers the greater part of the upper third of the thigh (Fig. 6-6). Its superior **base** is formed by the inguinal ligament, and its **sides** by the medial border of the sartorius laterally and the medial border of the adductor longus medially. The above muscles meet at the **apex** inferiorly, where a narrow intermuscular cleft, and **adductor canal,** continues distally. The **roof** of the femoral triangle is formed by skin and superficial and deep fasciae. It contains superficial inguinal lymph nodes and vessels, the femoral branch of the genitofemoral nerve, superficial branches of the femoral artery, and the upper end of the great saphenous vein. The iliopsoas, adductor longus, and pectineus muscles form the **floor** of the triangle. Its **contents** include the **femoral artery** and **vein** passing from the base to the apex; the **deep external pudendal, profunda femoral,** and **lateral** and **medial femoral circumflex arteries;** the **femoral branch** of the genitofemoral, the **lateral femoral cutaneous** and **femoral nerves,** and several **inguinal lymph nodes.**

FEMORAL VEIN CANNULATION

The ease of access to the femoral vein makes it a preferred site for cannulation, especially during cardiopulmonary resuscitation (CPR) when, because of chest compression, areas in the head and neck are difficult to stabilize. To find the vein the femoral artery is located as it extends into the thigh at the mid-point of the inguinal ligament. Placing your fingers on the pulsating artery, the needle may be introduced into the vein that lies medial to the artery as they traverse the femoral triangle.

Adductor Canal

At the apex of the femoral triangle, the femoral artery and vein continue inferiorly in the **adductor canal,** a muscular cleft between the vastus medialis and the adductors longus and magnus. It is deep to and covered by the sartorius muscle as it crosses the thigh obliquely. At the lowermost extent of the adductor canal, the posterior wall, formed by the adductor magnus, presents a deficiency, or opening, the **adductor hiatus,** which leads into the popliteal fossa. In addition to the femoral artery and vein, the canal contains the saphenous nerve and branches of the femoral nerve to the vastus medialis muscle.

Muscles

The muscles of the anterior aspect of the thigh are the iliopsoas, the sartorius, and the quadriceps femoris (Table 6-3). The latter two muscles are contained within the **extensor (anterior) compartment,** which is limited by the lateral and medial intermuscular septa. The **sartorius** acts as a flexor of the leg and, together with the rectus femoris and iliopsoas, also flexes the thigh. The **iliopsoas** also acts as a medial rotator of the thigh. The bulky **quadriceps,** one of the largest and most powerful muscles in the body, consists of the **rectus femoris** and the **vastus lateralis, medialis,** and **intermedius.** As powerful extensors of the leg, the four muscles form an aponeurotic and tendinous insertion on the tibia, with the patella interposed as a sesamoid bone. Muscles in the extensor compartment are innervated by the femoral nerve and receive their blood supply from the femoral artery (Fig. 6-7).

Arteries and Nerves

The **femoral artery,** the continuation of the external iliac, passes deep to the inguinal ligament to enter the femoral triangle. At the apex of the triangle it passes deep to the sartorius muscle to traverse the adductor canal. It becomes the **popliteal artery** at the adductor hiatus. Branches within the femoral triangle include the **superficial**

Table 6-3. Muscles of the Anterior Compartment of the Thigh

Muscle	Origin	Insertion	Action	Nerve
Iliopsoas: Compound muscle formed from iliacus and psoas major				
Iliacus	Iliac fossa and lateral portion of sacrum	Lesser trochanter of femur by way of iliopsoas tendon	Flexes and rotates thigh medially	Femoral
Psoas major	Lumbar vertebrae	Lesser trochanter of femur by way of iliopsoas tendon	Flexes and rotates thigh medially	Second and th lumbar
Sartorius	Anterior superior iliac spine	Upper part of medial surface of tibia	Acts on both hip and knee joints; mainly flexes leg	Femoral
Quadriceps: Consists of rectus femoris and vasti muscles, the four muscles combining into an aponeuroti tendinous insertion into tibial tuberosity, with patella interposed as sesamoid bone				
Rectus femoris	Straight head from anterior inferior iliac spine; reflected head from postero-superior aspect of rim of acetabulum	Tibial tuberosity	Extends leg and flexes thigh	Femoral
Vastus lateralis	Intertrochanteric line, greater trochanter, linea aspera, and lateral intermuscular septum	Tibial tuberosity	Extends leg	Femoral
Vastus medialis	Intertrochanteric line, spiral line, and medial intermuscular septum	Tibial tuberosity	Extends leg	Femoral
Vastus intermedius	Upper two-thirds of shaft of femur and distal one-half of lateral intermuscular septum	Tibial tuberosity	Extends leg	Femoral
Articularis genus	Variable slip of muscle on deep aspect of vastus intermedius that, in extension of pulls synovial membrane out of the way of the articular surfaces			

epigastric artery, which pierces the femoral sheath to ascend superficial to the inguinal ligament and courses toward the umbilicus; the **superficial iliac circumflex,** which penetrates the femoral sheath to run in the subcutaneous tissue toward the anterior superior iliac spine; the **superficial external pudendal,** which supplies skin and muscles in the inguinal region and gives anterior scrotal or labial branches to the skin of the external genitalia, and the **deep external pudendal,** which lies on and supplies the pectineus and adductor longus muscles, and sends twigs to the external genitalia.

The large **profunda branch** of the femoral artery arises from its posterolateal aspect and passes inferiorly, posterior to the adductor longus. It gives rise to the medial and lateral circumflex arteries, then courses deeply to give origin to the **three perforat-**

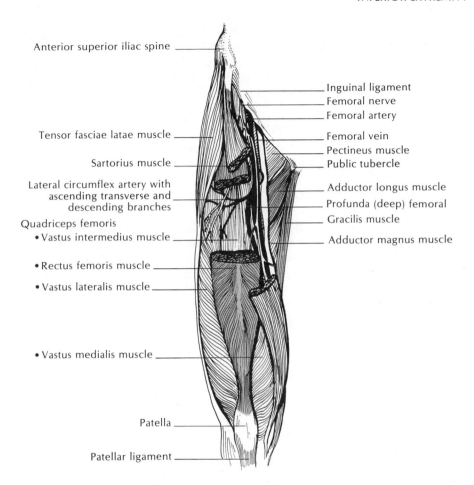

Anterior superior iliac spine

Tensor fasciae latae muscle

Sartorius muscle

Lateral circumflex artery with ascending transverse and descending branches

Quadriceps femoris
• Vastus intermedius muscle

• Rectus femoris muscle

• Vastus lateralis muscle

• Vastus medialis muscle

Patella

Patellar ligament

Inguinal ligament
Femoral nerve
Femoral artery
Femoral vein
Pectineus muscle
Public tubercle
Adductor longus muscle
Profunda (deep) femoral
Gracilis muscle
Adductor magnus muscle

Fig. 6-7. Anterior and medial views of thigh. Sartorius muscle removed to reveal contents of adductor canal; rectus femoris muscle resected to show principal arteries of anterior compartment.

ing branches, and terminates as the **fourth perforating artery.** These perforating vessels supply the adductor and the flexor muscle groups. The **medial femoral circumflex artery** courses between the pectineus and the iliopsoas, then around the neck of the femur, where it divides into a **superficial** and a **deep branch** to supply muscles in the region. The **lateral femoral circumflex** passes laterally, deep to the sartorius and rectus femoris, where it divides into ascending, transverse, and descending branches. The **ascending branch** distributes to the gluteal region. The **transverse branch** encircles the femur to anastomose with the medial femoral circumflex, while the **descending branch** passes inferiorly to terminate in the anastomosis around the knee joint.

The **femoral nerve,** the largest branch of the lumbar plexus, emerges from the lateral side of the psoas major muscle just below the iliac crest to descend between the psoas and the iliacus muscles. It passes deep to the inguinal ligament in the lateral neuromuscular compartment to enter the thigh and divide into a number of branches in the femoral triangle. Its terminal and longest branch, the **saphenous nerve,** passes through the adductor canal to become superficial at the adductor hiatus and course between the sartorius and gracilis muscles to supply the skin of the leg and the medial side of the foot. Branches of the femoral nerve in the abdomen include the **nerve to the iliacus** and the **nerve to the psoas.** In the thigh, it gives **muscular branches** to the sartorius, rectus femoris, and vasti muscles, and twigs to the pectineus, as well as the **intermediate** and **medial femoral cutaneous branches.**

ADDUCTOR COMPARTMENT OF THE THIGH

Muscles (Table 6-4)

The adductor muscles, arranged in three layers, occupy the medial (adductor) compartment which is limited by the medial and posterior intermuscular septa (Fig. 6-8). The **pectineus** and **adductor longus** are the most superficial; the **adductor brevis** lies intermediate, and the **adductor magnus** occupies the deepest stratum. The **gracilis** (also an adductor) lies superficially on the medial aspect of the thigh. The two divisions of the obturator nerve are interposed between the three muscular layers and supply the muscles in the adductor compartment. All three adductors act in adduction and lateral

Table 6-4. Muscles of the Adductor Compartment of the Thigh

Muscle	Origin	Insertion	Action	Nerve
Adductor longus	Body of pubis immediately below pubic crest	Linea aspera of femur	Adducts, flexes, and rotates thigh laterally	Obturator
Adductor brevis	Body of pubis below origin of adductor longus	Between lesser trochanter and linea aspera and upper part of linea aspera	Adducts, flexes, and rotates thigh laterally	Obturator
Adductor magnus	Side of pubic arch and ischial tuberosity	Extensive into linea aspera, medial supracondylar ridge, and adductor tubercle	Adducts, flexes, and rotates thigh laterally; distal fibers aid in extension and medial rotation of thigh	Obturator; distal portion by sciatic
Pectineus	Pectineal line and pectineal surface of pubis	Posterior aspect of femur between lesser trochanter and linea aspera	Adducts and assists in flexion of thigh	Obturator and femoral
Gracilis	Lower half of body of pubis	Upper part of medial surface of tibia	Adducts thigh; flexes knee joint and rotates leg medially	Obturator
Obturator externus	Margins of obturator foramen and obturator membrane	Posterior aspect of intertrochanteric fossa of femur	Flexes and rotates thigh laterally	Obturator

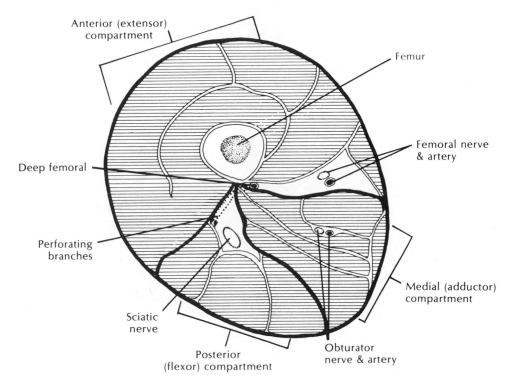

Fig. 6-8. Compartments of the thigh.

rotation of the thigh. The pectineus and gracilis, in addition to adduction, act as flexors. The pectineus aids in flexion of the thigh, and the gracilis in flexion of both the thigh and the leg. The gracilis also assists in medial rotation of the leg.

Arteries and Nerves

The **obturator artery** is a branch of the internal iliac that arises in the pelvis and accompanies the obturator nerve through the obturator canal. Here it divides into anterior and posterior branches that follow the margins of the obturator foramen deep to the obturator externus (Fig. 6-9). The **anterior branch** supplies the obturator externus, the pectineus, the adductors, and the gracilis; the **posterior branch** is distributed to muscles attaching to the ischial tuberosity and gives off a branch to supply the head of the femur as it passes deep to the transverse acetabular ligament.

The **obturator nerve** originates in the abdomen as a branch of the lumbar plexus. It emerges from the medial surface of the psoas major muscle to pass along the lateral wall of the pelvic cavity, through the obturator canal, and divide into anterior and posterior divisions as it enters the thigh. The divisions pass to either side of the adductor brevis muscle. The **anterior division** lies between the adductors longus and brevis, supplies these muscles, and sends branches to the gracilis and pectineus muscles. The **posterior division** passes between the adductors brevis and magnus, supplying these mus-

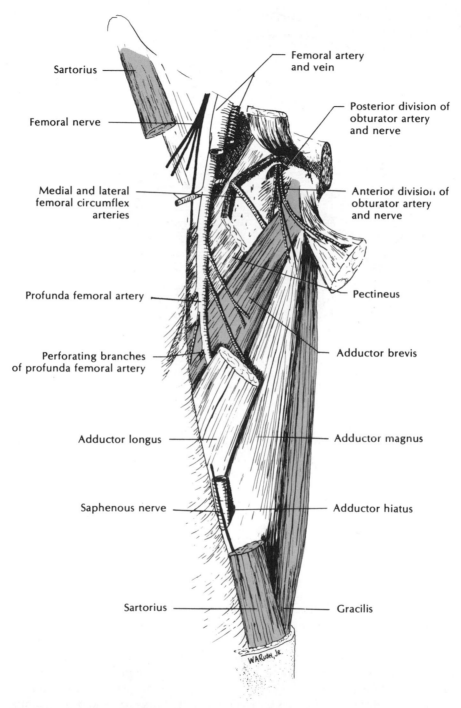

Sartorius

Femoral nerve

Medial and lateral
femoral circumflex
arteries

Profunda femoral artery

Perforating branches
of profunda femoral artery

Adductor longus

Saphenous nerve

Sartorius

Femoral artery
and vein

Posterior division of
obturator artery
and nerve

Anterior division of
obturator artery
and nerve

Pectineus

Adductor brevis

Adductor magnus

Adductor hiatus

Gracilis

Fig. 6-9. Adductor region of the thigh.

cles as well as the obturator externus. The pectineus and the adductor magnus each have dual innervations, with the pectineus receiving additional innervation from the **femoral nerve,** and the adductor magnus acquires a dual nerve supply from the **tibial division** of the sciatic nerve.

POSTERIOR COMPARTMENT OF THE THIGH

Muscles

The posterior compartment of the thigh is located between the posterior and lateral intermuscular septa and contains the flexor, or hamstring, muscles (Table 6-5). All the flexors are innervated by the sciatic nerve, with the short head of the biceps receiving its nerve supply from the common peroneal division and the remaining muscles from the tibial division. The **biceps** arises by a long and short head; the **semimembranosus** has a long membranous origin, and the **semitendinosus** a long tendinous insertion. The long head of the biceps, the semimembranosus, and the semitendinosus span both hip and knee joint, and all act as flexors of the leg and extensors of the thigh. They also act in rotation of the leg and thigh, with the biceps rotating laterally and the semimembranosus and semitendinosus rotating medially.

Arteries and Nerves

No major artery courses through the posterior compartment (Fig. 6-10). The blood supply is derived from the previously described **perforating branches** of the large profunda femoris branch of the femoral artery.

The **sciatic nerve,** the principal branch of the sacral plexus, is the largest nerve of the body. It enters the thigh inferior to the piriformis muscle, passing through the

Table 6-5. Muscles of the Posterior Compartment of the Thigh

Muscle	Origin	Insertion	Action	Nerve
Biceps femoris	Long head, common tendon with semitendinosus from ischial tuberosity; short head, linea aspera and upper half of supracondylar ridge of femur	Common tendon into head of fibula	Flexes knee; rotates leg laterally; long head extends hip	Long head by tibial portion, short head by peroneal portion of sciatic
Semitendinosus	In common with long head of biceps from ischial tuberosity	Upper part of medial surface of tibia	Flexes knee; rotates leg medially; extends hip joint	Tibial portion of sciatic
Semimembranosus	Ischial tuberosity	Medial condyle of tibia	Extends hip joint; flexes knee; rotates leg medially	Tibial portion of sciatic

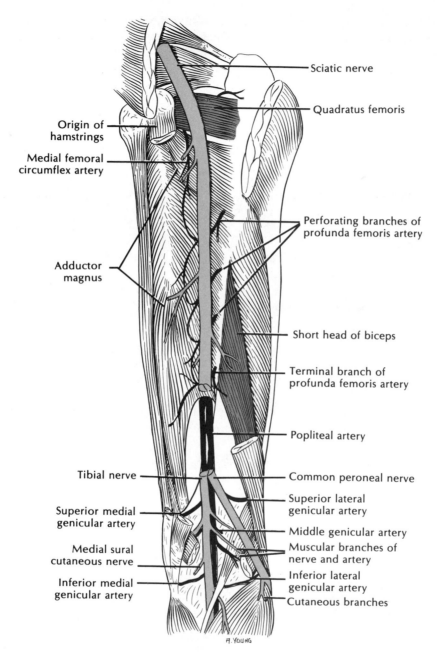

Fig. 6-10. Posterior region of the thigh. (Hollinshead WH: Textbook of Anatomy, 3rd ed. New York, Harper & Row, 1974)

greater sciatic foramen to descend between the gluteus maximus muscle, and the gemelli muscles, the obturator internus tendon, and the quadratus femoris muscle. As it continues distally through the thigh, it lies on the adductor magnus and is crossed obliquely by the long head of the biceps femoris. It terminates by dividing into **tibial** and **common peroneal nerves.** This division may occur anywhere from its origin in the pelvis to the popliteal fossa, but usually takes place as the sciatic nerve enters the popliteal fossa. Branches from the common peroneal division supply the short head of the biceps. The tibial division supplies the long head of the biceps, the semimembranosus, the semitendinosus, and the adductor magnus muscles (Table 6-6).

Table 6-6. Nerve Distribution to Thigh

Nerve	Origin	Course	Distribution
Genitofemoral	Anterior divisions of lumbar plexus	Pierces and runs on anterior surface psoas major; genital branch passes through superficial inguinal ring, femoral branch courses with femoral artery in femoral sheath	Genital branch supplies cremaster and skin of scrotum (labium majus); femoral branch-skin over femoral triangle
Ilioinguinal nerve	Anterior divisions of lumbar plexus	Passes in cleft between internal abdominal oblique and transversus abdominis to traverse inguinal canal	Skin over femoral triangle, scrotum (labium majus), and pubic symphysis
Lateral femoral cutaneous	Posterior divisions of lumbar plexus	Crosses posterior abdominal wall to pass deep to inguinal ligament 2 to 3 cm medial to the anterior superior iliac spine	Skin of lateral aspect of thigh
Medial and intermediate femoral cutaneous	Femoral	Arise in femoral triangle to course in superficial fascia	Skin of medial and anterior aspect of thigh
Femoral	Posterior divisions of lumbar plexus	Crosses posterior abdominal wall to pass deep to inguinal ligament in neuromuscular compartment	Muscles of anterior (extensor) compartment
Obturator	Anterior divisions of lumbar plexus	From deep to psoas major, crosses lateral wall of pelvic cavity to traverse obturator canal	Muscles of medial (adductor) compartment, twigs to pectineus, and skin over medial aspect of thigh
Tibial	Anterior divisions of sacral plexus as component of sciatic	Traverses greater sciatic foramen below level of piriformis, lies on adductor magnus as it passes through thigh	Semimembranosus, semitendinosus, long head of biceps and twigs to adductor magnus
Common peroneal	Posterior divisions of sacral plexus as component of sciatic	Traverses greater sciatic foramen below level of piriformis, lies on adductor magnus as it passes through thigh	Short head of biceps

HERNIATED DISC

A protrusion of the nucleus pulposus and concomitant pressure on a spinal nerve is the most common cause of pain in the lower back. The pain is due to a spasm of the intrinsic muscles of the back. The muscle spasm will result in a deviation of the vertebral column toward the affected side. The level of the disc involved may be ascertained by the deficit of sensation to the involved dermatome of the lower extremity or by loss of specific muscle reflexes.

POPLITEAL FOSSA

The **popliteal fossa** is a diamond-shaped area on the posterior aspect of the knee joint, extending from the lower third of the femur to the upper part of the tibia (Fig. 6-11). With the knee flexed this area presents a depression, but with the joint fully extended forms a slight posterior bulge. The fossa is **bounded** superiorly and laterally by the biceps femoris; superiorly and medially by the semitendinosus, semimembranosus, gracilis, sartorius, and adductor magnus, and inferiorly by the converging heads of the gastrocnemius and the laterally placed plantaris muscle. The **floor** is formed from above downward by the popliteal surface of the femur, the oblique popliteal ligament, the popliteus muscle, and the upper part of the tibia. The fossa contains the common peroneal and tibial nerves, the popliteal artery and vein, the posterior femoral cutaneous nerve, the small saphenous vein, lymph nodes, and numerous synovial bursae.

The **popliteal artery** is the direct continuation of the femoral at the adductor hiatus. It descends through the popliteal fossa to terminate at the lower border of the popliteus muscle by dividing into the **anterior** and **posterior tibial arteries.** Branches of the popliteal artery within the fossa consist of several **genicular branches** that join in the anastomosis around the knee joint.

Leg

In the leg, as in the thigh, the muscles are located in fascial compartments. The **lateral (peroneal) compartment** is limited by the **anterior** and **posterior intermuscular septa;** the **anterior (extensor) compartment** is between the **anterior intermuscular septum** and the **tibia;** and the **posterior (flexor) compartment** is bounded by the **tibia** and the **posterior intermuscular septum.** Passing between the tibia and the fibula, the deeply lying **interosseous membrane** separates the extensor from the flexor compartment (Fig. 6-12). The latter is further divided by two **transverse intermuscular septa** into a deep subdivision containing the tibialis posterior muscle, a superficial subdivision containing the gastrocnemius and soleus muscles, and an intermediate compartment containing the remaining muscles, the posterior tibial artery, and the tibial nerve. Each of the three major compartments transmits a major nerve that supplies muscles

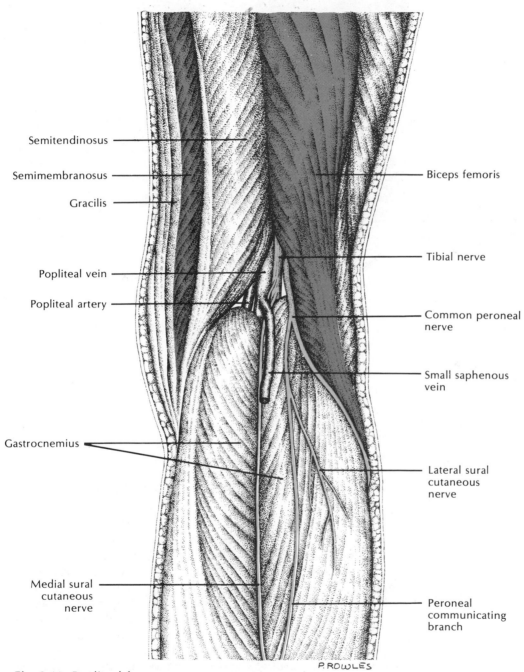

Semitendinosus

Semimembranosus

Gracilis

Popliteal vein

Popliteal artery

Gastrocnemius

Medial sural
cutaneous
nerve

Biceps femoris

Tibial nerve

Common peroneal
nerve

Small saphenous
vein

Lateral sural
cutaneous
nerve

Peroneal
communicating
branch

P. ROWLES

Fig. 6-11. Popliteal fossa.

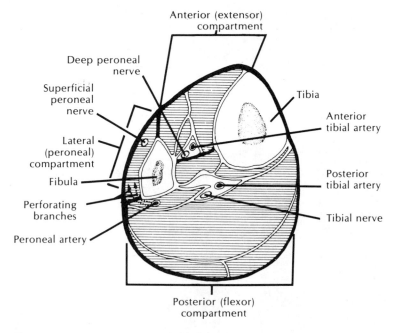

Fig. 6-12. Compartments of the leg.

within the compartment, but, as in the thigh, one compartment, the lateral, has no major artery coursing through it.

ANTERIOR COMPARTMENT OF THE LEG

Muscles (Table 6-7)

Within the anterior compartment (Fig. 6-13) the superficially located **tibialis anterior muscle** lies along the lateral side of the tibia, adjacent to the interosseous membrane. It acts in dorsiflexion and inversion of the foot and, with the tibialis posterior and the peroneus longus, functions to maintain the longitudinal arch of the foot. The long, thin **extensor digitorum longus** is situated along the fibula and acts to extend the toes and secondarily to dorsiflex and evert the foot. Usually, the **peroneus tertius** can be distinguished from the extensor digitorum longus only by its insertion into deep fascia of the foot or to the bases of the fourth and fifth metatarsals. The **extensor hallucis longus,** deeply located between the tibialis anterior and the extensor digitorum longus, extends the great toe and aids in dorsiflexion of the foot. Muscles in the anterior compartment are innervated by the deep peroneal (anterior tibial) nerve and receive their blood supply from the anterior tibial artery. The tibialis anterior usually receives additional innervation directly from the common peroneal.

Arteries and Nerves

At the bifurcation of the popliteal artery, its **anterior tibial branch** passes distally between the heads of origin of the tibialis posterior muscle. It then pierces the proxi-

Table 6-7. Muscles of the Anterior Compartment of the Leg

Muscle	Origin	Insertion	Action	Nerve
Tibialis anterior	Lateral condyle of upper two-thirds of tibia and interosseous membrane	First cuneiform and first metatarsal	Dorsiflexes and inverts foot	Deep peroneal
Extensor digitorium longus	Lateral condyle of tibia, upper three-fourths of fibula, and interosseous membrane	By four tendons that form membranous expansions over metatarsophalangeal joints of second to fifth toes	Extends toes; continued action dorsiflexes and everts foot	Deep peroneal
Peroneus tertius	Distal one-fourth of fibula and interosseous membrane	Fifth metatarsal or deep fascia of foot	Dorsiflexes and everts foot	Deep peroneal
Extensor hallucis longus	Middle half of fibula and interosseous membrane	Base of distal phalanx of great toe	Extends great toe; aids in dorsiflexion and inversion of foot	Deep peroneal

mal portion of the interosseous membrane to course inferiorly on the anterior surface of the membrane passing between the tibialis anterior and the extensor digitorum longus proximally, and the tibialis anterior and the extensor hallucis longus distally. The anterior tibial artery terminates under the inferior extensor retinaculum as the **dorsalis pedis artery.** In addition to supplying the muscles of the anterior compartment, its branches include the **posterior tibial recurrent,** which passes superiorly between the popliteal ligament and the popliteus muscle to join the genicular anastomosis, and the **anterior tibial recurrent,** which crosses the interosseous membrane to pass superiorly and join the genicular anastomosis. At the ankle the **medial anterior malleolar branch** passes deep to the tendon of the tibialis anterior to anastomose with branches of the posterior tibial artery, whereas the **lateral anterior malleolar** winds around the lateral malleolus deep to the extensor digitorum longus and peroneus tendons to join the malleolar branches of the posterior tibial and dorsalis pedis arteries.

The **deep peroneal** (anterior tibial) **nerve,** a branch of the common peroneal, courses around the head of the fibula to pass deep to the peroneus longus muscle, pierces the extensor digitorum longus, and descends on the interosseous membrane in the anterior compartment. It terminates under the inferior extensor retinaculum as **medial** and **lateral digital branches.** It gives twigs to the peroneus longus, supplies muscles in the anterior compartment, the extensor digitorum brevis on the dorsum of the foot, and its most distal branch supplies the skin of the contiguous sides of the great and second toe.

FOOT DROP

Disruption of the nerve supply to the extensor muscles in the leg will result in a "foot drop." Upon examination, a patient will be unable to

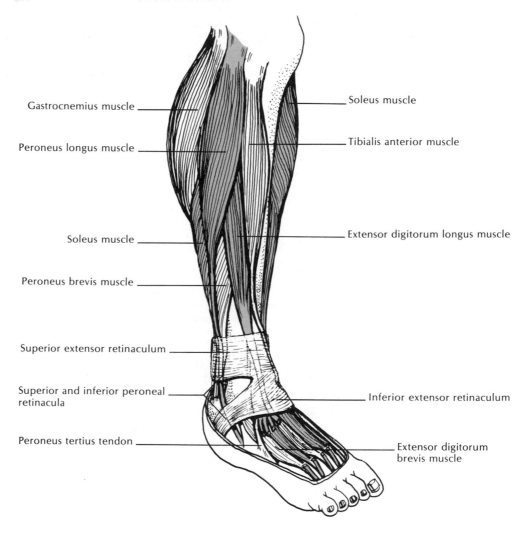

Gastrocnemius muscle

Peroneus longus muscle

Soleus muscle

Peroneus brevis muscle

Superior extensor retinaculum

Superior and inferior peroneal retinacula

Peroneus tertius tendon

Soleus muscle

Tibialis anterior muscle

Extensor digitorum longus muscle

Inferior extensor retinaculum

Extensor digitorum brevis muscle

Fig. 6-13. Anterior and lateral compartments of the leg.

extend his foot against resistance. Usually this is due to a herniated disc or to trauma to the common peroneal nerve. The superficial position of the common peroneal as it winds around the neck of the fibula makes it vulnerable to injury. It is the most frequently damaged nerve in the lower extremity.

LATERAL COMPARTMENT OF THE LEG

Muscles (Table 6-8)

Two muscles, the **peroneus longus** and the **peroneus brevis,** are located in the lateral compartment of the leg (see Fig. 6-13). Both are supplied by the superficial pero-

Table 6-8. Muscles of the Lateral Compartment of the Leg

Muscle	Origin	Insertion	Action	Nerve
Peroneus longus	Lateral condyle of tibia, head and upper two-thirds of fibula	First metatarsal and first cuneiform	Everts and aids in plantar flexion of foot	Superficial peroneal
Peroneus brevis	Lower two-thirds of fibula	Base of fifth metatarsal	Everts and aids in plantar flexion of foot	Superficial peroneal

neal nerve, with the peroneus longus frequently receiving additional innervation from the common peroneal. Both muscles act as evertors and aid in plantar flexion of the foot. The tendons of the peroneus longus and the tibialis anterior form a sling by their manner of insertion and give support for the longitudinal arch of the foot.

Arteries and Nerves

No major artery courses in the lateral compartment of the leg. The muscles are supplied by perforating twigs of the **peroneal branch** of the posterior tibial artery located within the posterior compartment.

The **superficial peroneal** (musculocutaneous) **nerve,** a branch of the common peroneal, passes between the extensor digitorum longus and the peronei muscles to descend in the lateral compartment. Within the compartment it supplies the peroneus longus and peroneus brevis muscles. In the distal third of the leg it becomes superficial to supply skin on the anterior surface of the leg and the dorsum of the foot.

POSTERIOR COMPARTMENT OF THE LEG

Muscles (Table 6-9)

The deeply placed **tibialis posterior muscle** is the principal invertor of the foot and is separated from the **flexor digitorum longus** and the **flexor hallucis longus muscles** by the deeper of two transverse intermuscular septa (Fig. 6-14). The tibialis posterior aids in plantar flexion of the foot, and both flexors assist in inversion of the foot. The **gastrocnemius** and the **soleus** are separated from the intermediate flexor muscles by the more superficial transverse intermuscular septum. Inserting by the common **calcaneal tendon** (tendon of Achilles) into the calcaneus, the gastrocnemius and the soleus are the strongest of the plantar flexors of the foot.

CALCANEAL BURSITIS

Calcaneal bursitis is an inflammation of the bursa located between the calcaneal tendon and the upper part of the calcaneus. It is fairly common in long distance runners.

Table 6-9. Muscles of the Posterior Compartment of the Leg

Muscle	Origin	Insertion	Action
Gastrocnemius	Lateral head, lateral condyle of femur; medial head, popliteal surface and medial condyle of femur	With soleus through calcaneal tendon into posterior surface of calcaneus	Plantar flexes foot and flexes knee
Soleus	Upper one-third of fibula, soleal line on tibia	With gastrocnemius through calcaneal tendon into posterior surface of calcaneus	Plantar flexes foot
Plantaris	Popliteal surface of femur above lateral head of gastrocnemius	Into medial side of calcaneal tendon	Plantar flexes foot
Popliteus	Popliteal groove, lateral condyle of femur	Tibia above soleal line	With knee fully extended, rotates femur laterally
Flexor digitorum longus	Middle one-half of tibia below soleal line	Tendon divides into four tendons, which insert into distal phalanges of four lateral toes	Flexes phalanges of four lateral toes; continued action plantar flexes and inverts foot
Flexor hallucis longus	Lower two-thirds of fibula and intermuscular septa	Base of distal phalanx of great toe	Flexes great toe; continued action plantar flexes and inverts foot
Tibialis posterior	Interosseous membrane and adjoining tibia and fibula	Into tuberosity of navicular with slips to cuneiforms, cuboid, and bases of second, third, and fourth metatarsals	Principal invertor of foot; plantar flexes foot

Two small muscles within the popliteal region complete the flexor group of muscles, namely, the **popliteus,** which performs the important initial "unlocking action" in flexion of the knee, and the relatively unimportant **plantaris,** which sends its long slender tendon to insert into the medial side of the calcaneal tendon.

TRICK KNEE

In the normal gait the final action in extension of the leg (knee) is a medial rotation of the femur that "locks" the femur and tibia so that the extremity becomes a rigid, weight-bearing pillar. Intermittently in some patients following knee injuries, this action is not finalized; the knee "gives way" when weight is placed on it, and the individual falls.

Arteries and Nerves

The **posterior tibial artery,** the larger terminal branch of the popliteal, passes distally in the intermediate portion of the posterior compartment of the leg on the superficial aspect of the flexor digitorum longus. At the ankle it passes deep to the flexor

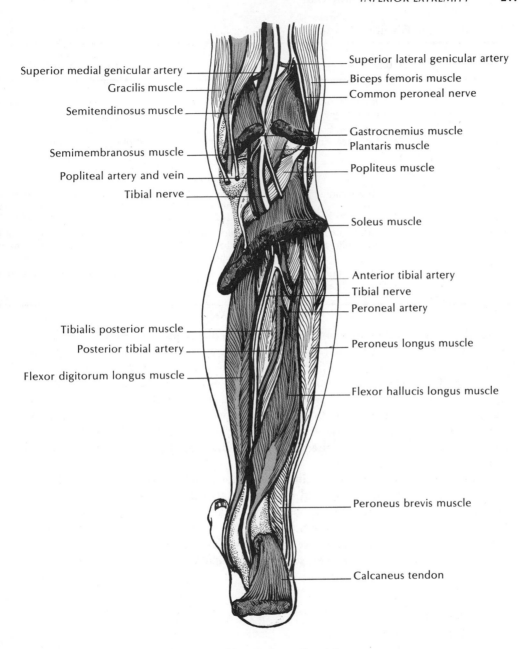

Superior medial genicular artery

Gracilis muscle

Semitendinosus muscle

Semimembranosus muscle

Popliteal artery and vein

Tibial nerve

Tibialis posterior muscle

Posterior tibial artery

Flexor digitorum longus muscle

Superior lateral genicular artery

Biceps femoris muscle

Common peroneal nerve

Gastrocnemius muscle

Plantaris muscle

Popliteus muscle

Soleus muscle

Anterior tibial artery

Tibial nerve

Peroneal artery

Peroneus longus muscle

Flexor hallucis longus muscle

Peroneus brevis muscle

Calcaneus tendon

Fig. 6-14. Posterior compartment of leg and popliteal fossa.

Table 6-10. Nerve Distribution to Leg

Nerve	Origin	Course	Distribution
Saphenous	Continuation of femoral in femoral triangle	Passes deep to sartorius to reach leg	Skin of medial aspect leg
Sural	Derived from both common peroneal and tibial	Tibial component passes between heads of gastrocnemius, joins common peroneal component in middle third of leg	Skin of posterior and lateral aspect of leg
Tibial	Branch of sciatic in popliteal fossa	Courses through leg in intermediate portion of posterior compartment	Muscles in posterior (flexor) compartment, tributes to sural
Superficial peroneal	Branch of common peroneal in popliteal fossa	Winds around neck of fibula to course deep to peroneus longus	Muscles of lateral (peroneal) compartment, skin on distal third of leg
Deep peroneal	Branch of common peroneal in popliteal fossa	Courses through leg deep to extensor digitorum longus	Muscles of anterior (extensor) compartment

retinaculum where it divides into the **lateral** and **medial plantar arteries.** At the knee it gives off the **fibular branch,** which passes laterally toward the head of the fibula, giving twigs to the peroneus longus and peroneus brevis muscles as it ascends to join the genicular anastomosis. The large **peroneal branch** courses laterally between the flexor hallucis longus and the fibula to give a nutrient branch to the fibula, a communicating branch to the posterior tibial, and perforating branches to the peroneal compartment. Its **branch to the tibia** is the largest nutrient artery in the body. At the ankle joint **medial** and **posterior malleolar** and **medial calcaneal branches** join, respectively, the malleolar and calcaneal anastomoses.

From the popliteal fossa the **tibial** (posterior tibial) **nerve** passes over the tendinous arch of the soleus to descend into the intermediate division of the posterior compartment of the leg. Initially, it lies on the tibialis posterior, then on the flexor digitorum longus as it descends to divide under the flexor retinaculum to become the **medial** and **lateral plantar nerves.** It gives **muscular branches** to all the muscles of the posterior compartment as well as a **communicating branch to the sural nerve** (Table 6-10).

Foot

Plantar Aponeurosis

The plantar aponeurosis is a sheet of deep fascia of great strength and importance. It is divided into medial, intermediate, and lateral portions, differentiated by their density and demarcated superficially by two shallow grooves that extend longitudinally along the foot. At the divisional lines septa pass from the deep aspect of the plantar aponeurosis to the deeper structures of the foot. The posterior end of the intermediate portion is narrow and attaches to the medial tubercle of the calcaneus. It widens as it extends forward and, near the heads of the metatarsals, divides into five slips, which pass to each of the five toes. This portion forms a strong tie, especially for the great toe, between the calcaneus and each proximal phalanx.

Muscles (Table 6-11)

The muscles within the sole of the foot function basically as a group and are important in posture, locomotion, and support of the arches of the foot (Fig. 6-15). The **plantar muscles** are described in **four layers** and arranged in **three groups,** a medial group for the great toe, a lateral group for the small toe, and an intermediate group for the remainder of the digits.

Table 6-11. Muscles of the Foot

Muscle	Origin	Insertion	Action	Nerve
Extensor digitorum brevis (only muscle on dorsum of foot)	Dorsal surface of calcaneus	Divides into four tendons which insert into tendons of extensor digitorum longus of four medial toes	Dorsiflexes toes	Deep peroneal
Abductor hallucis	Medial tubercle of calcaneus	With medial belly of flexor hallucis brevis into proximal phalanx of great toe	Abducts and aids in flexion of great toe	Medial plantar
Flexor digitorum brevis	Medial tubercle of calcaneous and plantar fascia	Divides into four tendons that enter flexor sheath and split to admit passage of long flexor tendons before inserting into middle phalanx of lateral four toes	Flexes lateral four toes	Medial plantar
Abductor digiti minimi	Medial and lateral tubercles of calcaneus	Lateral side of proximal phalanx of little toe	Abducts little toe	Lateral plantar
Quadratus plantae	Medial head, medial side of calcaneus and plantar fascia; lateral head, lateral margin of calcaneus and plantar fascia	Tendons of flexor digitorum longus	Assists flexor digitorum longus	Lateral plantar
Lumbricals (4)	Tendons of flexor digitorum longus	Medial side of base of proximal phalanx of lateral four toes and extensor expansion	Aid interossei in flexion of metatarsophalangeal joints; extend distal two phalanges	First by medial plantar; lateral three by lateral plantar

The **superficial layer** includes the **abductor hallucis, abductor digiti minimi,** and the **flexor digitorum brevis** muscles. In the **second layer** the **tendon of the flexor hallucis longus** grooves the under surface of the talus bone (sustentaculum tali) to pass medially under the tendons of the flexor digitorum longus and insert into the base of the terminal phalanx of the great toe. The **tendons of the flexor digitorum longus** enter the fibrous flexor sheaths at the middle of the foot, perforate the tendons of the flexor digitorum brevis, and insert into the bases of the terminal phalanges. In addition to the above tendons, the **quadratus plantae** and **lumbricales** are components of this layer.

The **third layer** is made up of the **flexor hallucis brevis,** the **adductor hallucis,** and the **flexor digiti minimi.** The flexor hallucis brevis lies on the first metatarsal along the lateral side of the abductor hallucis and is grooved by the tendon of the flexor hallucis

Table 6-11. Cont.

Muscle	Origin	Insertion	Action	Nerve
Flexor hallucis	Cuboid and third cuneiform	Divides into two tendons; medial, into base of proximal phalanx of great toe with abductor hallucis; lateral, with adductor hallucis	Flexes great toe	Medial and sometimes twigs from lateral plantar
Adductor hallucis Oblique head	Anterior end of plantar ligament and sheath of peroneus longus	With lateral belly of flexor hallucis brevis into proximal phalanx of great toe	Adducts and flexes great toe	Lateral plantar
Transverse head	Capsule of lateral four metatarsophalangeal joints	Joins oblique head to insert as above	Acts as tie for heads of metatarsals; adducts great toe	Lateral plantar
Flexor digiti minimi	Base of fifth metatarsal and plantar fascia	Lateral side of base of proximal phalanx of little toe	Flexes small toe	Lateral plantar
Plantar interossei (3)	Medial side of third, fourth, and fifth metatarsals	Medial side of base of proximal phalanges of third, fourth, and fifth toes	Adduct lateral three toes toward second toe; flex proximal, and possibly extend distal phalanges	Lateral plantar
Dorsal interossei (4)	Lie in intermetatarsal space and arise from adjacent bones	Tendons pass forward as above and insert into proximal phalanges on either side of second toe and into lateral side of third and fourth toes	Abduct second, third, and fourth toes from midline of second toe; flex proximal, and possibly extend distal, phalanges	Lateral plantar

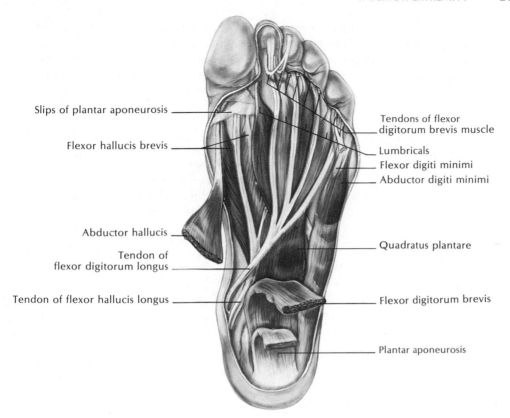

Slips of plantar aponeurosis

Flexor hallucis brevis

Abductor hallucis

Tendon of
flexor digitorum longus

Tendon of flexor hallucis longus

Tendons of flexor
digitorum brevis muscle

Lumbricals

Flexor digiti minimi

Abductor digiti minimi

Quadratus plantare

Flexor digitorum brevis

Plantar aponeurosis

Fig. 6-15. Sole of left foot.

longus to form two partially separated bellies. The adductor hallucis has two separate heads, the oblique and transverse, which may act as two distinct muscles.

Both muscles and tendons are present in the **fourth** and **deepest layer.** The **tendon of the peroneus longus** crosses the sole of the foot obliquely from the lateral to the medial side, and slips of the **tendon of the tibialis posterior** cross the sole in the opposite direction. These tendons form a sling for the foot to help maintain both the longitudinal and transverse arches. The muscles in this layer, the **three plantar** and **four dorsal interossei,** are thin and flattened, with the medial muscles more deeply situated. Both groups are more easily seen from the plantar than from the dorsal surface and lie between, and arise from, the metatarsal bones. The line of reference for the action of abduction and adduction for these muscles passes through the center of the second toe.

Arteries and Nerves

The **dorsalis pedis artery** is the continuation of the anterior tibial artery as it courses under the extensor retinaculum. It continues on the dorsum of the foot to the base of the first interosseous space, where it terminates by dividing into the **deep plantar** and the **first dorsal metatarsal arteries.** In its course it gives rise to **lateral** and **medial tarsal branches** supplying the extensor digitorum brevis, cutaneous and osseous twigs in the ankle region, and the **arcuate artery** near its termination. The latter passes laterally across the bases of the metatarsal bones and give rise to the second, third, and fourth **dorsal metatarsal arteries,** each of which subsequently divides into two **dorsal digital branches** to the sides of the toes. The deep plantar branch passes deeply through the first interosseous space to join with the lateral plantar artery to form the **deep plantar arch.**

DORSALIS PEDIS PULSE

The dorsalis pedis artery lies in the subcutaneous tissue on the dorsum of the foot, between the tendons of the extensor digitorum longus and the extensor hallucis longus. It is easily palpated and may be used to determine the pulse rate if the radial pulse is not accessible.

The smaller of two terminal branches of the posterior tibial, the **medial plantar artery,** arises under cover of the flexor retinaculum. It courses deep to the abductor hallucis, then between the abductor hallucis and the flexor digitorum brevis to supply branches to the medial side of the great toe and gives muscular, cutaneous, and articular twigs along its course. The larger of the two terminal branches of the posterior tibial is the **lateral plantar.** It passes forward between the first and second layers of muscles in the sole giving calcaneal, cutaneous, muscular, and articular branches as it continues toward the base of the fifth metatarsal. At the metatarsophalangeal joint it passes medially to join the deep plantar branch of the dorsalis pedis and together they form the deep plantar arch.

The **deep plantar arch** runs across the bases of the metatarsals to give a **plantar metatarsal artery** to each interosseous space, which subsequently divides into **plantar digital branches** to the adjacent sides of each toe. Separate branches supply the medial side of the great and the lateral side of the little toe. Perforating branches pass deeply in the interosseous space to join corresponding dorsal metatarsal vessels from the arcuate artery (Fig. 6-16).

The **medial plantar nerve,** the larger of the two terminal branches of the tibial nerve, arises under the cover of the flexor retinaculum. It passes forward between the abductor hallucis and the flexor digitorum brevis lateral to the medial plantar artery. It supplies the above muscles and sends **cutaneous branches** to the medial side of the sole. It terminates in four **digital nerves** that supply muscular branches to the flexor hallucis brevis, the first lumbricalis, and cutaneous branches to the adjacent sides of the four medial toes and the medial side of the foot (Table 6-12).

The **lateral plantar nerve** passes laterally in the foot to course anteriorly medial to the lateral plantar artery. It gives **muscular branches** to the quadratus plantae and abductor digiti minimi, and cutaneous branches to the lateral aspect of the sole. It terminates in superficial and deep branches. The **superficial branch** passes forward to supply the flexor digiti minimi, the interossei of the fourth intermetatarsal space, and skin on the lateral side of the sole, the fifth toe, and the lateral side of the fourth toe. The **deep branch** accompanies the lateral plantar artery and supplies the adductor hallucis, the remaining interossei, and the three lateral lumbricales muscles (Fig. 6-17).

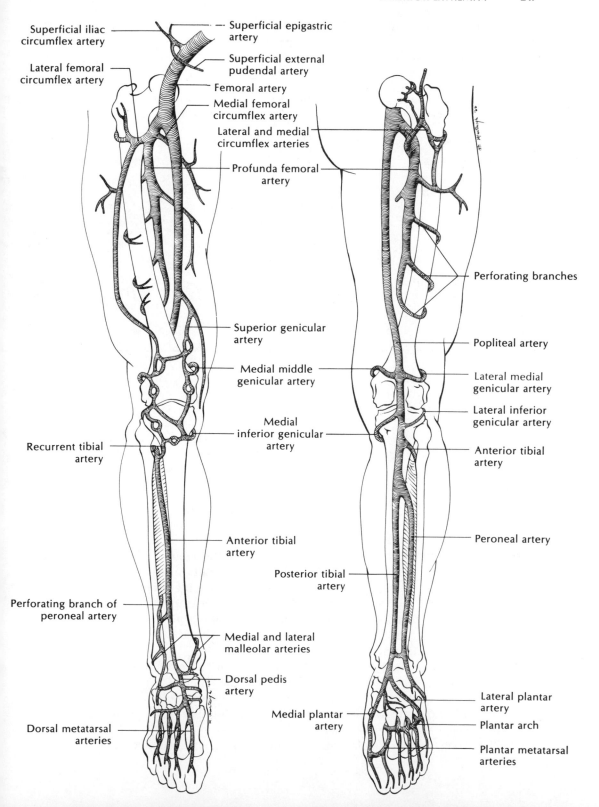

Superficial iliac circumflex artery

Superficial epigastric artery

Superficial external pudendal artery

Femoral artery

Medial femoral circumflex artery

Lateral femoral circumflex artery

Lateral and medial circumflex arteries

Profunda femoral artery

Perforating branches

Superior genicular artery

Popliteal artery

Medial middle genicular artery

Lateral medial genicular artery

Lateral inferior genicular artery

Medial inferior genicular artery

Recurrent tibial artery

Anterior tibial artery

Anterior tibial artery

Peroneal artery

Posterior tibial artery

Perforating branch of peroneal artery

Medial and lateral malleolar arteries

Dorsal pedis artery

Lateral plantar artery

Medial plantar artery

Plantar arch

Dorsal metatarsal arteries

Plantar metatarsal arteries

Table 6-12. Nerve Distribution to Foot

Nerve	Origin	Course	Distribution
Saphenous	Femoral in thigh	Through superficial fascia of leg and anterior to medial malleolus	Skin on medial aspect of fo
Superficial peroneal	Common peroneal	Becomes superficial in distal third of leg	Skin on dorsum of foot
Deep peroneal	Common peroneal	Passes deep to extensor retinaculum to enter foot	Extensor digitorum brevis a skin on contiguous sides of great and second toe
Medial plantar	Tibial	Passes deep to flexor retinaculum to enter foot and course in medial compartment	Abductor hallucis, flexor di torum brevis, flexor halluci brevis, first lumbrical; skin plantar surface medial to a splitting the fourth toe (ho ogous to median of hand)
Lateral plantar	Tibial	Passes deep to flexor retinaculum to enter foot and course in lateral compartment	Supplies all muscles not su plied by medial plantar; sk lateral to a line splitting fo toe (homologous to ulnar i hand)

Joints of the Inferior Extremity

HIP JOINT

The hip joint is the best example of an **enarthrodial** (ball-and-socket) type of joint in the body (Fig. 6-18). The spheroidal **head** of the femur fits into a cuplike cavity, the **acetabulum,** in the innominate bone. The head of the femur is covered by articular cartilage, except for a small central area, the **fovea capitis femoris,** where the **ligamentum teres** attaches. Articular cartilage on the acetabulum forms an incomplete horseshoe-shaped marginal ring, the **lunate surface,** with a central circular depression devoid of cartilage. In the fresh state, this is occupied by a mass of fat covered by the synovial membrane. A fibrocartilaginous rim, the **acetabular labrum,** deepens the articular cavity. It is triangular in cross section, with its base attached to the margin of the acetabulum and its free apex extending into the cavity.

The extensive **synovial membrane** of the hip joint passes from the margin of the articular cartilage of the head to cover the neck of the femur internal to the articular capsule. It reflects back onto the capsule and covers the acetabular labrum and the synovial fat pad, and ensheathes the ligamentum teres. It sometimes communicates with the bursa deep to the iliopsoas tendon.

The strong, dense, articular **capsule** of the hip joint is attached to the innominate

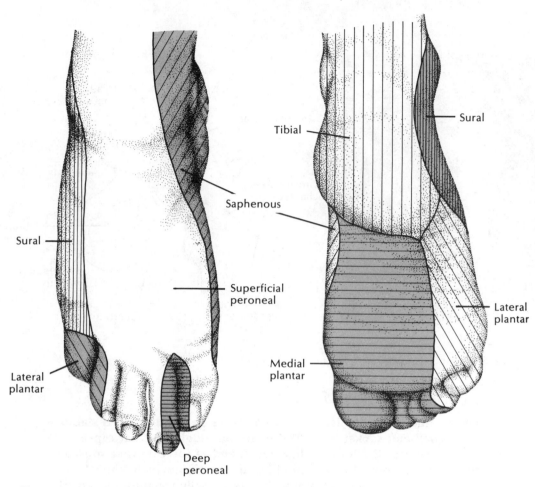

Fig. 6-17. Distribution of cutaneous innervation of the foot.

bone just beyond the periphery of the acetabular labrum. Its femoral attachments are the intertrochanteric line anteriorly, the base of the neck superiorly, and just above the intertrochanteric crest posteriorly. The capsule is composed of longitudinal collagenous fibers that are reinforced by accessory ligaments anteriorly and superiorly, and circular fibers (zona orbicularis) at the inferior and posterior aspect. The latter form a sling around the neck of the femur. The **iliofemoral ligament** (Y-shaped ligament of Bigelow) is a band of great strength at the anterior aspect of the articular capsule. It is intimately associated with the capsule and is attached superiorly to the anterior inferior iliac spine; inferiorly it divides into two bands that attach to the upper and the lower parts of the intertrochanteric line. The triangular **ischiofemoral ligament** reinforces the posterior aspect of the capsule. It attaches superiorly to the ischium, below and behind to the acetabular margins, and inferiorly blends with the circular fibers of the capsule.

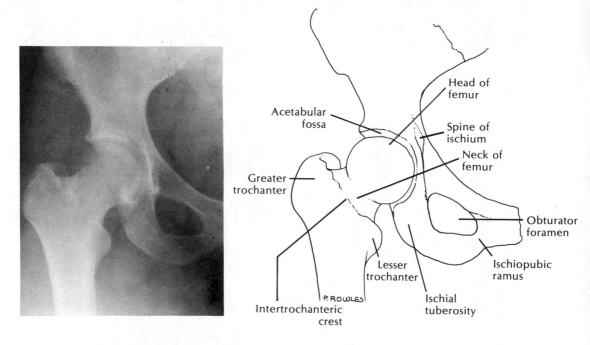

Fig. 6-18. Anteroposterior radiograph of hip joint.

DISLOCATION OF THE HIP

Dislocation of the hip is quite rare because of: 1) the marked stability of the ball-and-socket joint, 2) the strong, tough, articular capsule, 3) the strength of the intrinsic ligaments, and 4) the extensive musculature over the joint. In congenital hip dislocation, which is most common in females, the cartilaginous lips (especially the upper lip) of the acetabular fossa allows the head of the femur to slip out of the fossa onto the gluteal surface of the ilium. Since the head of the femur in the fossa promotes normal development of the acetabulum, the hip dislocation must be reduced and a cast or traction is used to prevent recurrent dislocation.

The **ligamentum capitis femoris** (round ligament of the femur) is a flattened triangular band covered by synovial membrane attached by its apex to the fovea capitis femoris, and by its base by two bands to either side of the **acetabular notch.** In the interval between the osseous attachments of the round ligament to the acetabulum, the ligament blends with the transverse acetabular ligament as the latter extends across the acetabular notch to convert it into a foramen.

KNEE JOINT

The knee joint is usually classified as a ginglymus (hinge) joint, but is actually much more complex in function (Fig. 6-19 A & B). It combines the actions of three types of diarthrodial joints, namely, a **ginglymus** (hinge), a **trochoid** (pivot), and an **arthrodial** (gliding). Furthermore, it may be described as three separate articulations, the **femoropatellar** and the two **tibiofemoral joints.** The articular surfaces of the medial and lateral condyles of the femur, covered by articular cartilage and separated by a groove, diverge posteriorly and articulate with the two entirely separate condyles of the tibia. Each of the condyles is deepened by a meniscus and separated by the intercondylar crest. The **medial** and **lateral menisci** are two crescentic cartilaginous lamellae that cover the peripheral two-thirds of the articular surface of the tibia. They are triangular in cross section with their peripheral bases attaching by the **coronary ligaments** to the articular capsule and the tibia. Their free apices project into the articular cavity.

ASPIRATION OF THE KNEE

Aspiration of the knee joint may be necessary to relieve pressure, to evacuate blood, or to obtain a fluid sample for laboratory studies. In this procedure, the needle is inserted somewhat proximal and lateral to the patella through the tendinous part of the vastus lateralis muscle and is directed towards the middle of the joint. Through the same route the joint cavity may be irrigated and steroids are administered in the treatment of knee pathology.

The **synovial membrane** lining the knee joint is the largest and most extensive in the body. It begins at the superior border of the patella, sends a blind sac deep to the quadriceps femoris muscle, and frequently communicates with the large **suprapatellar bursa** between the quadriceps femoris and the femur. Inferiorly the synovial membrane lies on either side of the patella, deep to the aponeurosis of the vasti muscles, and is separated from the patellar ligament by the **infrapatellar fat pad.** At the medial and lateral aspects of the patella reduplications of the synovial membrane pass into the interior of the joint as the **alar folds.** At the tibia the synovial membrane attaches to the peripheries of the menisci and reflects over and ensheathes the cruciate ligaments.

KNEE INJURIES

The integrity of the knee joint depends upon the strength of its femoral-tibial ligaments and the tonus of the muscles playing over the joint, especially the quadriceps femoris. This muscle is capable of functionally preserving the joint, even if one or more of its ligaments is disrupted.

The most common football knee injury is the rupture of the me-

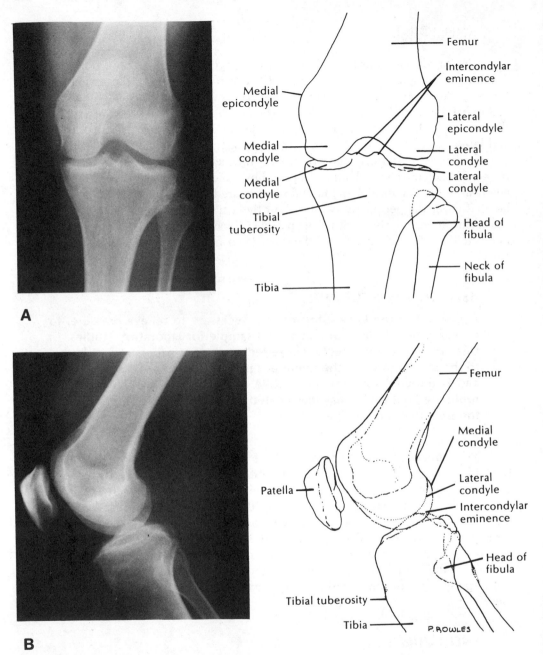

Fig. 6-19. A. Anteroposterior radiograph of knee joint. **B.** Lateral radiograph of knee joint.

dial collateral ligament, which is often accompanied by a tearing of the medial meniscus (medial semilunar cartilage). Since the ligament is firmly attached to the medial meniscus, these injuries usually occur together. (Injuries are less common to the lateral side of the knee because such an attachment is not present.) As the medial meniscus is torn, it may become wedged between the articular surfaces of the femur and tibia causing a "locking" of the joint.

A "clipping" injury is caused by a blow to the lateral side of the knee, damaging the anterior and posterior cruciate ligaments. Thus, in examining a patient for a knee injury, the physician is immediately concerned about the functional integrity of the three C's (collateral ligaments, cartilages, and cruciate ligaments).

The osseous arrangement of the knee joint as a weight-bearing structure is intrinsically unstable. The stability of this joint is realized by compensating mechanisms of surrounding muscles and tendons, a strong articular capsule, strong internal and external ligaments, and modification of the joint surface by the menisci. The **patellar ligament** (ligamentum patellae), the central portion of the tendon of the quadriceps femoris, attaches superiorly to the apex and adjacent margins of the patella, and inferiorly to the tuberosity of the tibia. The posteriorly situated **oblique popliteal ligament** attaches superiorly to the femur above the condyles and to the upper margin of the intercondylar fossa; inferiorly it attaches to the posterior margin of the head of the tibia. This ligament forms part of the floor of the popliteal fossa, and the popliteal artery passes inferiorly on its surface. The **arcuate popliteal ligament** arches inferiorly from the lateral condyle of the femur to the posterior surface of the capsule, where two converging bands attach it to the styloid process of the fibula. The **tibial (medial) collateral ligament,** a broad, flat, membranous band, attaches superiorly to the medial condyle of the femur and inferiorly at the medial condyle of the tibia. In its lower part it is crossed by tendons of the sartorius, gracilis, and semitendinosus muscles; covers part of the semimembranosus tendon, and is intimately adherent to the medial meniscus. The **fibular (lateral) collateral ligament,** a strong, rounded fibrous cord, is attached superiorly to the lateral condyle of the femur. Inferiorly it splits the inserting tendon of the biceps femoris to attach to the head of the fibula. It has no attachment to the lateral meniscus.

Strong internal cruciate ligaments cross like the limbs on an X, are named from their position and attachment to the tibia, and are ensheathed by synovial membrane. The **anterior cruciate ligament** attaches in a depression anterior to the intercondylar eminence of the tibia and passes superiorly, posteriorly, and laterally to attach to the medial and posterior aspect of the lateral condyle of the femur. The **posterior cruciate** is stronger, shorter, and less oblique than the anterior. It attaches posterior to the intercondylar eminence and to the posterior extremity of the lateral meniscus. It passes superiorly, anteriorly, and medially to the medial condyle of the femur. Another internal structure, the posterior **menisco-femoral ligament (of Wrisberg),** a strong fasciculus arising from the posterior attachment of the lateral meniscus, attaches to the medial condyle of the femur immediately behind the attachment for the posterior cruciate ligament. The **transverse ligament** of the knee connects the anterior margins of the menisci.

A large number of synovial bursae surround the knee joint. Anteriorly four bursae are present: one between the patella and the skin, another between the upper part of the tibia and the patellar ligament, one between the lower part of the tuberosity of the tibia and the skin, and a large bursa between the deep surface of the quadriceps femoris and the lower part of the femur. Laterally four bursae are noted: one between the lateral head of the gastrocnemius and the articular capsule, another between the fibular collateral ligament and the biceps tendon, a third between the fibular collateral ligament and the popliteus, and the last between the tendon of the popliteus and the lateral condyle of the femur. Five bursae are medially located: the first between the medial head of the gastrocnemius and the capsule; the second superficial to, and interposed between, the tibial collateral ligament and the sartorius, gracilis, and semitendinosus tendons; the third deep to, and interposed between, the tibial collateral ligament and the semimembranosus; the fourth between the semimembranosus and the head of the tibia; and the last between the semimembranosus and the semitendinosus. Any of these bursae may communicate with the synovial cavity of the knee joint.

ANKLE JOINT

The ankle presents two joints, the tibiofibular and the talocrural. The strength of the ankle joint is largely dependent upon the integrity of the **tibiofibular syndesmosis,** a strong fibrous union between the distal ends of the tibia and the fibula. Strong interosseous ligaments connect the roughened surfaces of adjacent portions of the bones to each other and are strengthened anteriorly and posteriorly by the **anterior** and **posterior tibiofibular ligaments.** The **talocrural joint** is between the tibia, the fibula, and the trochlea of the talus. The distal ends of the tibia and fibula form a deep socket, wider anteriorly than posteriorly, to receive the upper part of the talus. It is surrounded by the joint capsule, which is greatly thickened medially and laterally. The medial reinforcement, the **deltoid ligament,** is roughly triangular in shape and attaches superiorly to the medial malleolus and inferiorly to the talus, navicular, and calcaneus bones. Laterally three discrete ligaments, often referred to as the **lateral ligament,** include the **anterior talofibular ligament** passing from the lateral malleolus to the neck of the talus, the **posterior talofibular ligament** from the malleolar fossa to the posterior tubercle of the talus, and, between the two talofibular ligaments, the **calcaneofibular ligament** passing from the lateral malleolus to the lateral surface of the calcaneus.

POTT'S FRACTURE

Pott's fracture is a serious fracture in which the lower part of both the tibia and fibular are broken. If only one of the bones is fractured the unbroken bone acts as a splint resulting in minimal displacement of the limb. A complication of this injury is necrosis of the distal end of the fibula because of its poor blood supply. If this occurs, the stability of the ankle joint is impaired.

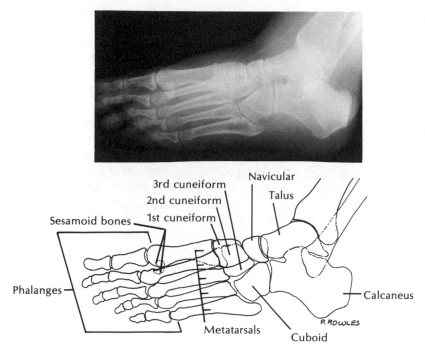

Fig. 6-20. Bones of the foot.

JOINTS OF THE FOOT

Tarsal and metatarsal bones are bound together by ligaments to form the longitudinal and transverse arches of the foot (Fig. 6-20). These arches are maintained partly by the shape of the bones, partly by the tension of the ligaments and the plantar aponeurosis, but most importantly by the bracing of muscle tendons attached to the foot. The **longitudinal arch** presents a greater height and wider span on the medial side of the foot, with the talus forming the "keystone" at the summit of the arch. The short, solid **posterior pillar** is formed by the calcaneus; the much longer **anterior pillar** is supported by a medial and a lateral column. The **medial column** is formed by the navicular, three cuneiforms, and three medial metatarsal bones; the **lateral column** by the cuboid and lateral two metatarsal bones. The weight of the body is transmitted to the talus at the summit. The important ligaments that prevent flattening of the arch lie in the plantar concavity. They include the **plantar calcaneonavicular (spring) ligament** which completes the socket formed by the navicular and the calcaneus; the **long plantar ligament** attaching to most of the plantar surface of the calcaneus and passing forward to attach to the tuberosity of the cuboid; the **plantar calcaneocuboid (short plantar) ligament,** placed deep to the long plantar and consists of a strong wide band 2 to 3 cm passing from the anterior part of the plantar surface of the calcaneus to a ridge behind the groove in the cuboid; and the **plantar aponeurosis,** which passes from the anterior to the posterior pillars and acts as a tie beam.

ARCHES OF THE FOOT

The entire body weight is supported by the feet. Normally, longitudinal and transverse arches of the foot result in contact at three points, the heel, and the first and fifth metatarsophalangeal joints. The integrity of these arches is maintained by the shape of tarsal bones, longitudinal and transverse ligaments, and leg muscles. If the arches are impaired by injury, excessive fatigue, illness, overweight, or congenital defects, the muscles are weakened, the ligaments are stretched, the arches flattened and pain is produced. Foot ailments that produce symptoms usually result from changes in the integrity of these arches or an excessive strain on the muscles or ligaments. The shape of the foot at birth may be flat foot (pes planus), normal arch, or claw foot (pes cavus), but these congenital foot shapes usually do not produce symptoms by themselves. The common flat foot is a foot with a flattened medial longitudinal arch, laterally displaced toes, and an everted foot. Claw foot describes a foot with an unusually high medial longitudinal arch that is caused by the powerful pull of the short intrinsic muscles of the sole of the foot. Postural strain, which usually involves such things as fatigue, overweight or poorly fitting shoes, usually produces pain in the foot or lower leg.

The **talocalcaneonavicular joint** lies anterior to the tarsal canal and forms part of the **transverse tarsal joint.** The head of the talus fits into a pocket formed by the navicular above and the calcaneous below. The considerable interval between the navicular and the calcaneus is occupied by the spring ligament. The **calcaneocuboid joint** completes the transverse tarsal joint. The **bifurcate ligament** reinforces the capsule, attaches in a depression on the upper part of the calcaneus, where it then divides to pass to the navicular and cuboid bones.

The **tarsometatarsal joints** are formed by the cuneiforms and the metatarsals. Individually there is little movement at the above joints, but working together they give elasticity and allow for twisting of the foot.

HAMMER TOE

Hammer toe is a common deformity of the second or third toe. It may be congenital, or acquired by wearing poorly fitting shoes. In the typical case, the metatarsophalangeal joint and the distal interphalangeal joint are hyperextended but the proximal interphalangeal joint is acutely flexed. A painful callus and small bursa often develop over the flexed joint.

The **metatarsophalangeal** and **interphalangeal joints** are similar. Between the heads of the metatarsal and the base of the proximal phalanx, the joint capsule is strengthened by **collateral ligaments.** The plantar portion of the capsule is thickened as the **plantar ligament** and is firmly fixed to the bases of the phalanges to allow the flexor tendons to glide freely over them. Plantar ligaments are connected by strong transverse fibers of the **deep transverse metatarsal ligament.** Aided by the transverse head of the adductor hallucis, this ligament helps to hold the heads of the metatarsals together with the tendons of the interossei passing deep, and the lumbricals superficial to this ligament.

7

head and neck

Surface Anatomy of Head and Neck

Viewed from the front, the most prominent midline surface feature of the neck is the bulging, laryngeal prominence (Adam's apple), which is formed by the V-shaped thyroid cartilage of the **larynx.** At its superior border the laryngeal notch is palpable, while above, the **hyoid bone** can be felt. Immediately below the larynx is an important landmark, the **cricoid cartilage,** which is joined inferiorly by the **trachea.** The latter is reinforced by a series of palpable, horseshoe-shaped cartilaginous bars with their open ends directed posteriorly.

At the base of the neck the superior border of the manubrium, located between the **sternal heads** of the sternocleidomastoid muscles, forms an obvious concave depression, the **suprasternal space,** or **jugular notch.** Here the clavicles can be seen projecting laterally towards the tips of the shoulders, which are formed by the acromion processes of the scapulae. The insertion of the sternocleidomastoid muscle is the large, rounded **mastoid process** easily palpated behind the ear.

The chin is formed by the **mental protuberance** of the mandible, while at the posterior border of the bone the prominent **angle** is continued **superiorly** as the **ramus.** When the jaw is clenched, the **masseter muscle** is easily demonstrable superficial to the angle and the ramus. The bony prominence of the cheek, formed by the zygomatic bone, is continuous anteriorly and posteriorly with the zygomatic processes of the maxillary and temporal bones, respectively, and superiorly with the zygomatic process of the frontal bone. The freely movable **cartilaginous portion** of the nose can be followed superiorly to the stationary **nasal bones,** which form the bridge of the nose. At the lips, the skin of the face and the mucous membrane of the oral cavity are continuous.

The **rima palpebrarum,** a slitlike orifice at the free margins of the eyelid, fuses laterally and medially, as the **inner** and **outer canthi.** The skin of the lid is continuous at the free margin with the conjunctivum covering the inner surface of the eyelid. The eyelashes curve outward from the free border, while the ducts of the **tarsal (meibomian) glands** open onto the free surface of the lid. **Supraciliary** (brow) **ridges,** covered partially by the eyebrows, indicate the anterior bulging of the frontal air sinuses. The external portion of the ear, consisting of the **auricula,** or **pinna,** and the **external acoustic meatus,** will be described with the ear.

The nostrils, or **external nares,** are oval openings separated from each other by the lower movable cartilaginous part of the nasal septum. They have stiff hairs (vibrissae) projecting into the vestibule that screen particulate matter carried into the nasal cavity by air currents.

Fasciae

The **superficial fascia of the head** and neck is continuous with the superficial fascia of the pectoral, deltoid, and back regions (Fig. 7-1). The muscles of facial expression are embedded within this layer.

The **deep fasciae** consist of the external investing layer, the middle cervical layer, and the visceral and the prevertebral fasciae. The **external investing layer (external**

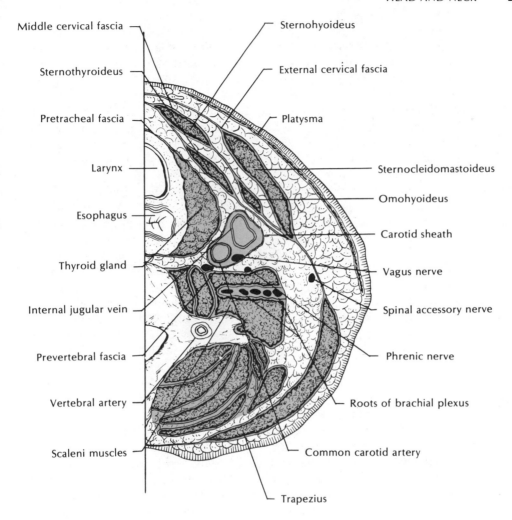

Fig. 7-1. Fasciae of the neck.

cervical fascia) completely invests the neck like a stocking and extends from the clavicle over the mandible to the zygomatic bone, where it blends with the fascia enclosing the masseter muscle. Posteriorly this layer fuses with the ligamentum nuchae and attaches superiorly to the external occipital protuberance and the superior nuchal line. In the anterior triangle this layer is bound to the hyoid bone and is subdivided into the suprahyoid and infrahyoid portions. The **suprahyoid portion** attaches to the inferior margin of the mandible, covers the submandibular gland, and sends a strong membranous process deep to the gland that attaches to the hyoid bone and angle of the mandible. This deep extension separates the submandibular and parotid glands and then splits to enclose the latter.

The **infrahyoid portion** of the external investing layer splits inferiorly to attach to the anterior and posterior aspect of the manubrium, where it forms the **suprasternal space.** This space is limited by the sternal heads of the sternocleidomastoideus and contains the lower portion of the anterior jugular veins, their communications across the midline, and some lymph nodes. Passing laterally this fascial layer opens to invest the sternocleidomastoideus, then fuses as it crosses the posterior triangle of the neck, separates again to ensheath the trapezius, and then fuses posteriorly in the midline with the ligamentum nuchae.

The **middle cervical fascia** is composed of two layers, with the stronger, more superficial layer enclosing the sternohyoideus and omohyoideus, and fusing superficially with the outer investing layer. The delicate deeper layer encloses the thyrohyoideus and sternothyroideus, and contributes to the formation of the carotid sheath before it fuses laterally with the external investing fascia.

The **prevertebral fascia** covers the anterior aspect of the cervical vertebrae, passes laterally to enclose the longus colli and scaleni muscles, then continues posteriorly to surround the levator scapulae muscle. Between the anterior and middle scaleni, this layer is prolonged into the axilla as the **cervicoaxillary sheath,** which surrounds the brachial plexus and the axillary artery. The prevertebral fascia is continuous inferiorly with the endothoracic fascia of the thoracic cavity. At the cervicothoracic aperture this fascial layer expands over the apex of the lung as the **cervical diaphragm** or **Sibson's fascia.** A potential cleft between the prevertebral fascia and the visceral fascia, the **retropharyngeal space,** is limited superiorly by the base of the skull and laterally by the attachment of the prevertebral fascia to the middle cervical fascia. Inferiorly the retropharyngeal space communicates with the posterior mediastinum.

The **visceral compartment** of the neck is located between the prevertebral fascia and the middle cervical fascia. It contains the major arteries and nerves within the neck, the cervical portions of the digestive and respiratory systems, and the thyroid and parathyroid glands. The **visceral (pretracheal) fascia** is a tubular prolongation into the neck of the visceral fascia of the mediastinum, where it is continuous with the fibrous pericardium. It encloses the esophagus, trachea, pharynx, and larynx, and contributes laterally to the formation of the carotid sheath. That portion covering the superior constrictor muscles is called the **buccopharyngeal fascia** and is attached superiorly to the pharyngeal tubercle at the base of the skull, and anteriorly to the pterygoid hamulus and the pterygomandibular raphe.

Neck

ANTERIOR TRIANGLE OF THE NECK

The **anterior triangle** of the neck is bounded posteriorly by the anterior border of the sternocleidomastoideus and anteriorly by the midline of the neck; the base is formed by the lower border of the mandible, and the apex is at the sternum. For descriptive purposes it is subdivided into three paired and one common triangle (Fig. 7-2). The **muscular triangle,** delineated by the superior belly of the omohyoideus, the sternocleidomastoideus, and the midline, contains the infrahyoid muscles and the thyroid gland. The superior belly of the omohyoid, the posterior belly of the digastric

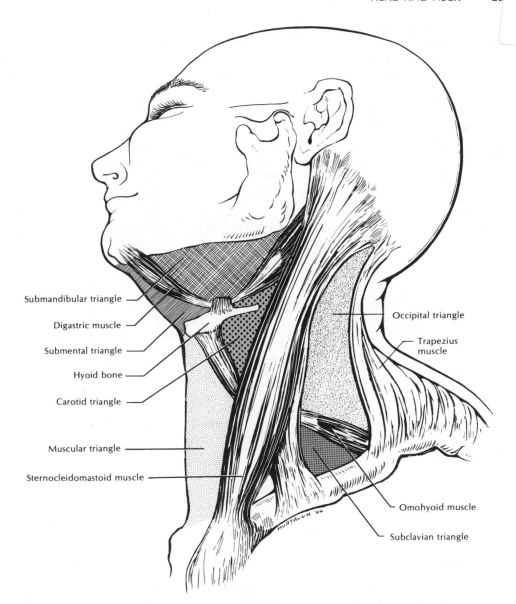

Fig. 7-2. Triangles of the neck.

muscle, and the sternocleidomastoideus bound the **carotid triangle,** which contains the vagus and hypoglossal cranial nerves, the common, external, and internal carotid arteries, the internal jugular vein, and the hyoid bone. The **submandibular** (digastric) **triangle** is limited by the two bellies of the digastric muscle and the mandible. It contains the submandibular gland, nerves to the anterior belly of the digastric and mylohyoideus, the hypoglossal nerve, and the lingual and facial arteries. The unpaired **submental triangle** is between the anterior bellies of the digastric muscles and the hyoid bone. It contains the anterior jugular veins.

Table 7-1. Muscles of the Anterior Triangle of the Neck

Muscle	Origin	Insertion	Action	Nerve
Platysma	Superficial fascia of upper pectoral and deltoid regions	Skin and facial muscles overlying mandible and border of mandible	Depresses mandible and lower lip; tenses and ridges skin of neck	Facial
Sternocleidomastoideus	Manubrium and medial one-third of clavicle	Mastoid process and lateral half of superior nuchal line	Singly rotates and draws head to shoulder; together flex cervical column	Spinal accessor and C_3
Omohyoideus	Medial lip of suprascapular notch	Lower border of body of hyoid	Steadies hyoid; depresses and retracts hyoid and larynx	C_2 and C_3 from ansa cervicalis
Digastric	Mastoid notch of temporal bone	Mandible near symphysis	Raises hyoid and base of tongue; steadies hyoid; depresses mandible	Posterior belly facial; anterior by mandibular sion of trigemi
Mylohyoideus	Mylohyoid line of mandible	Median raphe and hyoid bone	Elevates hyoid and base of tongue; depresses mandible; raises floor of mouth	Mylohyoid bra of inferior alve
Geniohyoideus	Genial tubercle of mandible	Body of hyoid	Elevates hyoid and base of tongue	C_1 and C_2 cour with hypogloss nerve
Sternohyoideus	Posterior surface of manubrium and medial end of clavicle	Lower border of body of hyoid	Depresses hyoid and larynx	C_1, C_2, and C_3 ansa cervicalis
Sternothyroideus	Posterior surface of manubrium	Oblique line of thyroid cartilage	Depresses thyroid cartilage	C_1, C_2, and C_3 ansa cervicalis
Thyrohyoideus	Oblique line of thyroid cartilage	Lower portion of body and greater horn of hyoid	Depresses hyoid; elevates thyroid cartilage	C_1 and C_2 cour with hypogloss nerve

Muscles

An extensive thin sheet of muscle, the **platysma,** covers the entire anterior aspect of the neck from the lower border of the mandible to the clavicle (Table 7-1, Fig. 7-3). This superficial muscle of facial expression attaches to the skin and mandible. It is innervated by the facial nerve and acts to tense the skin over the neck. The **sternocleidomastoideus,** extending obliquely from the sternoclavicular joint to the mastoid process, divides the neck into anterior and posterior triangles. It has two origins, a **sternal head,** arising as a rounded tendon from the sternoclavicular joint, and a flattened **clavicular head,** from the superior aspect of the medial third of the clavicle. It lies superficial to the great vessels of the neck and the cervical plexus, with cutaneous branches of the latter (the lesser occipital, greater auricular, transverse cervical, and supraclavicular nerves), all emerging at about the midpoint of its posterior border. This muscle is in-

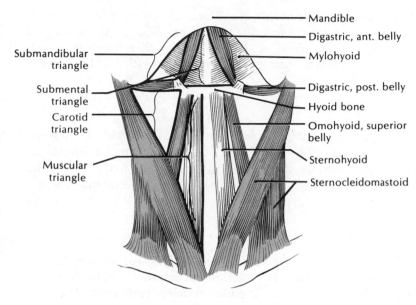

Submandibular triangle

Submental triangle

Carotid triangle

Muscular triangle

Mandible

Digastric, ant. belly

Mylohyoid

Digastric, post. belly

Hyoid bone

Omohyoid, superior belly

Sternohyoid

Sternocleidomastoid

Fig. 7-3. Muscles of the anterior triangle of the neck. (Hollinshead WH: Textbook of Anatomy, 3rd ed. New York, Harper & Row, 1974)

nervated by the spinal accessory nerve and twigs from the second and third cervical nerves.

A relatively extensive flat sheet of muscle, the two **mylohyoid** muscles, forms the floor of the submental and part of the digastric triangles, as well as the muscular floor of the mouth. From either side of the neck, these muscles meet in the midline to insert into the median raphe. Situated deep to the mylohyoideus, the small bandlike **geniohyoideus** muscle passes from the genial tubercle on the internal aspect of the mandible to the hyoid bone. The mylohyoideus is innervated by the mandibular division of the trigeminal nerve, and the geniohyoideus by twigs from the first loop of the cervical plexus passing with the hypoglossal nerve. Both muscles act in depressing the mandible and elevating the hyoid bone.

The **digastric muscle,** consisting of two bellies attached to an **intermediate tendon,** demarcates with the inferior border of the mandible, the digastric subdivision of anterior triangle. The intermediate tendon serves as a focal point in the relationships of the anterior triangle and is bound by a slip of fascia to the hyoid bone. The digastric is innervated by two nerves, the facial to the posterior belly and the mandibular division of the trigeminal to the anterior belly. It acts to elevate the hyoid bone and assists in depressing the mandible.

The four straplike muscles attaching either to the hyoid bone or the thyroid cartilage make up the **infrahyoid muscles.** The superficially placed **sternohyoideus** passes from the sternum to the hyoid bone. It lies superficial to the **thyrohyoideus.** The latter extends from the oblique line of the thyroid cartilage to the hyoid bone and may be considered the superior extension of the **sternothyroideus.** The **omohyoideus** consists of two bellies with an intermediate tendon. The muscle forms a wide V in passing from the superior notch of the scapula to the hyoid bone. The intermediate tendon attaches by deep fascia to the manubrium and the first rib or costal cartilage. All the infrahyoid muscles are innervated by the ansa cervicalis except for the thyrohyoideus, which receives a small branch from cervical components traveling with the hypoglossal nerve.

Arteries and Nerves

Arteries within the anterior triangle include the common carotid, the internal carotid, and the external carotid and its branches (Fig. 7-4). The **common carotid artery** courses superiorly from behind the sternoclavicular articulation to the level of the superior border of the thyroid cartilage, where it bifurcates into the internal and external carotid arteries.

CAROTID SINUS AND CAROTID BODY

At the bifurcation of the common carotid the internal carotid artery is dilated as the carotid sinus. At this location the internal carotid has abundant nerve endings supplied by the glossopharyngeal nerve. The carotid sinus is sensitive to changes in blood pressure and serves as a reflex pressoreceptor that assists in the regulation of blood pressure going to the brain. Posterior to the bifurcation in the wall of the inter-

Fig. 7-4. Lateral dissection of neck.

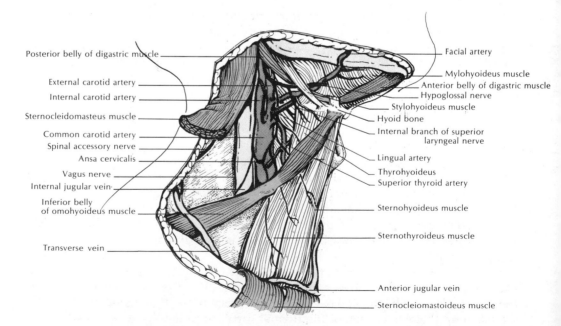

Posterior belly of digastric muscle

External carotid artery

Internal carotid artery

Sternocleidomasteus muscle

Common carotid artery

Spinal accessory nerve

Ansa cervicalis

Vagus nerve

Internal jugular vein

Inferior belly
of omohyoideus muscle

Transverse vein

Facial artery

Mylohyoideus muscle

Anterior belly of digastric muscle

Hypoglossal nerve

Stylohyoideus muscle

Hyoid bone

Internal branch of superior
laryngeal nerve

Lingual artery

Thyrohyoideus

Superior thyroid artery

Sternohyoideus muscle

Sternothyroideus muscle

Anterior jugular vein

Sternocleiomastoideus muscle

nal carotid, there is a small ovoid structure, the carotid body. It is a highly vascular, epithelioid body that is also well supplied by the glossopharyngeal nerve. It is sensitive to chemical changes in the blood, such as hypoxia. The carotid body acts as a chemoreceptor and reflexly causes an increase in heart rate, blood pressure, and respiratory rate, to eliminate the hypoxia.

The **internal carotid** has no branches in the neck but ascends within the carotid sheath with the internal jugular vein and the vagus nerve, to pass into the skull through the carotid canal as the principal blood supply to the brain and orbital cavity.

Six branches arise from the **external carotid artery** in the anterior triangle. Its two terminal branches, the superficial temporal and internal maxillary, bifurcate at the neck of the mandible. The six branches in the neck radiate from the area of the intermediate tendon of the digastric muscle to the various structures they supply.

LIGATION OF THE EXTERNAL CAROTID ARTERY

Ligation of the external carotid artery controls bleeding from a surgically inaccessible branch of this vessel, such as in the lingual, ascending pharyngeal, middle meningeal, or sphenopalatine. Collateral circulation is quickly established by anastomoses between the superior thyroid with the inferior thyroid arteries, and by branches crossing the midline of the face.

The **superior thyroid branch** arises from the anterior aspect of the external carotid opposite the thyrohyoid membrane and arches inferiorly to supply the thyroid gland. In its course it is accompanied by the external laryngeal nerve that innervates the cricothyroideus muscle. The **lingual artery** also arises from the anterior aspect of the external carotid, opposite the greater horn of the hyoid bone, and passes deep to the posterior belly of the digastric and mylohyoideus to supply the tongue. In its course it is crossed superficially by the hypoglossal nerve. The **facial (external maxillary) artery,** also from the anterior aspect of the external carotid, follows a sigmoid course to pass deep to the posterior belly of the digastric. It grooves the posterior and superior borders of the submandibular gland, then curves over the inferior border of the mandible onto the face. Its distribution will be considered with the discussion of the face. From the medial aspect of the external carotid the **ascending pharyngeal branch** ascends, with the internal carotid, to supply prevertebral muscles, the pharynx, and the palatine tonsil. Arising posteriorly, the **occipital branch** follows the inferior border of the posterior belly of the digastric muscle to pass superiorly and posteriorly to the mastoid process, where it ramifies on the back of the head. The **posterior auricular branch** courses along the superior border of the posterior belly of the digastric to pass to the notch between the external auditory meatus and the mastoid process. It supplies the area behind the ear. Deep to the parotid gland, the external carotid terminates by bifurcating into the **superficial temporal artery,** which ascends anterior to the ear and supplies the side of the face and head, and the **maxillary (internal maxillary) branch,** which will be described with the infratemporal fossa.

COLLATERAL CIRCULATION

Collateral circulation of the face and neck is extensive as many arteries of these regions anastomose freely with each other. In the neck, these channels are especially plentiful between arteries supplying the neck muscles and the thyroid gland. Anastomoses may also occur between the terminal branch of the facial artery and the ophthalmic artery. This establishes communication between the external and the internal carotid.

Nerves and their branches encountered in the anterior triangle of the neck include the glossopharyngeal, vagus, hypoglossal, cervical plexus, and cervical sympathetic chain (Table 7-2).

The **vagus nerve,** the longest of the cranial nerves, leaves the skull through the jugular foramen in company with the glossopharyngeal and the spinal accessory nerves, and the internal jugular vein. In the neck it is enclosed within the carotid sheath and enters the thoracic cavity by passing anterior to the subclavian artery. It has two **sensory ganglia,** the **superior** (jugular) at the jugular foramen and the **inferior** (nodose) 2 to 3 cm below. Initial branches of the vagus include the **recurrent meningeal,** which re-enters the cranial cavity by way of the jugular foramen to supply dura in the posterior cranial fossa; the **auricular branch,** cutaneous to the posterior aspect of the pinna of the ear and the floor of the external auditory meatus; the **pharyngeal branches,** which join branches of the glossopharyngeal and superior cervical ganglion to form the pharyngeal plexus, and the **nerve to the carotid body.**

The **superior laryngeal branch** of the vagus arises at the inferior ganglion, passes deep and medial to the internal and external carotid arteries, and divides into an internal and an external branch. The **internal laryngeal nerve** together with the laryngeal branch of the superior thyroid artery, pierces the thyrohyoid membrane to supply sensory fibers to the mucous membrane of the larynx above the level of the vocal folds, and parasympathetic fibers to glands of the epiglottis, base of the tongue, and the upper larynx.

The **external laryngeal nerve,** a long slender branch accompanies the superior thyroid artery. It passes deep to the sternothyroideus muscle and the thyroid gland to supply the cricothyroideus muscle.

INJURY TO THE EXTERNAL LARYNGEAL NERVE

Injury to the external laryngeal nerve may occur during thyroidectomy. It is a branch of the superior laryngeal nerve that lies next to the superior thyroid artery, often ligated in thyroid surgery. If the nerve is included in the ligature for the superior thyroid artery, the nerve supply to the cricothyroid muscle is interrupted and the vocal cord cannot be lengthened. With this loss of tension on one cord the voice becomes weak, hoarse, and easily fatigued.

Two or three **superior cardiac branches** arise at the level of the inferior ganglion and may join a sympathetic cardiac branch from the superior cervical ganglion as a

Table 7-2. Nerves in Anterior Triangle

Nerve	Origin	Course	Distribution
Vagus nerve (X)	Brain stem	Traverses jugular foramen to pass through neck in carotid sheath with carotid and internal jugular	Branches in neck include superior and inferior cardiac branches to cardiac plexus; superior and inferior laryngeal nerves to larynx; pharyngeal branches
Superior laryngeal	Vagus (X)	Divides into internal and external branches; internal laryngeal pierces thyrohyoid membrane to reach inside of larynx; external laryngeal follows superior thyroid artery	Internal laryngeal sensory to larynx above level of vocal folds; external laryngeal supplies cricothyroid
Inferior (Recurrent) laryngeal	Vagus (X)	On right side, hooks around subclavian artery; on left side, hooks around arch of aorta; nerves then run in tracheoesophageal groove to reach larynx	Motor to intrinsic muscles of larynx except cricothyroid (by external laryngeal), sensory below level of vocal folds
Ansa cervicalis	First and second loops of cervical plexus	Twig from first loop joins hypoglossal nerve, most fibers leave hypoglossal near intermediate tendon as superior limb of ansa; inferior limb from second loop of cervical plexus; limbs join to form loop (ansa) on superficial aspect of internal jugular	From ansa branches pass to sternohyoid, sternothyroid and omohyoid muscles
Nerve to thyrohyoid	Element of ansa cervicalis complex leaves hypoglossal just distal to superior limb of ansa	Passes directly to thyrohyoid muscle	Thyrohyoid
Nerve to geniohyoid	Element of ansa cervicalis complex leaves hypoglossal nerve in floor of oral cavity	Passes directly to geniohyoid muscle	Geniohyoid
Hypoglossal (XII)	Brain stem	Leaves cranial cavity through hypoglossal canal, courses deep to posterior belly and intermediate tendon of digastric; at intermediate tendon, passes forward on hyoglossus and mylohyoid to reach floor of oral cavity	All intrinsic and extrinsic tongue muscles except palatoglossus

common nerve to the heart. The **inferior cardiac branch** arises either from the vagus or from the recurrent laryngeal nerve at the root of the neck. All vagal cardiac branches from the right side and the superior cardiac branches from the left side pass behind the major vessels of the heart to terminate in the deep cardiac plexus. The inferior cardiac branch from the left side descends anterior to the arch of the aorta to end in the superficial cardiac plexus. Stimuli from cardiac branches of the vagus nerve decrease the rate and force of the heart beat.

Arising low in the neck, the **inferior (recurrent) laryngeal nerve** is sensory to the larynx below the level of the vocal folds and motor to all the intrinsic muscles of the larynx, except the cricothyroideus. On the right side of the inferior laryngeal nerve hooks around the subclavian artery, on the left around the arch of the aorta. Both nerves then pass posteriorly to ascend in the tracheoesophageal groove. At the larynx they penetrate the cricothyroid membrane to reach the interior of the larynx.

INFERIOR LARYNGEAL NERVE INJURY

In performing a thyroidectomy, this nerve is vulnerable to injury because of its intimate relation to the thyroid and the inferior thyroid artery. The surgeon must carefully identify and dissect the nerve away from the thyroid before removing the gland. Permanent postoperative hoarseness results from damage or division of this nerve. If only one nerve is involved, speech is not greatly affected because the other vocal cord is still functional. Bilateral cutting of the nerves causes the vocal cords to become fixed in a neutral abduction-adduction position, resulting in loss of speech and impaired breathing.

The **glossopharyngeal nerve** leaves the cranium through the jugular foramen together with the vagus and spinal accessory nerves, and the internal jugular vein. Initially, it courses between the internal jugular vein and internal carotid artery, then passes forward between the internal carotid and the styloid process. It presents two **sensory ganglia,** the small **superior** and the larger **inferior (petrosal) ganglion.** From the inferior ganglion the **tympanic nerve** passes through the temporal bone to join the tympanic plexus on the promontory of the middle ear cavity. It carries parasympathetic fibers destined for the parotid gland, and sensory fibers to the middle ear cavity. Filaments from the **promontory plexus** are reconstituted to form the **minor petrosal nerve,** which passes through a small canal in the temporal bone to emerge into the middle cranial fossa and then continues through the fissure between the temporal and the sphenoid bones to reach the otic ganglion. A second branch of the glossopharyngeal, the **nerve to the carotid sinus,** descends from the main trunk at the jugular foramen and joins with fibers from the vagus to pass to the anterior surface of the internal carotid and supply pressor receptor fibers to the carotid sinus. Three or four **pharyngeal branches** join with branches from the vagus and the superior cervical sympathetic ganglion. The **nerve to the stylopharyngeus** is the only named muscular branch of the glossopharyngeal nerve. **Tonsillar branches** supply sensation to the palatine tonsil, soft palate, and fauces, and communicate with the lesser palatine nerve. Two **lingual branches** supply the posterior third of the tongue with both general and special (taste) sensation.

Cervical Sympathetic Trunk

The **cervical portion of the sympathetic trunk** is embedded within the connective tissue posterior to the carotid sheath and anterior to the prevertebral muscles. Unlike the sympathetic chain in the thorax, where typically one ganglion occurs for each spinal nerve, only three ganglia are present. In the neck the first four cervical spinal nerves receive branches from the superior cervical ganglion, the fifth and sixth cervical nerves from the middle cervical ganglion, and the seventh and eighth cervical nerves from the inferior cervical ganglion. The latter frequently unites with the first thoracic ganglion to form the stellate ganglion. **No white rami communicantes are present in the cervical region.** Each of the cervical spinal nerves receives at least one gray ramus communicans from their respective fused ganglia.

The **superior cervical ganglion,** the largest ganglion of the sympathetic chain, is located between the internal carotid artery and the longus capitis muscle at the level of the second and third cervical vertebrae. It is spindle-shaped, about 2 to 3 cm long, and sends twigs superiorly along the internal carotid artery as the internal carotid plexus. It also sends **communicating twigs** to the glossopharyngeal, vagus, spinal accessory, and hypoglossal cranial nerves; **gray rami** to the first four cervical spinal nerves; **pharyngeal branches** to the pharyngeal plexus, and a **cardiac branch,** which on the left side passes to the superficial cardiac plexus, and on the right to the deep cardiac plexus.

HORNER'S SYNDROME

Horner's syndrome results from paralysis of the cervical sympathetic nerves. The injury may result from pressure on the cervical chain or ganglia by a malignant tumor in the neck or upper lung, surgery, or penetrating injuries to the neck. Symptoms include constriction of the pupil, narrowing of the palpebral fissure, recession of the eyeball (enophthalmos), drooping of the eyelid (pseudoptosis), loss of sweating (anhidrosis), and flushing of the face (vasodilation).

The smaller **middle cervical ganglion** is located opposite the summit of the loop of the inferior thyroid artery. It gives off **gray rami** to the fifth and sixth cervical nerves, a **thyroid branch,** which forms a plexus around the inferior thyroid artery, **cardiac branches** to the deep cardiac plexus, and the **ansa subclavia** as filaments that descend anteriorly to hook around the subclavian artery and join the inferior cervical ganglion.

The **inferior cervical ganglion** is small, irregular in shape, located behind the common carotid artery, and frequently joins with the ganglion of the first thoracic nerve to form the **stellate ganglion.** Its branches include **gray rami** to the seventh and eighth cervical spinal nerves, branches to the **ansa subclavia,** contributions to the **subclavian** and **vertebral plexuses,** and a **cardiac branch** to the deep cardiac plexus.

Cervical Plexus

The **cervical plexus** is formed by the **ventral rami** of the first four cervical nerves (Fig. 7-5). Each of these ventral rami divides into an **ascending** and a **descending division.** The ascending limb of the first cervical nerve passes into the skull to supply sen-

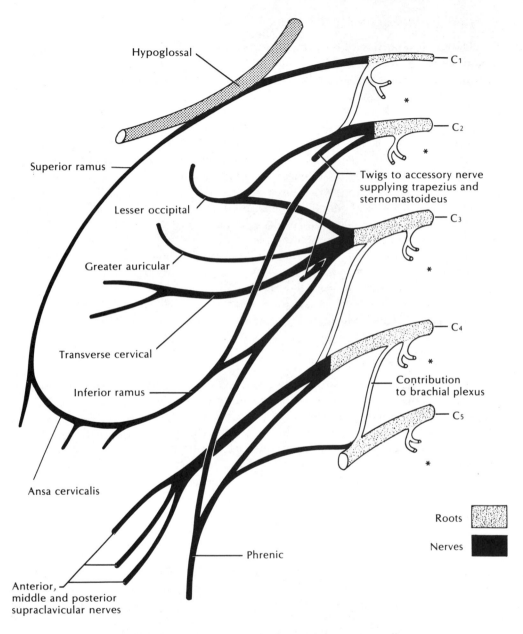

Hypoglossal

Superior ramus

Lesser occipital

Greater auricular

Transverse cervical

Inferior ramus

Ansa cervicalis

Anterior,
middle and posterior
supraclavicular nerves

Phrenic

C₁

C₂

Twigs to accessory nerve
supplying trapezius and
sternomastoideus

C₃

C₄

Contribution
to brachial plexus

C₅

Roots

Nerves

*Segmental branches to rectus capitis lateralis,
rectus capitis anterior, longus capitis, longus colli,
scaleni muscles and levator scapulae.

Fig. 7-5. Cervical plexus.

sation to the meninges. The descending limb of the fourth cervical nerve contributes to the brachial plexus. The remaining ascending and descending divisions unite to form **three loops** opposite the first four cervical vertebrae. The cervical plexus receives postganglionic sympathetic fibers from the superior cervical sympathetic ganglion. Branches of the plexus include the cutaneous nerves, to be described later with the posterior triangle (the **lesser occipital, greater auricular, transverse cervical,** and the three **supraclavicular nerves**), the ansa cervicalis, segmental muscular branches to the prevertebral muscles, and the phrenic nerve.

The ansa cervicalis, which innervates the infrahyoid muscles, is derived from the first and second loops of the cervical plexus. Twigs from the first loop (C_1 and C_2) pass to the hypoglossal nerve, travel with it for 1 or 2 cm, where most of the fibers leave the hypoglossal as the **superior ramus** (descendens hypoglossi) of the ansa cervicalis, and descends to form the ansa with a branch from the second loop (C_2 and C_3), the **inferior ramus** (descendens cervicalis). Fibers from the first and second cervical nerves, remaining with the hypoglossal nerve, branch off as independent twigs to supply the thyrohyoideus and geniohyoideus muscles. The superior ramus of the ansa cervicalis sends twigs to the superior belly of the omohyoideus. The inferior ramus supplies the inferior belly of the omohyoideus, and from the loop, branches pass to the sternohyoideus and sternothyroideus muscles.

Segmental muscular branches from the cervical plexus supply the prevertebral muscles. Twigs from the first loop of the cervical plexus pass to the rectus capitis lateralis, longus capitis, and rectus capitis anterior; from the second loop to the longus capitis and longus colli, and from the third loop to the middle scalenus and levator scapulae. Contributions from the cervical plexus to the spinal accessory nerve give additional innervation to the sternocleidomastoideus (C_2) and the trapezius (C_3 and C_4). The **phrenic nerve** is derived from the fourth cervical nerve with contributions from the third and fifth cervical nerves. It courses inferiorly to cross the scalenus anterior obliquely, passes deep to the transverse cervical and suprascapular arteries to pass through the thorax, where it innervates the diaphragm.

THE HICCUP

The hiccup is an involuntary, spasmodic contraction of the diaphragm which momentarily draws air into the lungs. At that instant, the glottis is closed causing the characteristic sound. The cause is unknown but may be due to the stimulation of nerve endings in the digestive tract or the diaphragm. Inhalation of air with about 5% CO_2 usually stops hiccuping. If the problem continues unabated, section of the phrenic nerve may be indicated.

PREVERTEBRAL REGION

The **prevertebral muscles** form a longitudinal muscular mass anterior to the vertebral column which includes the **anterior, middle,** and **posterior scaleni;** the **longus capitis** and **colli,** and the **rectus capitis anterior** and **lateralis muscles.** The scaleni,

Table 7-3. Prevertebral Muscles

Muscle	Origin	Insertion	Action	Nerve
Scalenus anterior	Transverse processes of third to sixth cervical vertebrae	Scalene tubercle on first rib	Bilaterally stabilize neck; unilaterally inclines neck to side	Twigs from ce plexus and C_5 through C_7
Scalenus medius	Transverse processes of lower five cervical vertebrae	Upper surface of first and second ribs	As above	Cervical plexu C_4 through C_8
Scalenus posterior	Transverse processes of fifth and sixth cervical vertebrae	Outer surface of second rib	As above	Twigs from C_7
Longus capitis	Transverse processes of third through sixth cervical vertebrae	Basilar portion of occipital bone	Flexes and rotates head	Twigs from C_1 through C_4
Longus colli	Transverse processes and bodies of third cervical to third thoracic vertebrae	Anterior tubercle of atlas, bodies of second to fourth cervical vertebrae, and transverse processes of fifth and sixth cervical vertebrae	Flexes and rotates head	Twigs from C_2 through C_6
Rectus capitis anterior	Lateral mass of atlas	Basilar portion of occipital bone	Flexes and rotates head	Twigs from C_1 C_2
Rectus capitis lateralis	Transverse process of atlas	Jugular process of occipital bone	Bends head laterally	Twigs from C_1 C_2

longus capitis, superior portion of the longus colli, and rectus capitis lateralis muscles all arise from transverse processes of the cervical vertebrae. The inferior portion of the longus colli arises from the bodies of cervical vertebrae, and the rectus capitis anterior from the lateral mass of the atlas. The scaleni insert into the first two ribs, the rectus capitis anterior and lateralis into the occipital bone, and the colli muscles into the bodies or transverse processes of the cervical vertebrae (Table 7-3). All the muscles are ensheathed by **prevertebral fascia,** are segmentally innervated, and act as a group in flexion and rotation of the head and neck.

Structures associated with the prevertebral region include the **vertebral artery, the carotid sheath** and its contents, the **cervical portion of the sympathetic trunk,** the **spinal nerves** emerging through the intervertebral foramina, and the **cervical plexus.**

POSTERIOR TRIANGLE OF THE NECK

The **posterior triangle** of the neck is bounded anteriorly by the posterior border of the sternocleidomastoideus and posteriorly by the anterior border of the trapezius (Fig. 7-6). The **apex** is at the junction of the above muscles on the superior nuchal line,

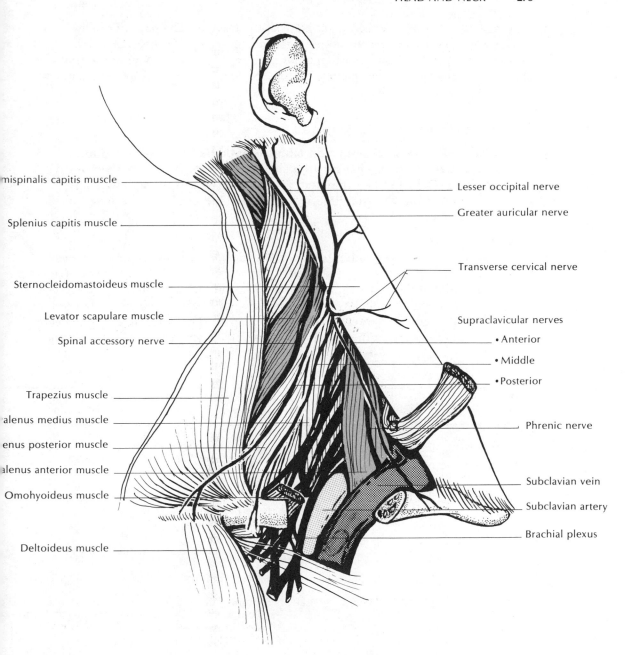

Semispinalis capitis muscle

Splenius capitis muscle

Sternocleidomastoideus muscle

Levator scapulare muscle

Spinal accessory nerve

Trapezius muscle

Scalenus medius muscle

Scalenus posterior muscle

Scalenus anterior muscle

Omohyoideus muscle

Deltoideus muscle

Lesser occipital nerve

Greater auricular nerve

Transverse cervical nerve

Supraclavicular nerves
• Anterior
• Middle
• Posterior

Phrenic nerve

Subclavian vein

Subclavian artery

Brachial plexus

Fig. 7-6. Contents of the posterior triangle of the neck.

and the **base** is formed by the middle third of the clavicle. The posterior belly of the omohyoideus muscle divides the posterior triangle into a small **subclavian** and a larger **occipital triangle.** From superior to inferior, six muscles form the **floor** of the triangle: the semispinalis capitis, the splenius capitis, the levator scapulae, and the anterior, middle, and posterior scaleni.

CENTRAL VENOUS PRESSURE

When lying on an examination table all patients have distended neck veins. If the patient changes to a sitting position, the central venous pressure (CVP) can be estimated. This is done by observing how many centimeters the external jugular vein is distended above the level of the clavicle.

Important structures within the posterior triangle include the spinal accessory, phrenic, lesser occipital, greater auricular, transverse cervical, and supraclavicular nerves; the roots of the brachial plexus; the subclavian, suprascapular, and transverse cervical arteries; and the subclavian and **external jugular veins** (Table 7-4). The latter, formed by the union of the retromandibular and posterior auricular veins, descends vertically from the angle of the mandible to pass obliquely across the sternocleidomas-

Table 7-4. Nerves in Posterior Triangle

Nerve	Origin	Course	Distribution
Spinal accessory (XI)	Brain stem and C_{1-5}	Exits cranial cavity via jugular foramen, crosses posterior triangle to reach trapezius	Sternocleidomastoid trapezius
Phrenic	Cervical plexus	Crosses anterior surface of scalenus anterior to enter inlet of thoracic cavity	Thoracic diaphragm
Lesser occipital	Second loop of cervical plexus	From midpoint of posterior border of sternocleidomastoid courses along posterior border to area behind ear	Skin behind ear
Greater auricular	Second loop of cervical plexus	From midpoint of posterior border of sternocleidomastoid crosses muscle to angle to mandible	Skin between sternoc mastoid and ramus o mandible
Transverse cervical	Second loop of cervical plexus	Crosses sternocleidomastoid transversely	Skin over anterior tri
Supraclaviculars	Third loop of cervical plexus	From midpoint of posterior border of sternocleidomastoid fan out inferiorly	Skin over shoulder a upper chest to secon tercostal space
Segmental branches	Ventral rami of spinal nerves	Pass directly to muscles supplied	Prevertebral muscles muscles in floor of p rior triangle

toideus, where it pierces the deep fascia about 2 to 3 cm above the clavicle and empties into the subclavian vein.

INTRAVENOUS CANNULATION

When other peripheral veins are collapsed or otherwise unavailable, the external jugular vein is readily available for cannulation, such as in old, obese, or chronically ill patients. The risk of complications is relatively small compared to subclavian or internal jugular entry. Intravenous cannulation is frequently used in a Code Blue (emergency) situation.

The **spinal accessory nerve** emerges from the posterior border of the sternocleidomastoideus to cross the triangle obliquely, dividing it into equal parts, and then disappears deep to the trapezius muscle that it supplies. The **phrenic nerve,** arising from the third, **fourth,** and fifth cervical nerves, crosses the scalenus anterior obliquely to descend into the thorax.

DROOPED SHOULDER

A superficial wound in the posterior triangle can sever the relatively superficial spinal accessory nerve. Injury to this nerve would denervate the trapezius muscle. This would result in asymmetry of the lateral slope of the neck from atrophy of the muscle and a downward displacement or drooping of the shoulder, on the affected side.

Inferior to the spinal accessory nerve, four cutaneous branches of the cervical plexus emerge from the posterior border of the sternocleidomastoideus. The **lesser occipital nerve** (C_2 and C_3) follows the posterior border of the sternocleidomastoideus cephalad to pierce the deep fascia near the mastoid process, where it supplies the scalp above and behind the ear, and the medial surface of the pinna. The **greater auricular nerve** (C_2 and C_3) passes obliquely across the sternocleidomastoideus to ascend behind the ear, where it gives a **mastoid branch** to the mastoid process, an **auricular branch** to both surfaces of the pinna, and **facial branches** to the skin in front of the ear and over the parotid gland. The **transverse cervical nerve** (C_2 and C_3) crosses the sternocleidomastoideus horizontally to divide into a **superior** and an **inferior branch,** which supply the skin over the anterior triangle of the neck. A large branch from C_3 and C_4 divides into the **medial, intermediate, and lateral supraclavicular nerves.** They pierce the platysma near the clavicle to supply the skin of the chest as low as the second intercostal space and the skin over the upper portion of the shoulder.

LOCAL ANESTHETIC BLOCK

Infiltration of a local anesthetic agent at the midpoint of the posterior border of the sternocleidomastoid can anesthesize most of the skin of

the neck. This injection blocks the four cutaneous branches of the cervical plexus that emerge behind the sternocleidomastoideus.

Within the posterior triangle, the **trunks of the brachial plexus** emerge through the **scalene gap** between the anterior and middle scaleni muscles to continue through the cervicoaxillary canal into the axilla.

SCALENUS ANTICUS SYNDROME (CERVICAL RIB SYNDROME)

This syndrome results from compression of the brachial plexus and subclavian artery. It may be due to a hypertrophy of the scalene muscles, which narrows the "scalene gap" through which these structures pass to enter the posterior triangle. It may also be caused by the presence of a supernumerary (cervical) rib. In this condition there may be pain, numbness, and weakness, if not corrected it can result in muscular atrophy and reflex disorders.

The **subclavian artery** crosses the first rib to pass through the scalene gap in company with, but anterior to, the brachial plexus, while the **subclavian vein** passes anterior to the scalenus anterior as it crosses the first rib. Two branches of the subclavian artery, the **suprascapular** and the **transverse cervical,** cross the scalenus anterior from medial to lateral to pin down the phrenic nerve as they course toward the scapular region.

SUBCLAVIAN VEIN CATHETERIZATION

The subclavian vein passes anterior to the scalenus anterior, which inserts into the scalene tubercle on the first rib. When using the supraclavicular approach to insert a needle or catheter into the subclavian vein, palpation of the scalene tubercle safeguards the underlying subclavian artery and brachial plexus. The latter structures course deep to the scalenus anterior as they pass through the scalene gap.

ROOT OF THE NECK

The **root of the neck** is the junctional area between the neck proper and the thorax and is limited laterally by the first rib, anteriorly by the manubrium, and posteriorly by the first thoracic vertebra (Fig. 7-7). It transmits all the structures passing between the neck and the thorax.

Thyroid Gland

The **thyroid** is an encapsulated, endocrine, butterfly-shaped gland, located in front and to the sides of the trachea at the level of the fifth through seventh cervical

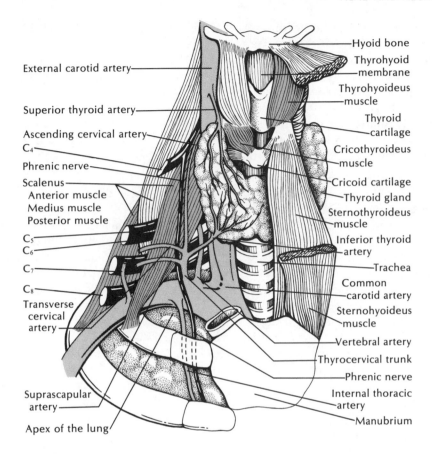

External carotid artery

Superior thyroid artery

Ascending cervical artery

C$_4$

Phrenic nerve

Scalenus
Anterior muscle
Medius muscle
Posterior muscle

C$_5$
C$_6$

C$_7$

C$_8$

Transverse
cervical
artery

Suprascapular
artery

Apex of the lung

Hyoid bone

Thyrohyoid
membrane

Thyrohyoideus
muscle

Thyroid
cartilage

Cricothyroideus
muscle

Cricoid cartilage

Thyroid gland

Sternothyroideus
muscle

Inferior thyroid
artery

Trachea

Common
carotid artery

Sternohyoideus
muscle

Vertebral artery

Thyrocervical trunk

Phrenic nerve

Internal thoracic
artery

Manubrium

Fig. 7-7. Root of the neck.

vertebrae. It consists of a **right** and a **left lobe** connected by the narrow **isthmus** of glandular tissue located at the level of the third or fourth tracheal ring. Each lobe is somewhat conical in form and consists of a base situated at the level of the fifth or sixth tracheal ring and an apex resting against the side of the thyroid cartilage extending to its superior border. Its full, rounded lateral (superficial) surface is covered by infrahyoid muscles, and the medial (deep) surface is molded by the structures on which it lies, namely, the cricoid and thyroid cartilages, and the cricothyroideus and inferior constrictor muscles superiorly, and the trachea and esophagus inferiorly. There may be a slender extension of the thyroid gland, the **pyramidal lobe,** passing either from the isthmus or the left lobe superiorly to the hyoid bone. A narrow slip of muscle, the **levator glandulae thyroideae,** is sometimes attached to this lobe.

THYROGLOSSAL DUCT

The thyroid gland develops from a downgrowth of a solid column of cells in the floor of the primitive pharynx at the foramen caecum of the tongue. During the caudal migration of the thyroid into the neck re-

gion, the cell column becomes canalized as the thyroglossal duct. The distal cells of the duct give rise to the thyroid. Although the fully developed thyroid has no duct, remnants of the duct may be found along its course. Cysts develop within such vestiges, especially in the sublingual region and in the midline above and below the thyroid cartilage. If they enlarge or become infected, these cysts must be removed surgically. Operative cure of a thyroglossal duct cyst requires removal of the cyst and the duct by removing the middle portion of the hyoid bone and ligating the duct close to its origin, at the base of the tongue.

The thyroid gland is a highly vascularized organ supplied by two pairs of relatively large arteries. The **superior thyroid arteries,** branches from the external carotids, supply the apices, and the **inferior thyroid arteries,** from the thyrocervical trunks, supply the base and the deep surface. A single, inconstant, small branch, the **thyroidea ima artery,** may arise from the brachiocephalic, the left common carotid, or the arch of the aorta. These vessels anastomose freely within each lobe but little communication occurs across the midline. **Venous drainage** is accomplished by three pairs of veins that form a superficial plexus; however, most of the blood is drained from the deep surface. The superior thyroid veins drain the upper portion of the gland and either cross the common carotids to join the internal jugular veins, or follow the superior thyroid arteries to end in the common facial veins. The middle thyroid veins arise near the lower portion of the gland and cross the common carotids to empty into the internal jugulars. The inferior thyroid veins are the largest veins draining the thyroid. They begin as a plexus over the isthmus and pass inferiorly to empty into the brachiocephalic veins.

Parathyroid Glands

Two or more pairs of small endocrine glands, the **superior** and **inferior parathyroids,** are usually embedded in the posterior aspect of the thyroid. The superior glands are more constant in position and usually lie at the level of the middle of the thyroid, while the inferior glands are situated near the base of the thyroid. Their vascular supply is usually derived from twigs of the inferior thyroid artery.

PARATHYROID TETANY

Parathyroid tetany results from the accidental removal of all or most of the parathyroids during a total thyroidectomy. Surgeons, therefore, are careful to identify and leave intact the parathyroid glands (usually four in number). The hormone of the parathyroid regulates calcium metabolism and plasma calcium concentrations. With removal of the parathyroids, the plasma calcium is decreased causing increased neuromuscular activity, such as muscular spasms, weakness, and nervous hyperexcitability, a condition called tetany. Tetany from removal of the parathyroid glands can cause death, unless adequate calcium or vitamin D is provided, or exogenous parathyroid hormone is administered.

Arteries in the Root of the Neck

The arteries in the root of the neck originate from the arch of the aorta and include the brachiocephalic (innominate) on the right side and the common carotid and subclavian arteries on the left. The **brachiocephalic** arises from the arch of the aorta to pass superiorly and divide behind the sternoclavicular joint into right common carotid and right subclavian arteries. Anteriorly the brachiocephalic artery is covered by the sternoclavicular joint and the sternohyoideus and sternothyroideus muscles; medially it rests on the trachea; laterally it is related to the right brachiocephalic vein, and posteriorly it is separated from the apical pleura by fat and connective tissue.

On the right side the **subclavian artery** originates behind the sternoclavicular joint as the terminal branch of the brachiocephalic; on the left it is a direct branch of the arch of the aorta and enters the root of the neck by passing behind the sternoclavicular joint. Both arteries arch laterally, groove the pleura and lungs, and pass between the anterior and middle scaleni muscles to become the axillary arteries at the lateral border of the first rib. The subclavian artery is arbitrarily divided into three parts by the anterior scalenus muscle. Branches from the first part (proximal to the muscle) include the vertebral, the thyrocervical trunk, and the internal thoracic arteries; the second part (deep to the muscle) gives off the costocervical trunk, while the dorsal scapular artery may arise as a single branch of the third part (distal to the muscle).

The **vertebral artery** arises from the posterosuperior aspect of the subclavian. It ascends vertically along the lateral border of the longus colli muscle to the level of the cricoid cartilage, to pass between the scalenus anterior and the longus colli muscles and enters the foramen intertransversarium of the sixth cervical vertebra. It traverses similar foramina in each of the cervical vertebrae to finally enter the cranial cavity through the foramen magnum. It contributes to the formation of the circle of Willis, which supplies the brain.

SUBCLAVIAN STEAL SYNDROME

The subclavian steal syndrome refers to a reversal of blood flow from the internal carotid through the basilar artery (that supplies the brain stem), down the vertebral to the subclavian artery. It can only occur when the subclavian artery is occluded proximal to the origin of the vertebral artery. When a patient with subclavian steal syndrome increases the demand for blood flow to the upper extremity by vigorous exercise, for example, the extra blood is "stolen" from the basilar-vertebral system, which can produce syncope or stroke.

Arising from the anterior aspect of the subclavian at the medial margin of the anterior scalenus muscle, the short, wide **thyrocervical trunk** gives origin to four branches. The largest branch, the **inferior thyroid artery,** follows an S-shaped course, ascending along the medial border of the scalenus anterior to pass medially between the carotid sheath and vertebral vessels at the level of the cricoid cartilage. It then descends along the posterior border of the thyroid gland, which it supplies from its deep aspect. The **ascending cervical artery** is a small, constant branch of the trunk which ascends obliquely across the scalenus anterior to give twigs to prevertebral muscles and the vertebral canal. Coursing laterally across the scalenus anterior, anterior to the phrenic nerve, the **transverse cervical branch** traverses the posterior triangle of the neck. It courses deep to the omohyoid muscle to reach the anterior border of the trapezius. It divides into a **superficial branch** (superficial cervical) ramifying on the deep surface of the trapezius and supplying it, and a **deep** (descending) **branch** that passes deep to the levator scapulae and rhomboids to follow the medial border of the scapula and to supply these muscles. The **suprascapular artery** follows the transverse cervical artery to pass to the scapular notch, where it meets the suprascapular nerve. At the notch the artery passes superficially and the nerve deep, to the transverse scapular ligament. Both structures divide into **supraspinatus** and **infraspinatus branches** to supply muscles in their respective fossae on the dorsum of the scapula.

The **internal thoracic artery,** arising from the inferior aspect of the subclavian, passes inferomedially to enter the thorax behind the first costal cartilage. It lies between the pleura and ribs and intercostal muscles and is usually crossed anteriorly by the phrenic nerve. This vessel continues inferiorly along the lateral border of the sternum to give **anterior intercostal branches** to the first six intercostal spaces, and then terminates by dividing into the **superior epigastric** and **musculophrenic arteries.**

The short **costocervical trunk** arises from the posterior aspect of the subclavian to pass superoposteriorly over the pleura to the neck of the first rib. Opposite the first intercostal space it gives a **deep cervical branch** that ascends between the neck of the first rib and the transverse process of the seventh cervical vertebra to supply deep muscles of the neck. The **superior intercostal branch** of the trunk courses inferiorly, anterior to the neck of the first rib, to give posterior intercostal branches to the first and second intercostal spaces. As a **variation** of the branches of the subclavian artery, the **dorsal scapular artery** arises, in about 50% of the cases, from the third part of the subclavian to course parallel to the medial border of the scapula. It supplies the levator scapulae and the rhomboids and replaces the deep (descending) branch of the transverse cervical artery from the thyrocervical trunk. When the dorsal scapular artery is present, the transverse cervical artery supplies only the trapezius muscle and is called the superficial cervical artery.

Deep Back

DEEP MUSCLES (TABLE 7-5)

The deep muscles of the back are arranged in two groups: a **longitudinal group** consisting of the **erector spinae,** or **sacrospinalis (iliocostalis, longissimus,** and **spinalis),** the **semispinalis,** and the **splenius;** and a **transverse group** including the **multifidi,** the **rotatores,** the **interspinales,** and the **intertransversarii** (Fig. 7-8). (Current terminology lists the term erector spinae for all the deep muscles of the back.)

The longitudinal group spans the interval from the spines of the vertebrae to the angles of the ribs and acts to extend the vertebral column. In width this group forms a

Table 7-5. Deep Muscles of the Back

Muscle	Origin	Insertion
Erector spinae	Series of muscles forming mass that extends from sacrum to skull. Acting unilaterally, they bend vertebral column to that side; bilaterally they extend vertebral column. They are segmentally innervated by dorsal rami of spinal nerves, as are all muscles of back listed below.	
Iliocostalis lumborum	Iliac crest and sacrospinal aponeurosis	Lumbodorsal fascia and tips of transverse processes of lumbar vertebrae and angles of lower six or seven ribs
Iliocostalis thoracis	Superior borders of lower seven ribs medial to angles	Angles of upper seven ribs and transverse process of seventh cervical vertebra
Iliocostalis cervicis	Superior borders at angles of third to seventh ribs	Transverse processes of fourth, fifth, and sixth cervical vertebrae
Longissimus thoracis	Sacrospinal aponeurosis, sacroiliac ligaments, transverse processes of lower six thoracic and first two lumbar vertebrae	Transverse processes of lumbar and thoracic vertebrae and inferior borders of ribs lateral to their angles
Longissimus cervicis	Transverse processes of upper five or six thoracic vertebrae	Transverse processes of second through sixth cervical vertebrae
Longissimus capitis	Transverse processes of first four cervical vertebrae and articular processes of last four cervical vertebrae	Mastoid process of temporal bone

Muscle	Origin	Insertion	Action
Spinalis thoracis	Spines of upper two lumbar and lower two thoracic vertebrae	Spines of second through ninth thoracic vertebrae	Extends vertebral column
Spinalis cervicis	Spines of upper two thoracic and lower two cervical vertebrae	Spines of second through fourth cervical vertebrae	Extends vertebral column
Semispinalis capitis	Transverse processes of upper six thoracic and seventh cervical vertebrae	Between superior and inferior nuchal lines	Extends and inclines head laterally
Semispinalis thoracis	Transverse processes of lower six thoracic verebrae	Spines of upper six thoracic and lower two cervical vertebrae	Extends and inclines head laterally
Semispinalis cervicis	Transverse processes of upper six thoracic vertebrae	Spines of second through sixth cervical vertebrae	Extends and inclines head laterally
Splenius cervicis	Spinous processes of third through sixth thoracic vertebrae	Transverse processes of first three cervical vertebrae	Inclines and rotates head and neck
Splenius capitis	Ligamentum nuchae and spinous processes of upper five thoracic vertebrae	Mastoid process and superior nuchal line	Inclines and rotates head and neck
Multifidus	Sacrum and transverse processes of lumbar, thoracic, and lower cervical vertebrae	Spinous processes of lumbar, thoracic, and lower cervical vertebrae	Abducts, rotates, and extends vertebral column

Table 7-5. Cont.

Muscle	Origin	Insertion	Action
Rotatores	Transverse processes of second cervical vertebra to sacrum	Lamina above vertebra of origin	Rotate and extend vertebral column
Interspinales	Superior surface of spine of each vertebra	Inferior surface of spine of vertebra above vertebra of origin	Extend and rotate vertebral column
Intertransversarii	Extend between transverse processes of cervical, lumbar, and lower thoracic vertebrae. Unilaterally, they bend vertebral column laterally; bilaterally they stabilize column.		

mass of muscle about as broad as the palm of the hand. These muscles extend vertically from the fourth segment of the sacrum to the mastoid process of the temporal bone and are placed side by side like three fingers.

The muscles of the transverse group extend from the spines of the vertebrae to the tips of the transverse processes and are present between the fourth sacral vertebra and the occipital bone. The muscles of the transverse group are placed one on top of the other, like the layers of a sandwich, and act primarily to twist the vertebral column. The deep muscles are innervated segmentally by dorsal rami of the spinal nerves, except for some of the intertransversarii, which are supplied by ventral rami.

Fascia of the deep muscles forms thin muscular envelopes, except in the lumbar region, where it thickens and is disposed in three anteroposterior layers, or lamina, as the **thoracolumbar fascia.** The thick, strong **posterior layer** covers the sacrospinalis and continues superiorly onto the thorax; inferiorly it attaches to the iliac crest and sacrum. Medially it fuses with the periosteum of the vertebral spines and superiorly with the ligamentum nuchae; laterally it joins the middle and anterior layers at the lateral border of the erector spinae. The **middle** and **anterior layers** sheathe the quadratus lumborum muscle. At the lateral border of this muscle they fuse with the posterior layer to give insertion to the three muscles of the anterolateral abdominal wall (Fig. 7-9.)

SUBOCCIPITAL TRIANGLE (TABLE 7-6)

The muscles that bound the small **suboccipital triangle** are medially, the **rectus capitis posterior major,** which overlies the **minor;** laterally, the **obliquus capitis superior,** and inferiorly, the **obliquus capitis inferior** (Fig. 7-10). The **semispinalis capitis** and the **longissimus capitis,** lying deep to the flattened, relatively extensive **splenius muscle,** form the roof of the triangle. The floor is formed by the posterior atlantooccipital membrane and the posterior arch of the atlas. The triangle contains the **vertebral artery** as it passes into the skull through the foramen magnum, and the **suboccipital nerve** (C_1), which supplies the muscles of the triangle. The **greater occipital nerve** (C_2) arches around the inferior border of the obliquus capitis inferior to supply sensation to the back of the scalp (Table 7-7). The muscles of the triangle may act as extensors and rotators of the head, but function chiefly as postural muscles.

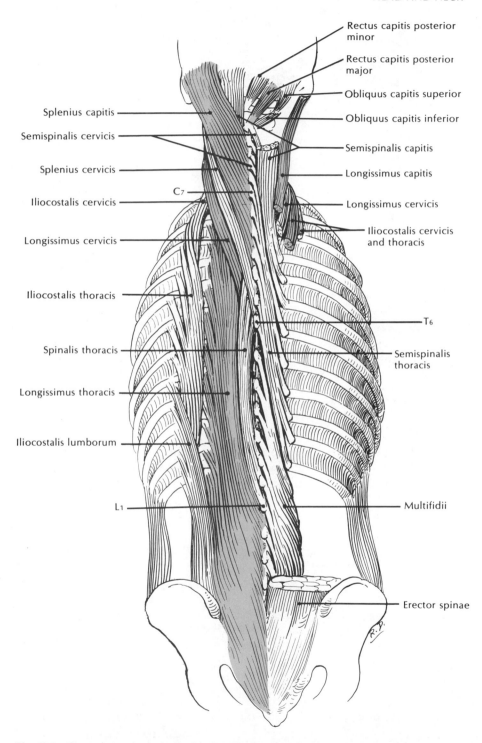

Splenius capitis

Semispinalis cervicis

Splenius cervicis

Iliocostalis cervicis

Longissimus cervicis

Iliocostalis thoracis

Spinalis thoracis

Longissimus thoracis

Iliocostalis lumborum

Rectus capitis posterior minor

Rectus capitis posterior major

Obliquus capitis superior

Obliquus capitis inferior

Semispinalis capitis

Longissimus capitis

Longissimus cervicis

Iliocostalis cervicis and thoracis

C₇

T₆

Semispinalis thoracis

L₁

Multifidii

Erector spinae

Fig. 7-8. Deep muscles of the back. (Hollinshead WH: Textbook of Anatomy, 3rd ed. New York, Harper & Row, 1974)

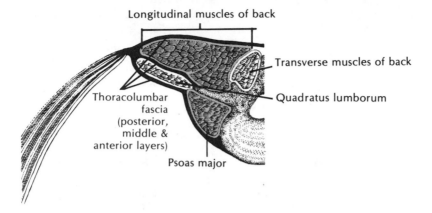

Fig. 7-9. Fasciae of back.

VERTEBRAL COLUMN

The **vertebral column** forms the central portion of the axial skeleton of the body. It presents two **primary curvatures,** the thoracic and the sacral, which are convex posteriorly, and two **secondary curvatures,** the cervical and the lumbar, which are concave posteriorly. It is composed of seven cervical, twelve thoracic, five lumbar, five fused sacral, and three to five fused coccygeal vertebrae.

SCOLIOSIS

This abnormal curvature of the spine is a condition in which a lateral bending of the vertebral column, usually in the thoracic region, is present. This is the most common of the abnormal vertebral curva-

Table 7-6. Muscles of the Suboccipital Triangle

Muscle	Origin	Insertion	Action	Nerve
Rectus capitis posterior major	Spine of axis	Lateral half of inferior nuchal line	Extends and rotates head	Dorsal ramus (suboccipital
Rectus capitis posterior minor	Posterior tubercle of atlas	Occipital bone below medial portion of inferior nuchal line	Extends head	As above
Obliquus capitis superior	Transverse process of atlas	Occipital bone between superior and inferior nuchal lines	Extends head	As above
Obliquus capitis inferior	Spine of axis	Transverse process of atlas	Rotates and extends head	As above

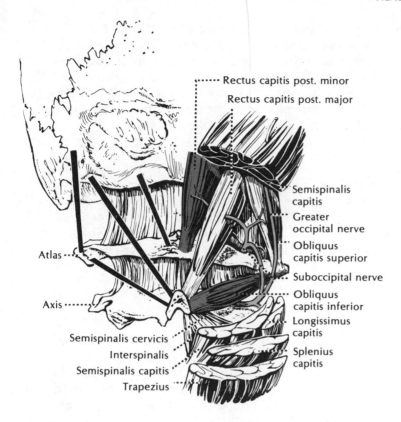

Rectus capitis post. minor
Rectus capitis post. major

Semispinalis capitis
Greater occipital nerve
Obliquus capitis superior
Suboccipital nerve
Obliquus capitis inferior
Longissimus capitis
Splenius capitis

Atlas
Axis
Semispinalis cervicis
Interspinalis
Semispinalis capitis
Trapezius

Fig. 7-10. Suboccipital triangle and contents. (Gardner E, Gray DJ, O'Rahilly R: Anatomy. Philadelphia, Saunders, 1975)

Table 7-7. Nerve Distribution to Suboccipital Triangle and Back

Nerve	Origin	Course	Distribution
Suboccipital	Dorsal ramus of C_1	Passes between skull and first cervical vertebra to reach suboccipital triangle	Muscles of suboccipital triangle
Greater occipital	Dorsal ramus of C_2	Emerges inferior to obliquus capitis inferior muscle and ascends to reach back of scalp	Skin over occipital bone
Least occipital	Dorsal ramus of C_3	Passes directly to skin	Dermatome of C_3
Dorsal rami	Spinal nerves	Pass segmentally to muscles and skin	Intrinsic muscles of back (erector spinae) and skin over back of lower neck and back

tures. It may be congenital due to an absence of the lateral half of a vertebra (hemivertebra) or acquired from a persistent severe sciatica. In the latter the trunk is usually bent away from the painful side, thus reducing the pressure on the sciatic nerve fibers as they emerge through the intervertebral foramina. Poliomyelitis may cause scoliosis by the paralysis of muscles on one side of the body, which produces a lateral deviation of the trunk towards the unaffected side.

KYPHOSIS

This abnormal curvature of the spine is an exaggeration of the convex curvature of the thoracic region (hunchback) of the vertebral column. In tuberculosis of the spine vertebral bodies may partially collapse causing an acute angular bending of the vertebral column, which is called gibbus. In the elderly, degeneration of the intervertebral discs leads to senile kyphosis. "Round shouldered" is an expression of mild kyphosis.

LORDOSIS

This abnormal curvature of the spine is an exaggerated convex curvature of the vertebral column in the lumbar region (sway back). It is present in congenital double dislocation of the hip, in which the support of the pelvis occurs posterior to the acetabulum. It may be caused from increased weight of abdominal contents, as in pregnancy or extreme obesity.

The component parts of a **typical vertebra** are the body and a number of processes that surround the centrally located vertebral foramen (Fig. 7-11). The massive **body** gives strength to the vertebral column and is separated from adjacent vertebral bodies by the intervertebral discs. The **transverse processes** project laterally from the junction of the pedicles and the laminae. In the cervical region they contain a foramen for the passage of the vertebral artery, and in the thoracic region they afford articulation for the ribs. The short, slightly rounded **pedicles** project posteriorly from the body and are joined by the flattened **laminae** to form the arch. The pedicles, laminae, and posterior surface of the body form the vertebral foramen. Adjacent vertebral foramina form the vertebral canal. At the junction of the laminae the **spinous processes** project posteriorly. At the junction of the pedicle with the lamina the **superior** and **inferior processes** bear articular facets that form synovial joints with adjacent vertebrae. The deep **vertebral notch** on the inferior border and the shallow notch on the superior border of each pedicle, together with the intervertebral discs, form the **intervertebral foramina** between adjacent vertebrae. They transmit spinal nerves, vessels, and meninges.

Intervertebral discs unite adjacent, unfused vertebral bodies. They consist of an outer layer of fibrocartilage, the **anulus fibrosus,** while centrally they present a relatively soft gelatinous mass, the **nucleus pulposus.**

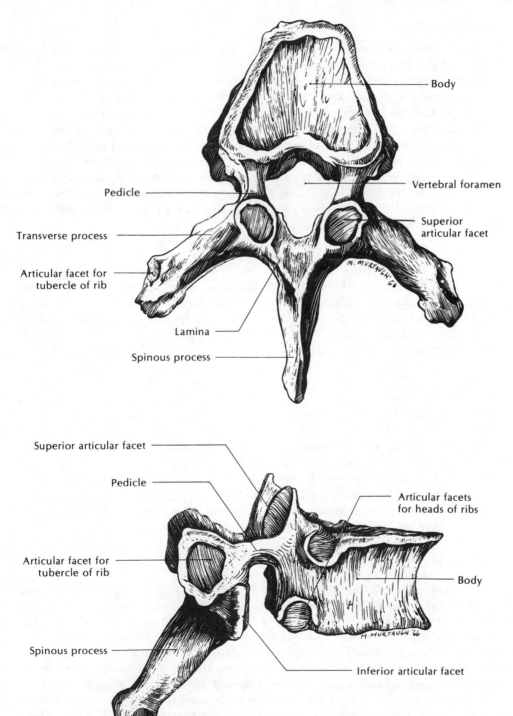

Body

Vertebral foramen

Superior
articular facet

Pedicle

Transverse process

Articular facet for
tubercle of rib

Lamina

Spinous process

Superior articular facet

Pedicle

Articular facets
for heads of ribs

Articular facet for
tubercle of rib

Body

Spinous process

Inferior articular facet

Fig. 7-11. Typical thoracic vertebra.

SLIPPED DISC

Slipped disc is a layman's term for a herniated intervertebral disc. Discs are subject to pathologic changes where the strain is greatest, for example, lower lumbar region. If the compression force is excessive, a rupture of the disc occurs, and the semisolid nucleus pulposus herniates partially or completely through the anulus fibrosus, to impinge on spinal nerves that emerge from the vertebral column adjacent to the discs in this region. Such nerve compression causes a painful neuralgia (sciatica) down the back, and lateral side of leg and into sole of foot. Traction, bed rest, and analgesia usually relieve the pain. If such conservative treatment is ineffective, then surgical decompression of the spinal nerves by laminectomy or removal of some of the nucleus pulposus may be necessary to relieve pain.

Regional characteristics differentiate vertebrae of the cervical, thoracic, and lumbar regions. Cervical vertebrae have foramina within their transverse processes, the **foramina transversaii,** which transmits the vertebral artery. The transverse processes present anterior and posterior tubercles, while the spinous processes of the third through sixth vertebrae are bifid. The first and second cervical vertebrae are atypical. The first, or **atlas,** lacks both a body and a spinous process. Instead of the usual articular processes, the lateral mass of the atlas articulates superiorly with the occipital condyles of the skull and inferiorly with the second cervical vertebra. These lateral masses are joined by the anterior and posterior arches, which present, in the midline, anterior and posterior tubercles. The **odontoid process** projects superiorly from the upper part of the body of the second cervical vertebra, or **axis,** and represents the transposed body of the first vertebra. **Thoracic vertebrae** present **articular facets for the ribs,** one on the body for articulation with the head of the rib, and one at the transverse process for the rib tubercle. The long slender spinous processes of thoracic vertebrae are directed inferiorly. The **lumbar vertebrae** are differentiated by their lack of foramina transversarii and rib facets. They present massive bodies, broad and somewhat quadrilateral horizontal spinous processes, and long slender transverse processes.

Ligaments of the Vertebral Column

The broad, thick **anterior longitudinal ligament** passes over the anterior surface of the vertebral bodies from the skull to the coccyx and is firmly attached to the bodies and the intervertebral discs. The **posterior longitudinal ligament** extends over the posterior surface of the vertebral bodies and, therefore, along the inner anterior surface of the vertebral canal. The **supraspinous ligament** attaches along the tips of the spinous processes, where it enlarges as the ligamentum nuchae in the cervical region. **Interspinous ligaments** pass between adjacent superior and inferior borders of spinous processes and fuse with the supraspinous ligament. **The ligamentum flavum** extends from the anterior aspect of one lamina to the posterior aspect of the lamina below. **Intertransverse ligaments** form small bands between adjacent transverse processes.

LUMBAR PUNCTURE

Lumbar puncture is a procedure performed to: 1) obtain a sample of cerebrospinal fluid for diagnostic studies, 2) relieve intracranial pressure, 3) administer drugs, or 4) induce anesthesia. Since the spinal cord, in the adult, does not extend inferior to the second lumbar vertebra, the interspace between L_4 and L_5 is usually chosen for the puncture. At this level there is no danger of damage to the spinal cord. In its course the needle pierces the skin, superficial fascia, supraspinous ligament, interspinous ligament, epidural space, dura, and arachnoid to enter the subarachnoid space.

SPINAL CORD AND MENINGES

As an extension of the brain stem, the **spinal cord** is located in the vertebral canal between the foramen magnum and the level of the first or second lumbar vertebra. Below this level nerve rootlets and the filum terminale, a prolongation of pia mater, form the **cauda equina,** which occupies the vertebral canal. The cylindric spinal cord is slightly flattened anteroposteriorly and has **cervical** and **lumbar enlargements** at the levels of origin of the nerves to the upper and lower extremities. The cord is grooved anteriorly by the **anterior median fissure,** and posteriorly by the **posterior median sulcus.** The designation of a **spinal cord segment** or **spinal level** refers to that portion of the cord associated with the origin of ventral and dorsal roots of a specific spinal nerve.

RELATION OF SPINAL CORD SEGMENTS TO VERTEBRAE

The spinal cord is considerably shorter than the vertebral column, extending inferiorly only to the first or second lumbar vertebra. Thus, the relationship between origin of the nerves from the spinal cord and their emergence through intervertebral foramina of the vertebral canal do not coincide. In the lower cervical and upper thoracic region the spinal cord segment is approximately two vertebrae proximal to the vertebral level. Thus, the spinous process of C_6 overlies spinal cord segment C_8; in the lower thoracic and upper lumbar regions, T_{11} and T_{12} spinous processes overlie the five lumbar spinal cord segments, while L_1 spinous process overlies the five sacral segments.

Such knowledge is critical to the surgeon in locating a tumor in a patient showing signs of spinal cord compression, for example, loss of sensation over the thumb, an area innervated by C_6. Therefore, to expose this spinal cord segment a laminectomy would be performed on vertebra C_4.

The **blood supply** of the spinal cord is from the **vertebral arteries,** supplemented by segmental **spinal branches** of deep cervical, intercostal, lumbar, and sacral arteries. As the vertebral artery enters the cranial cavity, just prior to its union to the artery of

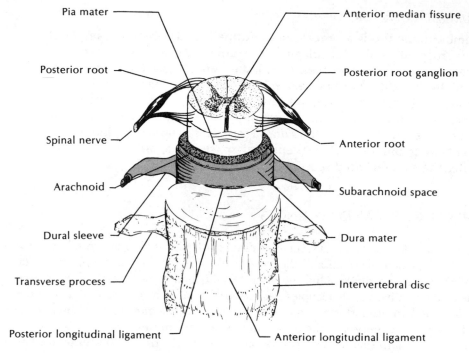

Fig. 7-12. Cross section of the spinal cord and meninges.

the opposite side to form the basilar artery, it gives off **anterior** and **posterior spinal branches.** The branches from each side unite to form a single trunk anteriorly while posteriorly paired trunks extend inferiorly. These arterial trunks course along the entire length of the cord and receive segmental contributions from the above-named intervertebral branches.

The spinal cord, bathed in cerebrospinal fluid, is invested by three membranes (meninges): the dura, arachnoid, and pia mater (Fig. 7-12). The outer, tough, dense **dura mater** is a fibrous tube that extends from the foramen magnum, where it is continuous with dura of the brain, to the coccyx. Below the level of the second sacral vertebra it narrows considerably and blends with the connective tissue covering the posterior aspect of the coccyx. As each spinal root emerges from the vertebral canal, it carries with it, for a short distance, a prolongation of the dura as the **dural sleeve.** The **epidural** space, between the dura and walls of the vertebral canal, contains fat and a plexus of thin-walled veins. The **subdural space** is a capillary interval between the dura and the arachnoid, containing a small quantity of fluid.

The avascular, delicate, transparent **arachnoid mater,** coextensive with the dura mater, is also continuous through the foramen magnum with the arachnoid covering the brain. Filamentous extensions pass through the subarachnoid space to fuse with the pia mater. Laterally this membrane is prolonged a short distance with the sheath for the spinal nerve. The **subarachnoid space,** a relatively extensive trabeculated interval between the arachnoid and the pia, contains the **cerebrospinal fluid** that serves as a protective liquid cushion for the cord.

CEREBROSPINAL FLUID PRESSURE

Cerebrospinal fluid pressure is measured by a manometer attached to a needle inserted into the subarachnoid space in a lumbar puncture procedure. With the patient in a lateral recumbent position, the manometer should register about 120 mm of water.

The **pia mater** is a delicate, vascular membrane intimately adherent to the spinal cord. From the termination of the cord at the first or second lumbar vertebra the pia mater continues inferiorly as the **filum terminale. Denticulate ligaments** are serrated lateral extensions from the pia that anchor and limit torsion or twisting of the cord within the dural sac. The medial edge of this small ligament has a continuous attachment to the cord midway between the anterior and posterior nerve roots, while the serrated lateral border attaches to the dura at intervals between emerging spinal nerves.

Face

Cutaneous Innervation

The cutaneous innervation to the face is supplied by all three divisions of the **trigeminal,** or fifth cranial, nerve (Fig. 7-13). The **ophthalmic division** (V_1) divides into three branches, the lacrimal, frontal, and the nasociliary nerves, which supply cutaneous innervation above the level of the eyes and on the dorsum of the nose. The **lacrimal nerve** supplies the conjunctiva and skin of the upper eyelid. The **frontal nerve** divides into a **supratrochlear branch** that supplies skin of the forehead and upper eyelid, and a **supraorbital branch** passing through the supraorbital foramen to supply the upper eyelid and the scalp as far posteriorly as the lambdoidal suture. The nasociliary nerve gives **long ciliary branches** to the eye that pierce the lamina cribrosa, and a **posterior ethmoidal** branch that supplies posterior and middle ethmoidal air cells. The nasociliary nerve then bifurcates to give an **anterior ethmoidal** branch to the anterior ethmoidal air cells, the anterosuperior aspect of the nasal cavity, and the skin over the dorsum of the nose, and an **infratrochlear** branch that supplies skin of the upper eyelid.

TRIGEMINAL NEURALGIA

Trigeminal neuralgia, or tic douloureux, is associated with excruciating pain over the face, especially areas innervated by the mandibular and maxillary divisions of the fifth cranial nerve. The etiology is unknown but it is most common in the middle aged and elderly. Alcohol injections into the trigeminal ganglion or around the divisions of the nerves as they emerge from the skull, may relieve trigeminal neuralgia.

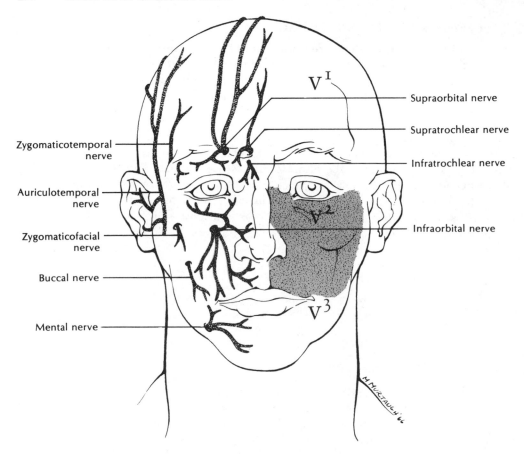

Fig. 7-13. Cutaneous innervation of the face.

The **maxillary division** (V₂) divides into the infraorbital and zygomatic nerves, which supply an oblique area between the mouth and the eyes. The **infraorbital nerve** appears on the face at the infraorbital foramen and divides into an **inferior palpebral branch** to the skin and conjunctiva of the lower lid, an **external nasal branch** to the side of the nose, and a **superior labial branch** to the upper lip. The **zygomatic nerve** terminates in the **zygomaticofacial** and **zygomaticotemporal branches** to the skin over the zygomatic bone and the temporal region.

The **mandibular division** (V₃) supplies the remainder of the face. Its cutaneous branches are derived from both the anterior and posterior divisions of the nerve. The **buccal branch,** from the anterior division, breaks into a series of twigs to supply the muscosa of the cheek. The **auriculotemporal nerve,** from the posterior division, divides into an auricular branch to the external auditory meatus, tympanic membrane, skin of the tragus, and the upper and outer part of the pinna, and a **superficial temporal branch,** which supplies most of the temporal region. A terminal branch of the mandibular division is the **inferior alveolar nerve.** It passes through the mandibular foramen to course in the mandibular canal. It emerges from the canal at

Table 7-8. Cutaneous Innervation of Face

Nerve	Origin	Course	Distribution
Frontal	Ophthalmic nerve (V_1)	Traverses orbit on superior aspect of levator palpebrae superioris; divides into supraorbital and supratrochlear branches, which traverse foramina or notches of same name	Skin of forehead and scalp to lambdoidal suture
Infratrochlear	Nasociliary nerve	Follows medial wall of orbit to upper eyelid	Skin of upper lid
Dorsal (External) nasal	Anterior ethmoidal nerve	Terminal branch in nasal cavity emerges on face between nasal bone and nasal cartilage	Skin on dorsum of nose
Zygomatic	Maxillary nerve (V_2)	Arises in floor of orbital cavity, divides into zygomaticofacial and zygomaticotemporal, which traverse foramina of same name	Skin over zygomatic arch and anterior temporal region
Infraorbital	Terminal branch of maxillary nerve (V_2)	In floor of orbit runs in orbital groove and canal to emerge at infraorbital foramen	Skin of cheek, lateral side of nose and upper lip
Buccal	Mandibular nerve (V_3)	From anterior division of V_3 in infratemporal fossa courses anteriorly to reach cheek	Skin of cheek
Auriculotemporal	Mandibular nerve (V_3)	From posterior division of V_3 passes between neck of mandible and external auditory meatus to course with superficial temporal artery	Skin in front of ear and temporal region
Mental	Terminal branch of inferior alveolar	Emerges from mandibular canal at mental foramen	Skin of chin and lower lip

the mental foramen to divide into the **mental branch** supplying the skin of the chin, and the **inferior labial branch** to the lower lip (Table 7-8).

NERVE BLOCK

Nerve block is a term signifying loss of sensation in a region, such as in local dental anesthesia. For dental procedures on the lower teeth, analgesia is obtained by depositing an anesthetic solution in the proximity of the inferior alveolar nerve as it enters the mandibular foramen. This effectively blocks all sensations carried centrally by this nerve. For anesthesia to the upper teeth, superior alveolar nerve endings are

Table 7-9. Muscles of the Face

Muscle	Origin	Insertion	Action	N
Frontalis	Epicranial aponeurosis	Skin of forehead	Raises the eyebrows, wrinkles forehead	F
Corrugator supercilii	Medial portion of supraorbital margin	Skin of medial half of eyebrow	Draws eyebrows downward and medialward	F
Orbicularis oculi	Medial orbital margin, medial palpebral ligament, and lacrimal bone	Skin and rim of orbit and tarsal plate	Sphincter of eyelids	F
Procerus	Lower part of nasal bone, upper part of lateral nasal cartilage	Skin between eyebrows	Wrinkles skin over bridge of nose	F
Nasalis	Canine eminence lateral to incisive fossa	Nasal cartilages	Draws alae of nostril toward septum	F
Depressor septi	Incisive fossa of maxilla	Posterior aspect of ala and nasal septum	Draws nasal septum inferiorly	F
Dilator nares	Margin of piriform aperture of maxilla	Side of nostril	Widens nostril	F
Orbicularis oris	Surrounds oral orifice, forming intrinsic muscle of lips. Interlaces with other muscles associated with lips.		Acts in compression, and protrusion of lips	F
Levator labii superioris	Frontal process of maxilla, infraorbital region, and inner aspect of zygomatic bone	Greater alar cartilage, nasolabial groove, and skin of upper lip	Elevates lip, dilates nostril, and raises angle of mouth	F
Caninus	Canine fossa of maxilla	Angle of mouth	Raises angle of mouth	F
Zygomaticus	Zygomatic arch	Angle of mouth	Elevates and draws angle of mouth backward	F
Depressor labii inferioris	Mandible between symphysis and mental foramen	Into orbicularis oris and skin of lower lip	Depresses and everts lower lip	F
Depressor anguli oris	Oblique line of mandible	Angle of mouth	Turns corner of mouth downward	F
Risorius	Fascia overlying masseter muscle	Angle of mouth	Retracts angle of mouth	F
Platysma	Superficial fascia of pectoral and deltoid regions	Mandible, skin of neck and cheek, angle of mouth, and orbicularis oris	Depresses lower jaw; tenses and ridges skin of neck	F
Buccinator	Pterygomandibular raphe, alveolar processes of jaw opposite molar teeth	Angle of mouth	Compresses cheek as accessory muscle of mastication	F
Mentalis	Incisive fossa of mandible	Skin of chin	Elevates and protrudes lower lip	F
Auricularis anterior, superior, and posterior (all rudimentary)	Temporal fascia, epicranial aponeurosis, and mastoid process	Front of helix, triangular fossa, and convexity of concha	May retract and elevate ear	F

**blocked by inserting the needle beneath the mucous membrane and
the anesthetic solution is then infiltrated slowly throughout the area of
the roots of the teeth to be treated.**

Muscles (Table 7-9)

The muscles of facial expression lie within the superficial fascia of the face (Fig.
7-14). They gain their origin from fascia or underlying bone and insert into the skin.
They are usually described by their action on, or relation to, orifices of the face; are

Fig. 7-14. Muscles of the facial expression and the facial nerve.

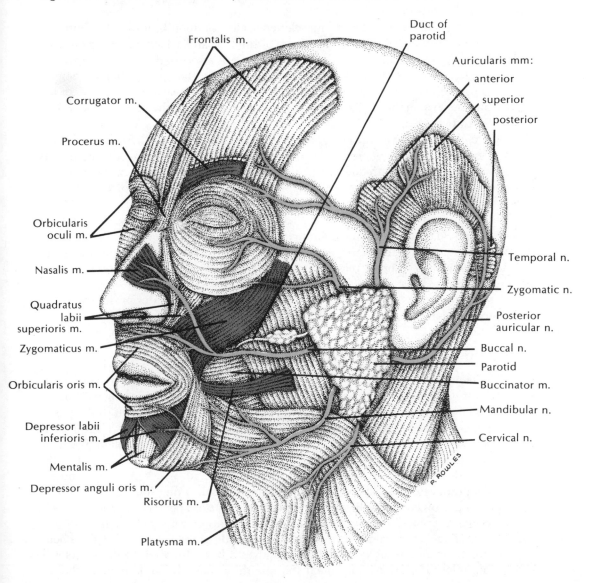

frequently fused with adjacent muscles, and are innervated by the facial, or seventh, cranial nerve. Muscles associated with or adjacent to the eye include the **frontalis,** which wrinkles the forehead and raises the eyebrows, as in registering surprise, or draws the skin of the scalp forward; the **corrugator** furrows and wrinkles the brow, and the **orbicularis oculi** acts in winking and blinking.

Muscles around the nose are the **procerus** (pyramidalis nasi) over the bridge of the nose, which wrinkles the skin between the eyebrows and draws the latter medially in registering a fierce expression; the **nasalis** (compressor nares), which compresses the nostril, drawing its margins toward the septum; the **dilator nares,** which enlarges the nostril, and the **depressor septi,** which draws the cartilaginous portion of the septum inferiorly and constricts the nostril.

The **orbicularis oris** ("kissing" muscle) encircles the mouth, forms the muscular bulk of the lips, interlaces with other muscles of the lips, and acts in compression, contraction, and protrusion of the lips. Three muscles are associated with the upper lip: the **levator labii superioris,** which functions in elevation of the lips, dilation of the nostril, and raising the angle of the mouth; the **caninus,** acting to elevate the corner of the mouth, and the **zygomaticus** ("laughing" muscle), which draws the angle of the mouth backward and upward.

Muscles of the lower lip are four in number: the **depressor labii inferioris** expresses terror or grief in depressing and everting the lip; the **depressor anguli oris** draws the angle of the mouth downward; the **risorius** retracts the angle of the mouth in grinning, and the **platysma,** interdigitating with muscles of the lower lip, depresses the lower lip and mandible and tenses or ridges the skin of the neck.

The **buccinator** muscle makes the cheek rigid in sucking or blowing. It acts as an accessory muscle of mastication by eliminating the space of the vestibule between the cheek and the jaws. The **mentalis** muscle of the chin elevates and protrudes the lower lip and, when well developed, may form with its member of the opposite side the dimple in the chin.

Three poorly developed muscles of the ear, the **auricularis anterior, superior,** and **posterior,** insert into the pinna and act in feeble movements of the ear.

Arteries and Nerves (Table 7-10)

The **facial nerve** follows a complicated intracranial course from its origin on the brain stem to its emergence from the cranium at the stylomastoid foramen. Here its branches include the **posterior auricular nerve,** with an **auricular branch** to the posterior auricularis muscle, and an **occipital branch** to the occipitalis muscle. The **digastric branch** supplies the posterior belly of the digastric muscle, and the **stylohyoid branch,** the stylohyoideus muscle. The facial nerve then turns anteriorly to pass lateral to the styloid process and enter the substance of the parotid gland, where it divides into an **upper (temporozygomatic)** and a **lower (cervicomandibular) division.**

The upper division supplies all the superficial muscles above the zygomatic arch, with a **temporal branch** innervating the anterior and superior auricularis, frontalis, orbicularis oculi, and corrugator muscles, and a **zygomatic branch** to the orbicularis oculi and zygomaticus muscles. The lower division gives a **buccal branch** to the muscles of expression below the orbit and around the mouth, a **mandibular branch** to the de-

Table 7-10. Terminal Branches of Facial Nerve

Nerve	Origin	Course	Distribution
Posterior auricular	Facial nerve (VII)	From stylomastoid foramen passes posteriorly to occipital region	Posterior and superior auricular muscles and occipitalis
Nerve to stylohyoid	Facial nerve (VII)	From stylomastoid foramen passes to styloid process	Stylohyoid and posterior belly of digastric
Branches to facial muscles	Facial nerve (VII)	From stylomastoid foramen facial turns anteriorly, cleaves parotid, and bifurcates into upper (temporozygomatic) and lower (cervicomandibular) divisions	Five terminal branches (*i.e.,* temporal, zygomatic, buccal, mandibular and cervical) pass to and supply facial muscles in respective regions

pressor labii inferioris and mentalis muscles, and a **cervical branch** to the platysma (Fig. 7-14).

PAROTIDECTOMY AND FACIAL PARALYSIS

In partial or complete surgical removal of the parotid gland the facial nerve may be damaged causing facial (Bell's) paralysis. The most important part of removal of the parotid is the identification, dissection, and preservation of the facial nerve. The safest procedure is to identify the nerve as it exits from the stylomastoid foramen and then follow its course through the parotid gland. A malignant parotid tumor may involve the facial nerve and cause unilateral facial palsy. In removing such a tumor it may be necessary to sacrifice the facial nerve or its branches. Nerve grafts may reestablish facial muscle innervation and unilateral motor paralysis of the face can be averted in some cases.

The **facial artery** appears on the face at about the middle of the inferior border of the mandible and passes from the anterior border of the masseter muscle toward the medial angle of the eye. In its course it gives off the **superior** and **inferior labial branches** to the upper and lower lips, respectively, and the **lateral nasal branch** to the ala and dorsum of the nose. It terminates as the **angular artery** at the medial canthus of the eye. The angular artery anastomoses with the dorsal nasal and palpebral branches of the ophthalmic artery to establish a communication between branches of the internal and external carotid arteries. Additional anastomoses are present between the facial artery and the infraorbital branches of the maxillary artery.

The **superficial temporal branch** of the external carotid artery arises in the substance of the parotid gland and courses superficially toward the temporal region. It gives off the **transverse facial artery,** which parallels the course of the parotid duct. It supplies the parotid gland and duct, the masseter and buccinator muscles, the skin over the cheek, and anastomoses with branches of the facial artery. The superficial temporal artery terminates by bifurcating into the **frontal (anterior) branch,** which anastomoses with branches of the ophthalmic artery, and the **parietal (posterior) branch,** which joins with the posterior auricular and occipital arteries.

Venous Drainage of Face

Under normal conditions the venous return of the face is essentially superficial. The forehead is drained by the **supraorbital** and **supratrochlear veins** which pass to the medial canthus to unite to form the **angular vein.** The continuation of the angular vein is the **facial vein.** It descends obliquely across the face and is accompanied by the facial artery. Just below the inferior border of the mandible it pierces the external investing layer of deep fascia to join the **internal jugular vein.**

Along its downward course the facial vein is joined by several additional tributaries draining specific regions of the face. These veins include the **palpebral** (from the eyelids), **external nasal** (about the nose), **labial** (from the lips), **deep facial** (draining part of the pterygoid venous plexus), **submental** (from the chin), **external palatine** (draining the tonsillar bed), **superficial temporal** (over the temporalis muscle), and the **retromandibular vein.** That part of the facial vein, inferior to its union with the retromandibular vein, is sometimes called the **common facial** vein before it empties into the internal jugular vein.

THE DANGER AREA OF THE FACE AND SCALP

The danger area of the face and scalp extends from the upper lip and the lower nasal regions to the midpoint of the scalp. This region is drained by the facial veins. It is a dangerous area because boils, pustules (pimples) and other skin infections commonly occur here. Manipulation of these infected structures may spread infection into the facial veins. These veins anastomose with the inferior ophthalmic veins, the pterygoid plexus, and the cavernous dural sinus. Clotting of the blood is a specific problem in the cavernous sinus because of the unusually slow movement of the blood. The extensive endothelial surface of this sinus encourages bacterial growth and precipitates the formation of thrombi. Septicemia is a very dangerous condition because of the probability of meningitis. Before antibiotic therapy such conditions had a mortality of over 90%.

PAROTID GLAND

The **parotid gland,** largest of the three major salivary glands, occupies the depression between the sternocleidomastoideus muscle and the ramus of the mandible. It is roughly quadrilateral in shape and somewhat flattened, with a **deep process** passing to the inner aspect of the mandible.

MUMPS

Mumps, or parotitis, is a common viral inflammatory lesion of the parotid gland that may spread to the testes and cause sterility.

The **superficial facial process** (accessory parotid) extends anteriorly beyond the ramus of the mandible, surrounds the proximal part of the large parotid (Stensen's) duct, and overlies the masseter muscle. The gland is enclosed by the external investing layer of deep cervical fascia. The thick-walled excretory **duct of the parotid** emerges from the facial process to continue across the masseter muscle and turns deeply to pass through the suctorial fat pad and buccinator muscle. It terminates inside the oral cavity opposite the upper second molar tooth.

CALCULUS

A calculus is an abnormal concretion of mineral salts, which usually occurs in excretory ducts of glands. For example, calculi may form in any of the three major salivary glands. Obstruction of the parotid duct by a calculus is often associated with chronic parotitis and is best treated by surgical removal.

The parotid gland lies in the concavity formed by the sternocleidomastoid muscle and the ramus of the mandible. It is related laterally to the skin and superficial fascia of the face, branches of the greater auricular nerve and parotid lymph nodes, and superiorly to the external auditory meatus. The posteromedial surface of the parotid overlies the mastoid process, the sternocleidomastoideus, posterior belly of the digastric and stylohyoideus muscles, and the styloid process. The anteromedial surface is molded to the posterior border of the ramus of the mandible and the structures attaching to it, that is, the masseter and medial pterygoideus muscles, and the temporomandibular ligament.

The most superficial structures passing through the substance of the gland are the terminal branches of the facial nerve. Anteriorly this nerve divides into upper and lower trunks within the gland. The retromandibular vein, the external carotid artery and some of its branches, and the auriculotemporal and posterior auricular nerves are related to the deep aspect of the parotid.

The above structures radiate from the periphery of the gland to pass to their destinations. The superficial temporal artery and vein, the auriculotemporal nerve, and the temporal branch of the facial nerve pass superiorly; the zygomatic, buccal, and mandibular branches of the facial nerve and the transverse facial artery pass anteriorly. The cervical branch of the facial nerve passes inferiorly, and the posterior auricular artery and nerve pass posteriorly to supply the area behind the ear.

TEMPORAL AND INFRATEMPORAL FOSSAE

The **temporal fossa** is an oval area on the lateral aspect of the skull. It is continuous inferiorly with the infratemporal fossa and contains the fan-shaped **temporalis muscle** of mastication.

The **infratemporal fossa** is the area bounded laterally by the ramus of the mandible, anteriorly by the body of the maxilla, superiorly by the infraorbital fissure, and inferiorly by the upper second and third molar teeth and their alveolar processes. The **medial wall** consists of the lateral or muscular plate of the pterygoid process, and the

roof is formed by the inferior surface of the greater wing of the sphenoid and part of the temporal bone. The medial and lateral pterygoid muscles, the pterygoid plexus of veins, the mandibular nerve, and the maxillary artery are the contents of the infratemporal fossa. The foramen ovale penetrates the roof of the fossa at the posterior border of the lateral pterygoid plate. The infraorbital fissure opens into it at right angles to the pterygomaxillary fissure, and the foramen spinosum is situated just posterolateral to the fossa.

Muscles (Table 7-11)

Located in the temporal and infratemporal fossae, the four muscles of mastication act in movements of the mandible and are innervated by the mandibular division of the fifth cranial nerve (Fig. 7-15). The fan-shaped **temporalis muscle** originates from the temporal fossa, inserts into the coronoid process, and elevates and retracts the mandible. The thick quadrilateral **masseter muscle** covers the lateral surface of the ramus, the angle, and the coronoid process of the mandible. It raises the lower jaw. In forcible clenching of the jaw this muscle can be felt as it bulges over the angle of the mandible. The **medial** and **lateral pterygoids** are both situated deep to the mandible in the infratemporal fossa. Both have two heads partially originating from respective surfaces of the lateral pterygoid plate and act to protrude and move the mandible from side to side. In addition, the medial pterygoid elevates, and the lateral pterygoid depresses, the mandible.

Arteries and Nerves

The **maxillary artery,** the larger of the two terminal branches of the external carotid, passes deep to the neck of the mandible and courses across the lower border of the lateral pterygoideus muscle to disappear into the pterygopalatine fossa (Fig. 7-15). It is divided into three parts by the lateral pterygoideus. The **first part,** proximal to the muscle, gives four branches: the **deep auricular artery** follows the auriculotemporal

Table 7-11. Temporal and Infratemporal Muscles

Muscle	Origin	Insertion	Action	Nerve
Temporalis	Temporal fossa and temporal fascia	Coronoid process and anterior border of ramus of mandible	Raises and retracts mandible	Mand
Masseter	Lower border and deep surface of zygomatic arch	Lateral surface of ramus and coronoid process of mandible	Raises and helps protract mandible	Mand
Lateral pterygoid	Intratemporal surface of sphenoid and lateral surface of lateral pterygoid plate	Neck of mandible and capsule of temporomandibular joint	Protrudes and depresses mandible; draws it toward opposite side	Mand
Medial pterygoid	Maxillary tuberosity and medial surface of lateral pterygoid plate	Mandible between mandibular foramen and angle	Raises and protrudes mandible; draws it toward opposite side	Mand

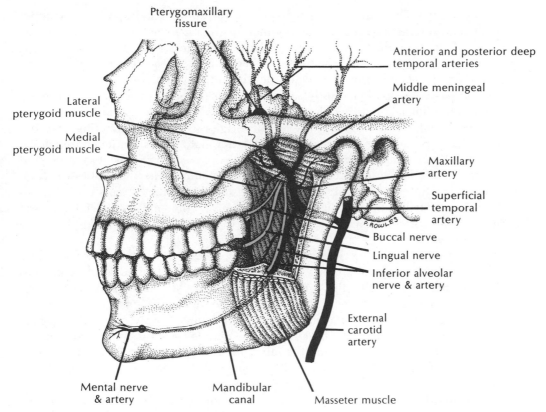

Pterygomaxillary fissure

Anterior and posterior deep temporal arteries

Middle meningeal artery

Lateral pterygoid muscle

Medial pterygoid muscle

Maxillary artery

Superficial temporal artery

P. ROWLES

Buccal nerve

Lingual nerve

Inferior alveolar nerve & artery

External carotid artery

Mental nerve & artery

Mandibular canal

Masseter muscle

Fig. 7-15. The infratemporal fossa.

nerve to supply the skin of the external auditory meatus and outer aspect of the tympanic membrane; the **anterior tympanic branch** passes behind the capsule of the temporomandibular joint to pass into the tympanic cavity by way of the petrotympanic fissure; the **middle meningeal artery** ascends medial to the lateral pterygoideus muscle to pass through the foramen spinosum to supply the meninges, and the **inferior alveolar** accompanies the inferior alveolar nerve as it passes through the mandibular foramen to traverse the mandibular canal and supply the lower teeth. The latter vessel terminates as the **mental branch** as it passes onto the face through the mental foramen. The **second part** of the maxillary artery gives **muscular branches** to the lateral and medial pterygoidei, the masseter, and the temporalis muscles. The terminal, or **third part** of the artery passes into the pterygopalatine fossa and will be described with that area.

The sensory portion of the **mandibular division** of the trigeminal nerve originates from the trigeminal ganglion and enters the infratemporal fossa by passing through the foramen ovale, where it is joined by the small **motor root** of the trigeminal. Two branches originate from the main trunk, one the **recurrent meningeal** (nervus spinosus), which reenters the skull through the foramen spinosum with the middle meningeal artery to supply sensory fibers to the meninges, and the other the **nerve to the medial pterygoideus,** which supplies motor fibers to the muscle and sends sensory twigs to the otic ganglion. The main trunk then divides into an anterior and a posterior division. The smaller **anterior division,** essentially motor, gives a branch to each of the four muscles of mastication: the **masseteric branch** to the masseter, the **deep temporal branch** to the temporalis, and the **nerves to the lateral** and **medial pterygoidei.** The buccal nerve is the only sensory branch from the anterior division.

The **posterior division,** mostly sensory, has three main branches: the auriculo-temporal, the lingual, and the inferior alveolar. The **auriculotemporal nerve,** which embraces the middle meningeal artery near its origin, sends **communicating branches** to the facial nerve and the otic ganglion. In the latter motor fibers continue through the ganglion without synapsing to innervate the tensor veli palatini and the tensor tympani muscles. An **anterior auricular branch** of the auriculotemporal nerve supplies sensation to the front of the ear. The tympanic membrane and skin of the external auditory meatus are innervated by the **external acoustic branch,** and the small **superficial temporal branch** of the auriculotemporal nerve supplies the parotid gland and skin over the temporal region.

The **lingual nerve** descends anteroinferiorly, medial to the lateral pterygoideus and anterior to the inferior alveolar nerve. Passing between the medial pterygoideus and the mandible, it reaches the submandibular region where it supplies general sensation to the mucous membrane of the floor of the mouth and the anterior two-thirds of the tongue. **Communicating branches** are given to the inferior alveolar and hypoglossal nerves, and the submandibular ganglion. The **chorda tympani branch** from the facial nerve joins the lingual nerve in the infratemporal fossa to supply special sensory (taste) fibers to the anterior two-thirds of the tongue and preganglionic parasympathetic fibers to the submandibular ganglion.

The **inferior alveolar nerve** descends with the inferior alveolar artery to enter the mandibular foramen and traverses the mandibular canal to emerge at the mental foramen as the mental nerve. The **mylohyoid branch** arises as the inferior alveolar nerve enters the mandibular foramen and supplies the anterior belly of the digastricus and the mylohyoideus muscles. Within the mandibular canal the inferior alveolar nerve sends twigs to all the lower teeth; its terminal branches, the **mental** and **inferior labial nerves,** supply the skin of the chin, and the skin and mucous membrane of the lower lip (Table 7-12).

Skull

Scalp

The skull is covered by the **scalp** which has three layers. The outer layer is the skin. Next to the skin is a layer of dense connective tissue through which nerves and blood vessels course. The **epicranium** is adjacent to the periosteum. It is a musculoaponeurotic sheet with the **occipitalis muscle** at its posterior end and the **frontalis muscle** attached anteriorly. The strong aponeurosis between these muscles is the **epicranial aponeurosis (galea aponeurotica).** All three layers are bound tightly together and are separated from the periosteum of the skull by a loose connective tissue space called the **danger area of the scalp** when infected.

Table 7-12. Distribution of Mandibular (V_3), Chorda Tympani and Minor Petrosal Nerves

Nerve	Origin	Course	Distribution
Branches to muscles of mastication	Anterior division or main trunk of (V_3)	Pass directly to muscles	Masseter, temporalis, medial and lateral pterygoids
Buccal	Anterior division of (V_3)	Passes anteriorly to cheek	Mucous membrane and skin of cheek
Lingual	Posterior division of (V_3)	Crosses lateral aspect of medial pterygoid to reach tongue	General sensation to anterior ⅔ of tongue
Chorda tympani	Facial nerve (VII)	From vertical portion of facial traverses middle ear cavity to reach infratemporal fossa and join lingual nerve	Taste to anterior ⅔ of tongue; conveys preganglionic parasympathetic fibers to submandibular ganglion. Postganglionics supply submandibular and sublingual glands
Inferior alveolar	Posterior division of (V_3)	Runs on lateral aspect of medial pterygoid, enters mandibular foramen, traverses mandibular canal, and terminates at mental foramen as mental nerve	Sensory to lower teeth, and skin of chin and lower lip
Nerve to mylohyoid	Inferior alveolar	At mandibular foramen, follows mylohyoid groove to reach superficial aspect of mylohyoid	Mylohyoid and anterior belly of digastric
Auriculotemporal	Posterior division of (V_3)	Passes between external auditory meatus and neck of mandible to course with superficial temporal artery	Skin of sideburn area; muscular twigs to tensors tympani and veli palatini
Minor petrosal	Glossopharyngeal (IX) through tympanic nerve and plexus	From tympanic plexus, leaves middle ear cavity via petrotympanic fissure to reach otic ganglion that hangs on auriculotemporal nerve	Conveys preganglionic parasympathetic fibers to otic ganglion. Postganglionics supply parotid gland

The blood supply to the scalp comes from **occipital, superficial temporal,** and **posterior auricular** branches of the external carotid. In addition, the internal carotid supplies the scalp by way of **supratrochlear** and **supraorbital** branches of the ophthalmic artery. All of these branches anastomose freely so scalp lacerations tend to bleed profusely.

The nerves to the scalp include the **supratrochlear** and **supraorbital** to the forehead and anterior scalp. The temporal region is supplied by the **auriculotemporal** nerve; the **lesser occipital** innervates the skin behind the ear, and the **greater occipital** nerve is distributed to the back of the head. The latter extends anteriorly to the lambdoidal suture.

THE DANGER AREA OF THE SCALP

The danger area of the scalp includes the subaponeurotic tissue cleft beneath the epicranial (galea) aponeurosis. Due to the looseness of the connective tissue in this area, blood and pus may spread over the skull, limited only by the attachments of the galea aponeurotica. Infection may spread through the valveless emissary veins, to skull bones, which may become infected and necrotic, or to the dural sinuses and result in meningitis. Septic emboli can form and pass into the dural sinuses to be disseminated throughout the venous system. An old surgical axiom states that if it were not for emissary veins, wounds of the scalp would lose half their significance.

The **skull** is the most complex osseous structure of the body (Fig. 7-16). It is adapted to house the brain and special sensory organs, and encloses the openings into the digestive and respiratory tracts. It is composed of twenty-two flattened, irregular bones that, except for the mandible, are joined by immovable, sutural-type articulations. For descriptive purposes, the skull is subdivided into the **cranium,** which houses the brain and special sense organs and is formed by eight bones, and the **facial skeleton,** composed of fourteen bones. There is no special demarcation of this subdivision but junctional areas contribute to the support of nasal, ocular, and auditory organs. The major **cranial sutures** are the **sagittal,** passing in the midline between the two parietal bones, and two transverse sutures, the **coronal** between the frontal and the parietal bones, and the **lambdoidal** between the occipital and parietal bones.

PREMATURE CRANIAL SYNOSTOSIS

Premature cranial synostosis is an early union of the bones of the skull before the brain has reached its normal size. Such premature closure reduces the size of the cranial vault substantially, thus preventing normal brain development. Tribasilar synostosis is the fusion in infancy of the three bones at the base of the skull. Mental retardation is a common sequela to any abnormal or early union of cranial bones.

Basic points of reference of the skull include the **nasion,** the midpoint of the nasofrontal sutures; the **bregma,** at the junction of the sagittal and coronal sutures; the **obelion,** that portion of the sagittal suture adjacent to the parietal foramina; the **lambda,** at the junction of the sagittal and lambdoidal sutures; the **inion,** or external occipital protuberance; the **asterion,** at the junction of the occipital, parietal, and temporal bones, and the **pterion,** at the junction of the frontal, sphenoid, parietal, and temporal bones.

THE SOFT SPOT

The "soft spot" of the infant's skull is a diamond-shaped membranous area where the sagittal and coronal sutures will eventually fuse.

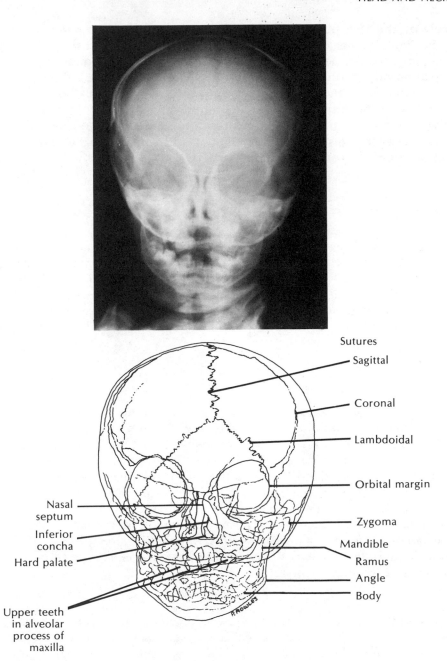

Fig. 7-16. X-ray and schematic drawing of skull.

Usually by the 18th postnatal month the sutures of the two parietal and the frontal bones unite and the anterior fontennelle (soft spot) is no longer palpable. Before closure, the physician can determine, by observation or palpation, increased intracranial pressure by a bulging of the fontennelle. Because the superior sagittal venous sinus lies in the midline of the anterior fontennelle, blood can be obtained by inserting a needle into the sinus. The posterior fontennelle occupies the interval between the sagittal and lamboidal sutures. This smaller gap closes during the second postnatal month.

CRANIAL CAVITY (TABLE 7-13)

Superiorly in the median plane of the internal aspect of the **calvarium,** or skull cap, a shallow groove lodges the superior sagittal venous sinus. Numerous small pits lie adjacent to this groove, which correspond to the position of the arachnoid granulations. Many grooves on the inner aspect of the calvarium are formed by the meningeal vessels. The largest and most prominent of these grooves, extending laterally over the inner surface from the foramen spinosum, lodges the middle meningeal artery and its branches.

CONTRECOUP FRACTURE

Contrecoup fracture is a term applied to a fracture of the skull at some distance from the point of contact of the blow. Since the osseous elements of the skull form a hollow but somewhat elastic shell, the force of a blow to the cranium can be transmitted to the opposite side where a fracture may be sustained. Blows to the top of the head may cause fractures of the bones of the floor of the cranial vault, often involving the sella turcica, one of the weakest parts of the base of the skull. The floor of the middle cranial fossa is especially involved because it is weakened by various foramina and fissures. Further, the strength of the petrous bone is decreased by the cavities of the internal ear, the carotid canal, and the jugular fossa.

The **floor** of the cranial cavity is subdivided into anterior, middle, and posterior cranial fossae (Fig. 7-17). The **anterior cranial fossa** is formed by portions of the ethmoid, sphenoid, and frontal bones and is adapted for the reception of the frontal lobes of the brain. Posteriorly the fossa is limited by the posterior border of the lesser wings of the sphenoid and the anterior margin of the chiasmatic groove. The **crista galli,** a midline process of the ethmoid bone, affords attachment for a longitudinal fold of dura mater, the **falx cerebri.** The **cribriform plate of the ethmoid,** at either side of the crista galli, admits passage of filaments of the olfactory nerve from the nasal mucosa to synapse in the olfactory bulb.

Table 7-13. Foramina of the Skull

Name	Bone	Position on bone	Structures passing through
Foramina Associated with Floor of Skull			
Cecum	Frontal	Between frontal and ethmoid (crista galli) at anterior end of crista galli in midline	Vein from nasal cavity to superior sagittal sinus
Olfactory	Ethmoid	In cribriform plate of ethmoid	Olfactory nerve branches and nasociliary nerve
Pterygoid (Vidian) canal	Sphenoid	Through root of pterygoid process	Nerve, artery, and vein of pterygoid canal
Sphenopalatine	Sphenoid and palatine	Between vertical part of palatine and ventral surface of sphenoid	Superior nasal nerve, nasopalatine nerves, and sphenopalatine vessels
Rotundum	Sphenoid	At junction of anterior and medial parts of sphenoid	Maxillary nerve
Vesalii	Sphenoid	Opposite root of pterygoid process and medial to foramen ovale	Small vein from cavernous sinus
Ovale	Sphenoid	Base of lateral pterygoid plate, in greater wing of sphenoid	Mandibular nerve, motor root of trigeminal, accessory meningeal artery, sometimes minor petrosal nerve
Spinosum	Sphenoid	Posterior angle of sphenoid, medial to spine of sphenoid	Middle meningeal vessels and recurrent branch from mandibular nerve
Lacerum	Sphenoid and petrous part of temporal	Bounded in front by sphenoid, behind by apex of petrous, medially by sphenoid and occipital bones	Internal carotid artery passes along upper part surrounded by nerve plexus, vidian nerve, meningeal branch of ascending pharyngeal artery
Jugular	Temporal and occipital	Behind carotid canal, between petrous part of temporal and occipital	Inferior petrosal sinus and bulb of internal jugular vein; ninth, tenth, and eleventh cranial nerves
Hypoglossal	Occipital	Above base of condyles	Hypoglossal nerve, meningeal branch of ascending pharyngeal artery
Magnum	Occipital	Center of posterior cranial fossa	Medulla oblongata and membranes, spinal accessory nerve, vertebral and spinal arteries
Foramina Associated with Orbit			
Anterior ethmoidal	Frontal and ethmoid	In frontoethmoidal suture on medial wall of orbit	Anterior ethmoidal vessels and nerve

Table 7-13. Cont.

Name	Bone	Position on bone	Structures passing thro
		Foramina Associated with Orbit	
Posterior ethmoidal	Frontal and ethmoid	2 to 3 cm posterior to anterior ethmoidal	Posterior ethmoidal ve sels
Optic	Sphenoid	Between upper and lower roots of lesser wing of sphenoid	Optic nerve and ophth mic artery
Superior orbital fissure	Sphenoid	Between greater and lesser wings of sphenoid	*Above* superior head o lateral rectus: trochlea frontal, and lacrimal nerves. *Between* heads lateral rectus: abducen oculomotor, and nasoc liary nerves; ophthalm veins
Inferior orbital fissure	Sphenoid and maxilla	Between greater wing of sphenoid and maxilla	Maxillary and zygomat nerves, infraorbital ves and veins to pterygoid plexus
		Foramina Associated with Mouth	
Mandibular	Mandible	Center of medial surface of ramus of mandible	Inferior alveolar nerve vessels
Greater palatine	Maxilla and palatine	At either posterior angle of hard palate	Greater palatine nerve descending palatine ve sels
Lesser palatine	Palatine	In pyramidal process of palatine bone (two or more)	Lesser palatine nerves
Incisive	Maxilla	Anterior end of median palatine suture just behind incisor teeth	Terminal branches of c scending palatine vess and nasopalatine nerv
Scarpa's	Maxilla	Incisive foramen in midline	Nasopalatine nerve
Stensen's	Maxilla	Lateral openings in incisive foramen	Terminal branch of de scending palatine arte
		Foramina Associated with Face	
Supraorbital	Frontal	Supraorbital margin of orbit	Supraorbital nerve anc tery
Infraorbital	Maxilla	Above canine fossa and below orbit	Infraorbital nerve and tery
Zygomaticoorbital	Zygoma	Anteromedial surface of orbital process	Zygomaticotemporal a zygomaticofacial nerve
Zygomaticofacial	Zygoma	Near center of deep surface of zygoma	Zygomaticofacial nerv and vessels
Zygomaticotemporal	Zygoma	Near center of temporal surface of zygoma	Zygomaticotemporal r
Mental	Mandible	Below second premolar tooth	Mental nerve and vess

Table 7-13. Cont.

Name	Bone	Position on bone	Structures passing through
		Foramina Associated with External Aspects of Skull	
Stylomastoid	Temporal	Between styloid and mastoid processes	Facial nerve and stylomastoid artery
Mastoid	Temporal	Near posterior border of mastoid process of temporal bone	Vein to transverse sinus, small branch of occipital artery to dura
Parietal	Parietal	Posterior aspect of parietal close to midline	Emissary vein to superior sagittal sinus

The floor of the **middle cranial fossa** is composed of the body and great wings of the sphenoid, and the squamosal and petrous portions of the temporal bones. Laterally it contains the temporal lobes of the brain. It is limited posteriorly by the superior angle of the petrous portion of the temporal bone and the dorsum sellae centrally. The **sella turcica,** the site of the hypophyseal fossa, is bounded by the tuberculum sellae anteriorly, the dorsum sellae posteriorly, and the anterior and posterior clinoid processes laterally. The anterior and posterior clinoid processes give attachment to a dural fold, the diaphragma sella, which covers the sella. A crescentic arrangement of the **foramina spinosum, ovale,** and **rotundum** and the **superior orbital fissure** is present in the floor of the middle cranial fossa. The **trigeminal impression** on the anterior surface of the petrous portion of the temporal bone lodges the trigeminal ganglion (semilunar or gasserian) of the fifth cranial nerve. The **tegmen tympani,** the lateral part of the anterior and superior surfaces of the petrous bone, forms the roof of the tympanic cavity, the mastoid antrum, and the internal acoustic meatus.

The **posterior cranial fossa** comprises the remainder of the cranial cavity. Its floor is formed by parts of the sphenoid, temporal, and occipital bones. The posterior cranial fossa contains the cerebellum, pons, and medulla oblongata, with a sheet of dura mater, the **tentorium cerebelli,** extending in a transverse plane to separate the cerebellum from the cerebrum. The basilar part of the occipital bone articulates with the sphenoid at the dorsum sellae and is related posteriorly to the pons and medulla of the brain stem. The **internal occipital protuberance** is located at the region of the confluence of the dural sinuses. The inferiormost portion of the posterior cranial fossa presents the large **foramen magnum,** through which the spinal cord passes. At either side of the foramen magnum, the **hypoglossal canal** transmits the hypoglossal nerve. Laterally the deep groove for the **lateral (transverse** and **sigmoid) dural sinus** extends from the internal occipital protuberance to the **jugular foramen.** At the foramen the glossopharyngeal, vagus, and spinal accessory nerves leave the cranial cavity, and the sigmoid sinus becomes the internal jugular vein. On the posterior surface of the petrous portion of the temporal bone, the prominent **internal acoustic meatus** transmits the seventh and eighth cranial nerves as well as branches of the basilar artery to the internal ear.

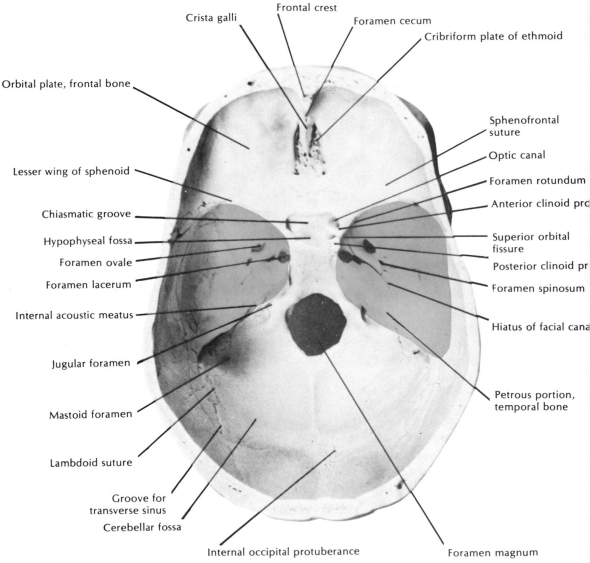

Fig. 7-17. Floor of the cranial cavity.

CRANIAL MENINGES

Three distinct connective tissue membranes (**meninges**), the dura, the arachnoid, and the pia mater cover the brain. The **dura mater** forms a tough outer covering and is composed of two closely adherent layers. The **outer layer** fuses to the periosteum lining the internal surface of the cranial cavity (endosteum) and is continuous through the foramina of the skull with the ectocranial periosteum.

EXTRADURAL HEMORRHAGE

An extradural hemorrhage, so named from the location of the hemorrhage in relation to the meninges, may result from a blow to the side of the head over the inferior part of the parietal bone causing fracture of the bone and rupture of the middle meningeal artery. The bleeding occurs extradurally stripping away the periosteum and dura from the internal surface of the cranium.

Within the outer layer **arachnoid granulations,** bulging cauliflowerlike masses, pit the inner surface of the parietal bones. The meningeal vessels, grooving the calvarium, also run in this layer. The **inner layer** forms four inward-projecting reduplicated folds, namely, the falx cerebri, the falx cerebelli, the tentorium cerebelli, and the diaphragma sellae, which partially divide the cranial cavities into compartments. Separations of the outer and inner layers create venous spaces called the **dural sinuses.**

The midline, sickle-shaped **falx cerebri** projects inward between the cerebral hemispheres. Anteriorly it is attached to the crista galli of the ethmoid bone; superiorly the convex upper border extends from the crista galli, along the midline of the inner surface of the calvarium, to the internal occipital protuberance, and is the site of the superior sagittal sinus. The lower concave border is free anteriorly and contains the inferior sagittal sinus. Posteriorly the lower border unites with the tentorium cerebelli where it forms the straight sinus. The **tentorium cerebelli** is situated between the cerebellum and the occipital and posterior portions of the temporal lobes of the cerebrum. The **falx cerebelli** is a slight fold, attached posteriorly to the internal occipital crest and the tentorium. Anteriorly it is free and projects between the cerebellar hemispheres. The occipital sinus is in its posterior attachment. The **diaphragma sellae** bridges over the sella turcica and covers the hypophysis. A central aperture admits passage of the stalk of the hypophysis.

SUBDURAL HEMORRHAGE

A subdural hemorrhage is a blood collection in the potential space between the dura and the arachnoid. It results from a rupture of the large veins that return blood from the surface of the brain to the superior sagittal sinus. It often results from a blow on the front or back of the head causing considerable anteroposterior movement of the brain within the cranium.

The **arachnoid** is separated from the dura mater by a capillary (subdural) space that contains just sufficient fluid to keep the adjacent surfaces moist. From the inner surface of the arachnoid, cobweblike trabeculae extend across the subarachnoid space to become continuous with the pia mater. Between the arachnoid and the pia the cerebrospinal fluid is contained within the relatively large **subarachnoid space.** The **arachnoid villi** or **granulations** project into the superior sagittal sinus to permit absorption of the cerebrospinal fluid.

SUBARACHNOID HEMORRHAGE

A subarachnoid hemorrhage may come from leakage of blood vessels that traverse the subarachnoid space. Aneurysms of the cerebral arteries often occur in or near the circle of Willis. Rupture of such an aneurysm bleeds into the subarachnoid space and is a common cause of a cerebrovascular accident in a young person. Blood detected in the cerebral spinal fluid, aspirated during a lumbar spinal puncture, may be due to a hemorrhage of this nature.

At certain areas around the base of the brain the arachnoid and the pia mater are widely separated as the **cerebellomedullary, pontine, interpeduncular, chiasmatic,** and **ambiens cisternae,** which contain large amounts of cerebrospinal fluid. The subarachnoid space communicates with the ventricular system of the brain through small apertures in the roof of the fourth ventricle (the midline **foramen of Magendie** and the lateral **foramina of Luschka**) and is continuous with the perineural space around nerves emerging from the brain and spinal cord.

The **pia mater** is a thin, highly vascular layer intimately adherent to the cortex of the brain and follows closely the contours of the brain.

DURAL SINUSES

The **dural sinuses** are venous channels that drain blood from the brain and meninges (Fig. 7-18). They contain no valves and are located between the inner and outer layers of the dura, except for the inferior sagittal and straight sinuses that are between the reduplications of the inner dural layer. The **superior sagittal sinus** is triangular in cross section, occupies the entire length of the attached superior portion of the falx cerebri, and increases in size as it passes posteriorly. At the internal occipital protuberance it usually continues as the right lateral sinus. Arachnoid granulations bulge into its lateral expansions, the **lacunae laterales.** The superior sagittal sinus receives several cerebral veins, diploic veins, and some drainage from the meningeal veins. The smaller **inferior sagittal sinus** occupies the free inferior edge of the falx cerebri. It receives adjacent cerebral veins and, at the junction of the falx cerebri with the tentorium cerebelli, receives the **great cerebral vein** (of Galen) to become the **straight sinus.** The latter is situated along the fusion of the falx cerebri to the tentorium cerebelli. At the internal occipital protuberance the straight sinus usually continues as the left lateral sinus. The **lateral sinuses** are continuations of either the superior sagittal or straight sinuses as noted above, but at their origin may form a common space, the confluence of sinuses. The lateral sinuses are subdivided into transverse and sigmoid portions. The **transverse sinus** occupies the attached portion of the tentorium cerebelli (between inner and outer layers of dura) and receives as tributaries the superior petrosal sinus, diploic veins, and adjacent cerebral and cerebellar veins. The **sigmoid sinus,** a continuation of the transverse, follows an S-shaped course internal to the junction of the petrous and mastoid portions of the temporal bone. It receives the occipital sinus and, at the jugular foramen, becomes the internal jugular vein.

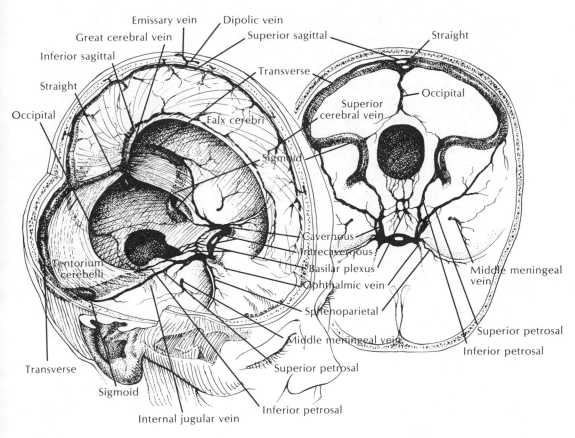

Fig. 7-18. Dural sinuses.

The **confluence of sinuses,** as noted above, is a dilatation commonly at the beginning of the right transverse sinus and forms a wide, shallow depression where the right and left transverse sinuses communicate. It may receive the superior sagittal and straight sinuses and may give origin to the **occipital sinus.** The latter occupies the attached border of the falx cerebelli and is variable in size. Inferiorly the occipital sinus bifurcates, partially encircles the foramen magnum, and ends in the sigmoid sinus. It receives cerebellar veins and communicates with the vertebral plexus of veins.

The **cavernous sinus** is an expanded, trabeculated dilatation at either side of the hypophyseal fossa. The oculomotor, trochlear, ophthalmic, and maxillary nerves are embedded in its lateral wall, while the internal carotid artery with its sympathetic plexus and the abducens nerve pass through the sinus close to its medial wall. It receives ophthalmic and cerebral veins, and the sphenoparietal sinus, and is drained by the superior and inferior petrosal sinuses. The **intercavernous (circular) sinus** connects the two cavernous sinuses.

The **sphenoparietal sinus** lies beneath the lesser wing of the sphenoid where it receives the anterior branch of the middle meningeal vein and diploic veins, and drains into the cavernous sinus. Within the attached margin of the tentorium cerebelli at the superior border of the petrous bone, the **superior petrosal sinus** bridges the trigeminal ganglion and drains the cavernous sinuses into the transverse sinus. The larger **inferior**

petrosal sinus occupies a groove between the petrous and basioccipital bones, where the abducens nerve passes. This sinus drains the cavernous sinus, the internal auditory and adjacent veins, and passes independently through the jugular foramen to empty into the internal jugular vein. The **basilar sinus** is on the posterior aspect of the dorsum sellae and the superior surface of the basioccipital bone. It drains both the cavernous and inferior petrosal sinuses and communicates inferiorly with the vertebral plexus of veins. All dural sinuses ultimately drain into the lateral sinuses and finally into the **internal jugular vein,** except for the inferior petrosal sinus, which empties directly into the internal jugular.

Diploic veins drain the diploë (the marrow-filled space between the inner and outer tables of the cranial bones) into dural sinuses adjacent to the bone. They have no accompanying arteries, the diploë being supplied by meningeal and superficial arteries of the scalp.

Emissary veins pass through foramina of the skull and connect the dural sinuses with the veins on the external surface of the skull. These vessels include a connection between the veins of the nose and the superior sagittal sinus through the foramen cecum, a communication between the veins of the scalp and the superior sagittal sinus through the parietal foramina, a connection (the largest of the emissary veins) between the posterior auricular vein and the sigmoid sinus through the mastoid foramen, and a communication between the suboccipital veins and the sigmoid sinus through the condylar foramen. Additional emissary veins include communications between the pterygoid plexus and the cavernous sinus through the foramen ovale; between supraorbital, ophthalmic, and facial veins and the cavernous sinus through the supraorbital fissure and the optic foramen, and between the pharyngeal venous plexus and the cavernous sinus by way of the carotid canal.

HYPOPHYSIS

The **hypophysis cerebri** is an endocrine gland situated in the sella turcica. Superiorly the sella turcica is roofed by the diaphragm sellae, which has a central aperture through which the **infundibulum,** or stalk, of the hypophysis connects the posterior lobe of the gland to the **tuber cinereum** of the hypothalamus. The cavernous sinuses are at either side of the sella turcica while anteriorly the diaphragma sellae separates the anterior lobe of the hypophysis from the optic chiasma.

CENTRAL NERVOUS SYSTEM

The central nervous system consists of the brain and the spinal cord. Subdivisions of the brain include the cerebrum, diencephalon, midbrain, pons, medulla oblongata, and the cerebellum.

Cerebrum

The **cerebrum** is the largest subdivision of the brain and fills most of the cranial cavity (Fig. 7-19). It is partially divided by a deep midline cleft, the **longitudinal fis-**

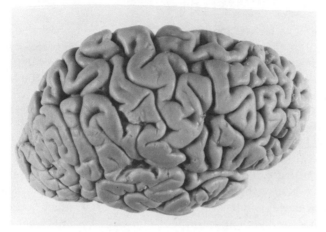

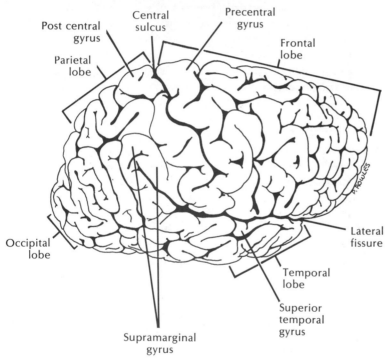

Post central gyrus

Central sulcus

Precentral gyrus

Parietal lobe

Frontal lobe

Occipital lobe

Lateral fissure

Temporal lobe

Superior temporal gyrus

Supramarginal gyrus

Fig. 7-19. Photo and schematic drawing of cerebrum.

sure, into bilaterally symmetrical hemispheres. At the depth of the fissure a thickened band of transverse fibers, the **corpus callosum,** unites the two hemispheres.

The surface of the cerebrum is corrugated in appearance. The furrows in this corrugation are designated as **sulci,** the ridges as **gyri.** Similar patterns of sulci and gyri are present in all brains though they may differ in detail. Certain sulci and fissures divide each of the cerebral hemispheres into **lobes.** The prominent central sulcus extends transversely at about the middle of the cerebral hemisphere and separates the frontal lobe from the parietal lobe. The **frontal lobe** lies subjacent to the frontal bone of the skull; the **parietal lobe** is internal to the parietal bone. Posteriorly the **parietooccipital fissure** divides the parietal lobe from the **occipital lobe.** The latter lies in the posterior portion of the cranial cavity, internal to the occipital bone, and rests on the superior surface of the **tentorium cerebelli.**

A deep cleft, the **transverse fissure,** separates the elongated **temporal lobe** from the frontal lobe. The temporal lobe is located in a pocketlike recess of the cranial cavity internal to the temporal bone.

The surface of the cerebrum, the **cerebral cortex,** is composed of gray matter that extends about 1 to 2 cm internally. The corrugation of the surface of the cerebrum is more extensive in man than in any other animal and greatly increases the surface area of the cortex. The cerebral cortex controls all conscious motor activity. Sensory impulses must reach the cortex for conscious interpretation. Distinct functional areas of the cortex have been localized. They include the **precentral gyrus,** which lies immediately in front of the central sulcus, and is the primary motor area; the **postcentral gyrus** located immediately behind the central sulcus, is the primary area for interpretation of general sensation; the posterior portion of the occipital lobe is the primary area of visual interpretation, while the temporal lobe is associated with interpretation of auditory, gustatory stimuli, and speech activity.

LESIONS OF THE CEREBRUM

Lesions of the cerebrum are many and varied, but their symptoms and signs are relatively few—vomiting, dizziness, headache, convulsions, and partial or complete paralysis. The location of the lesion is more important in eliciting specific symptoms than the nature of the pathological disturbance. For example, any lesion that blocks the motor pathway between the motor cortex of the cerebrum and the skeletal muscles where the nerves terminate, will cause paralysis whether the obstruction is due to a tumor, blood clot, inflammation, compression, or scar.

The underlying white matter of the cerebral cortex includes numerous **association fibers** that connect portions of the cortex with one another. For example, a great mass of fibers, the **corpus callosum,** which connects the cerebral hemispheres; the **optic radiation** that sweeps posteriorly to the occipital (visual) cortex, and fibers of the **internal capsule** that connect lower centers of the central nervous system to the cortex.

Diencephalon

The **cerebral hemispheres** cover the centrally located **diencephalon.** It lies to either side of the third ventricle and is subdivided into the **thalamus, hypothalamus** and **epithalamus.** The latter forms the roof of the third ventricle. The **thalamus,** an oval mass of cell bodies, is an important sensory center. It functions as a relay station, receiving impulses from sensory fibers ascending from the spinal cord, and transmitting impulses to the cerebral cortex. The thalamus is under the control of the cerebral cortex, and therefore, some of the impulses from the cerebral cortex to the spinal cord are relayed through this center. It also acts as an integrating center through its connections with **subcortical nuclei** (collections of nerve cell bodies situated deeply in the central nervous system) in the basal ganglia and the nuclei in the hypothalamus.

The **hypothalamus** occupies the floor and a portion of the lateral wall of the third ventricle. It serves as the great integrating center for the autonomic nervous system.

On the inferior aspect of the diencephalon a number of structures are visible. The most prominent of these is the midline **hypophysis** or **pituitary gland,** which, with the hypothalamus, functions to facilitate rapid and effective adjustments to sudden changes in the environment and is the chief regulator of the endocrine glands. Additional structures visible on the central aspect of the diencephalon include the **optic chiasma** and the paired, rounded **mammillary bodies.**

HYPOPHYSEAL TUMORS

Hypophyseal tumors cause erosion and enlargement of the sella turcica, which are evident on routine skull roentgenograms. The pituitary is in close proximity to the optic tracts, chiasma and nerves; thus, any enlargement of this gland may cause pressure on these vital structures causing partial blindness in one or both eyes. After surgical removal of the tumor, vision is usually improved and may return to normal.

Midbrain

The **midbrain** is the smallest subdivision of the brain stem. It is situated between the pons and the diencephalon as it surrounds the **cerebral aqueduct (of Sylvius),** which connects the third and fourth ventricles. Four rounded eminences, the **corpora quadrigemina,** are visible on the dorsal aspect of the midbrain. The two more superiorly situated bodies are called the **superior colliculi** and serve as synaptic stations in the visual pathway. The two **inferior colliculi** are synaptic areas in the auditory pathway.

Two large ropelike bundles of nerve fibers, the **cerebral peduncles** (crura cerebri), are visible on the ventral aspect of the midbrain. They are composed of motor fibers that extend from the cerebral cortex to the spinal cord. These fibers are cell processes from the initial (upper motor) neuron in the two-neuron pathway for motor impulses from the cerebral cortex to skeletal muscle fibers (Fig. 7-20).

Internally the midbrain includes the **oculomotor** and **trochlear nuclei** containing cell bodies for the fibers that form the third and fourth cranial nerves. The **red nucleus,** a part of the **reticular formation,** which appears pinkish in color in the fresh state, is also present in the midbrain.

Pons

When viewed from the ventral aspect of the brain stem the **pons** is seen as a prominent band of fibers that extends transversely toward the cerebellum. These fibers form the **middle cerebellar peduncle** (brachium pontis). Internally interspersed in white matter are the **pontine nuclei,** which lie ventral to the **reticular formation.** The

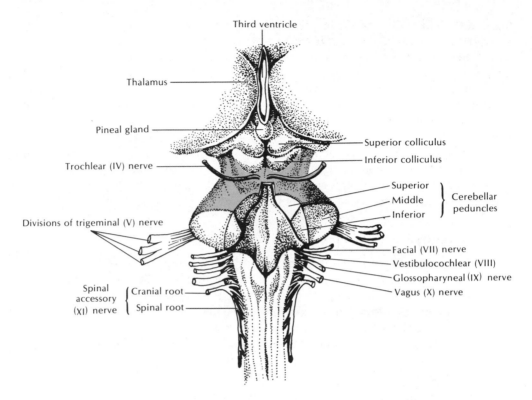

Fig. 7-20. Brain stem (posterior aspect).

nuclei of the trigeminal nerve are also present in the pons. Fiber tracts forming the **superior cerebellar peduncle** or **brachium conjunctivum** are also visible on the cross section of the pons. The most anterior part of the fourth ventricle extends into the pons.

Medulla Oblongata

The **medulla oblongata** is the caudalmost extent of the brain stem. It extends from the pons to the **foramen magnum** where it is continuous inferiorly as the spinal cord. Anteriorly the medulla oblongata rests on the basal portion of the occipital bone, while posteriorly it is covered over by the cerebellum. The extension of the cerebral aqueduct widens considerably in the midportion of the medulla as the **fourth ventricle.** Fiber tracts of the spinal cord extend into the medulla where they become rearranged and may decussate or cross to the opposite side.

A series of longitudinal ridges or elevations are visible on the ventral aspect of the medulla. The more prominent of these, located adjacent to the midline, are the two **pyramids** that contain fibers that will form the lateral corticospinal tracts of the spinal cord. The crossing of these tracts is visible as the **pyramidal decussation.** Laterally ridges corresponding to the **nucleus gracilis and nucleus cuneatus** are evident. These two nuclei contain cell bodies of **second order sensory neurons.** Two flattened masses, the **olives,** are present at the upper end of the medulla oblongata. Rootlets of origin of the ninth, tenth, and eleventh cranial nerves emerge in the more lateral of two parallel grooves on the ventral aspect of the medulla. Fascicles of origin of the **hypoglossal nerve** emerge along the more medial groove between the olive and the pyramid. At the junctional area between the pons and the medulla the **abducens, facial,** and **vestibulo-cochlear nerves** are visible as they emerge from the brain stem.

Nuclei of the above nerves are demostrable within the medulla as are the nuclei of the reticular substance and the olive. The **vestibular** and **cochlear nuclei** associated with impulses originating in the internal ear are also located in the medulla oblongata.

Cerebellum

The **cerebellum** is situated posteriorly and superiorly to the pons and medulla oblongata. It occupies the **posterior cranial fossa** inferior to the level of the **tentorum cerebelli.** It is divided into two lateral hemispheres that are united in the midline by a median portion, the **vermis.**

The surface of the cerebellum is covered by a superficial layer of gray matter, the **cerebellar cortex.** It appears laminated due to the presence of sulci and delicate gyri that are arranged in a parallel pattern. These laminae are referred to as **folia.** Deep to the cortex a number of nuclei are embedded in the white matter. The cerebellum functions to coordinate muscular activity of the body, particularly gross muscular movements.

Cerebellar peduncles connect the cerebellum to other components of the brain stem. These include the **superior cerebellar peduncle** (brachium conjunctivum), which connects the cerebellum to the midbrain; the **middle cerebellar peduncle** (brachium pontis), connecting the cerebellum to the pons; and the **inferior cerebellar peduncle** (restiform body), connecting the cerebellum to the medulla oblongata.

LESIONS OF THE CEREBELLUM

Lesions of the cerebellum may be from tumors, abscesses, cysts, or inflammation. The symptoms for all of these lesions are essentially similar—instability of equilibrium and locomotion with severe dizziness. Voluntary movements are impaired by a marked ataxia (incoordination of muscular activity) and nystagmus. Because of the ataxia, the patient tends to lean or fall towards the side of the lesion.

Ventricles

Some parts of the central nervous system are not solid structures but have cavities. In the brain and its subdivisions these cavities are termed **ventricles,** while in the spinal cord the cavity is called the **central canal** (Fig. 7-21). The cavity associated with each cerebral hemisphere is called the **lateral ventricle.** The **third ventricle** is the cavity of the diencephalon, which communicates with the two lateral ventricles through the **interventricular foramina (of Monro).** The **cerebral aqueduct (of Sylvius)** associated with the midbrain, extends between the third and **fourth ventricle.** The latter lies internal to the pons and medulla oblongata and is under cover of the cerebellum. Small openings in the fourth ventricle, the **foramen of Magendie** in the midline, and

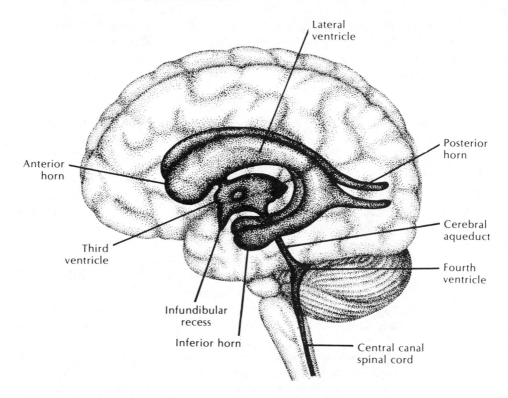

Fig. 7-21. Ventricles of the brain.

the two **foramina of Luschka** laterally, communicate with the **subarachnoid space.** The fourth ventricle continues distally as the small **central canal** of the spinal cord.

In the roof of the ventricles, specialized vascular structures, the **choroid plexuses,** elaborate **cerebrospinal fluid.** The latter circulates from the ventricles through the foramina of Magendie and Luschka into the subarachnoid space where it is reabsorbed into the vascular system at the arachnoid villi. The latter are specializations of the **arachnoid membrane** that project into the **dural venous sinuses.**

HYDROCEPHALUS

Hydrocephalus (water on the brain) is due to an abnormally large accumulation of cerebrospinal fluid within the ventricles of the brain. It may be caused by: 1) obstruction of some part of the ventricular system, 2) excessive production of fluid, or 3) interference with the absorption of the fluid. Obstruction is most commonly caused by scar tissue produced by inflammation of the meninges, which may close the foramina of the fourth ventricle. It may also be due to a tumor of the mesencephalon, which blocks the cerebral aqueduct. The resultant pressure dilates the ventricles and compresses the cerebral cortex against the unyielding skull. If this occurs in infants when the cranial bones are not yet united, the internal pressure greatly enlarges the skull as well as the ventricles of the brain. The cerebral cortex becomes markedly thinned-out and degenerates causing severe mental retardation.

Blood Vessels

The brain is supplied by two pairs of arteries, the internal carotid and the vertebral arteries, which form an anastomosis at the base of the brain called the **circulus arteriosus cerebri,** or the circle of Willis (Fig. 7-22). The **internal carotid** is subdivided for

Fig. 7-22. Circle of Willis and arterial supply to brain.

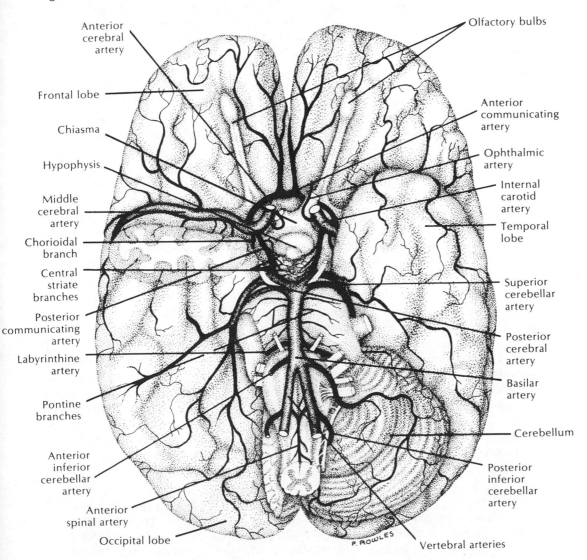

Anterior cerebral artery

Olfactory bulbs

Frontal lobe

Anterior communicating artery

Chiasma

Ophthalmic artery

Hypophysis

Internal carotid artery

Middle cerebral artery

Temporal lobe

Chorioidal branch

Central striate branches

Superior cerebellar artery

Posterior communicating artery

Posterior cerebral artery

Labyrinthine artery

Basilar artery

Pontine branches

Cerebellum

Anterior inferior cerebellar artery

Posterior inferior cerebellar artery

Anterior spinal artery

Occipital lobe

Vertebral arteries

P. ROWLES

descriptive purposes into four parts. The **cervical portion** begins at the bifurcation of the common carotid artery opposite the upper border of the thyroid cartilage and passes vertically upward anterior to the transverse processes of the upper three cervical vertebrae. In its course it is crossed by the twelfth cranial nerve, the digastric and stylohyoideus muscles, and the occipital and posterior auricular arteries. At the base of the skull the ninth, tenth, and eleventh cranial nerves pass between the artery and the internal jugular vein. **The internal carotid artery has no branches in the neck.** Its **petrous portion** follows an S-shaped course through the petrous portion of the temporal bone within the carotid canal. It is separated by a thin layer of bone from the tympanic cavity and the trigeminal ganglion and gives the **caroticotympanic branch** to the tympanic cavity. The **cavernous portion** within the cavernous sinus follows a second sigmoid course passing upward to the posterior clinoid process, then forward and superiorly again at the anterior clinoid process. Branches within the sinus include twigs to the cavernous sinus, the hypophysis, the trigeminal ganglion, and the meninges. The **ophthalmic artery** arises just as the internal carotid passes through the roof of the cavernous sinus.

Passing between the second and third cranial nerves, the **cerebral portion** of the internal carotid artery extends to the medial end of the lateral cerebral fissure, where it divides into terminal branches, the anterior cerebral and the much larger middle cerebral. Its **posterior communicating branch** courses posteriorly beneath the optic tract to join the posterior cerebral branch of the basilar artery. The internal carotid artery supplies the optic tract and cerebral peduncles and gives deep branches to the internal capsule and thalamus of the brain. Its small **choroidal branch** follows the optic tract and cerebral peduncles to the choroid plexus. The **anterior cerebral artery** passes anteromedially above the optic chiasma, then along the fissure over the corpus callosum to the parietooccipital sulcus, giving cortical branches to the medial surface of the hemisphere and deep central branches. The **anterior communicating branch** connects the anterior cerebral arteries of either side as they course over the corpus callosum. The direct continuation of the internal carotid, the **middle cerebral artery,** passes upward and laterally in the lateral cerebral fissure to spread out over the surface of the brain, giving orbital, frontal, temporal, and parietal cortical branches to the lateral surface of the hemisphere, and **central striate branches** (the arteries of cerebral hemorrhage), which pass deeply into the corpus striatum of the brain.

The **vertebral artery,** a branch of the subclavian, courses superiorly through the foramina intertransversarii of the cervical vertebrae to enter the cranial cavity by way of the foramen magnum. As it enters the cranium it gives off **anterior** and **posterior spinal arteries** that descend on either surface of the spinal cord. They anastomose with each other and, with segmental spinal branches, form longitudinal vessels along the length of the cord. The **meningeal branch** supplies the meninges of the posterior cranial fossa. The largest branch of the vertebral artery, the **posterior inferior cerebellar,** passes between the tenth and eleventh cranial nerves to supply the undersurface of the cerebellum.

The **basilar artery** is formed by the union of the two vertebral arteries. It passes in the midline of the inferior surface of the pons and gives the following branches: the **anterior inferior cerebellar** to the under surface of the cerebellum, the **pontine branches** to the pons, and the **labyrinthine artery,** which passes through the internal auditory meatus to the internal ear. Its **superior cerebellar branch** courses laterally just posterior to the third cranial nerve, supplies the superior surface of the cerebellum, and anastomoses with the anterior and posterior inferior cerebellar arteries. The **posterior cerebral branch** passes laterally to join the **posterior communicating vessel** from the internal carotid and contributes to the circle of Willis. The latter branch winds around the cerebral peduncle giving cortical branches to the occipital lobe, to the area adjacent to the parietooccipital fissure, and to the temporal lobe.

The **meningeal arteries** are periosteal arteries that groove the inner surface of the calvarium, lie in the outer layer of the dura mater, and supply the dura, the inner table of the skull, and the diploë. The **middle meningeal** is a branch of the internal maxillary artery and enters the cranial cavity through the foramen spinosum to divide into anterior and posterior branches, which supply most of the meninges. Other arteries to the meninges include branches from the anterior and posterior ethmoidal arteries to the anterior cranial fossa; the accessory meningeal and branches of the internal carotid to the middle cranial fossa; branches of the vertebral, the ascending pharyngeal, and the occipital artery to the posterior cranial fossa.

A STROKE OR CEREBROVASCULAR ACCIDENT (CVA)

A stroke or CVA is the rupture or occlusion of certain cerebral arteries, especially the striate arteries leading to the internal capsule of the brain. Such a lesion usually produces hemiplegia on the opposite side of the body. If the hemorrhage is massive, the area of the brain supplied by the ruptured artery will degenerate, and the neurologic deficits will be permanent. Fortunately, in many cases the motor and sensory functions partially or completely return with an early dissolution of the blood clot and resultant release of pressure on the cerebral cortex.

CRANIAL NERVES

Ten of the **twelve cranial nerves** originate directly from the brain stem (Table 7-14). Each has a superficial attachment as well as a deeply located nucleus of origin or termination. The olfactory (I) and optic (II) nerves are unusual in that they do not arise from the brain stem. The optic nerve is actually a brain tract rather than a nerve. The olfactory nerve begins in the olfactory mucosa and terminates in the olfactory bulb. Five of the twelve cranial nerves attach to the ventral aspect of the brain: the olfactory (I) at the olfactory bulb on the cribriform plate; the optic (II) at the anterolateral angle of the optic chiasma; the oculomotor (III) in a groove, the oculomotor sulcus, between the cerebral peduncle and the interpeduncular fossa; the abducens (VI) in a groove between the pons and the lateral aspect of the pyramid, and the hypoglossal (XII) as a row of rootlets in a groove between the pyramid and the olive. Six cranial nerves attach to the lateral aspect of the brain stem: the trigeminal (V) by two roots, a large sensory and a small motor at the side of the pons; the facial and vestibulocochlear (VII and VIII) in line with the trigeminal at the border of the pons and the inferior cerebellar

Table 7-14. Cranial Nerves and Their Components

Number	Name	Component	Cell bodies	Distribution
I	Olfactory	Sensory (smell)	Olfactory mucosa	Nasal mucosa of up part of septum and rior concha
II	Optic	Sensory (vision)	Ganglion layer of retina	Retina
III	Oculomotor	Somatic motor	Brain stem	Superior, inferior, ar medial rectus; inferi oblique and levator pebrae superioris m
		Parasympathetic	Preganglionic, brain stem; postganglionic, ciliary ganglion	Sphincter muscle of ciliary muscle
IV	Trochlear	Somatic motor	Brain stem	Superior oblique mu
V	Trigeminal	Sensory (general)	Semilunar ganglion	Skin and mucosa of meninges
		Motor	Brain stem	Muscles of masticati (masseter, temporali ternal and internal p goids), mylohyoid, a rior belly of digastric tensor tympani, and sor veli palatini
VI	Abducens	Somatic motor	Brain stem	Lateral rectus muscle
VII	Facial	Motor	Brain stem	Facial muscles, poste belly of digastric, sty hyoid, and stapedius
		Parasympathetic	Preganglionic, brain stem; postganglionic, pterygo-palatine ganglion	Nasal, palatine, and mal glands
			Preganglionic, brain stem; postganglionic, subman-dibular ganglion	Submandibular and lingual glands
		Sensory (general)	Geniculate ganglion	Skin of mastoid regi and external acousti atus
		Sensory (taste)	Geniculate ganglion	Anterior two-thirds tongue by way of ch tympani; soft palate way of major petros
VIII	Vestibulocochlear	Sensory (hearing)	Spiral ganglion	Organ of Corti
		Sensory (equilib-rium)	Vestibular ganglion	Semicircular canals, cle, and saccule
IX	Glossopharyngeal	Motor	Brain stem	Stylopharyngeus mu and pharyngeal mus
		Parasympathetic	Preganglionic, brain stem; postganglionic, otic gan-glion	Parotid gland
		Sensory (general)	Superior ganglion	Mucosa of pharynx, panic cavity, posteri one-third of tongue, carotid sinus

Table 7-14. Cont.

Number	Name	Component	Cell bodies	Distribution
		Sensory (taste)	Inferior ganglion	Posterior one-third of tongue
X	Vagus	Motor	Brain stem	Muscles of pharynx, larynx, levator veli palatini, palatoglossus
		Parasympathetic	Preganglionic, brain stem; postganglionic, on, in, or near viscera	Thoracic and abdominal viscera
		Sensory (general)	Superior ganglion	Skin of external acoustic meatus
		Sensory (general)	Inferior ganglion	Pharynx, larynx, and thoracic and abdominal viscera
		Sensory (taste)	Inferior ganglion	Epiglottis and base of tongue
XI	Spinal accessory cranial root	With motor branches of vagus	Brainstem	
	Spinal root	Somatic motor	Spinal cord	Sternocleidomastoid and trapezius
XII	Hypoglossal	Somatic motor	Brain stem	Intrinsic and extrinsic muscles of tongue

peduncle, and the glossopharyngeal, vagus, and spinal accessory (IX, X, and XI) as a row of rootlets in a narrow groove along the entire lateral side of the medulla. This origin of the spinal accessory is the cranial or accessory portion of that nerve; the spinal portion is derived from the upper five cervical segments of the cord. Only one cranial nerve, the trochlear (IV), attaches to the dorsum of the brain stem, emerging immediately posterior to the inferior colliculus of the midbrain at the superior medullary velum (Fig. 7-23).

The **olfactory nerve (I)** is entirely sensory. It arises in the olfactory mucosa of the nasal cavity as bipolar neurons and is limited in origin to the mucous membrane covering the superior nasal concha and the adjacent nasal septum. Numerous filaments from this distribution area pierce the cribriform plate to synapse with secondary neurons in the **olfactory bulb.** The **olfactory tract** passes posteriorly from the olfactory bulb to the olfactory trigone of the brain.

LESION

Olfactory (I): **If the lesion is unilateral, anosmia (loss of smell) results on the affected side.**

The **optic nerve (II)** is also entirely sensory and originates in the **ganglionic cells** of the retina, which are tertiary neurons in the visual pathway. The orbital portion of

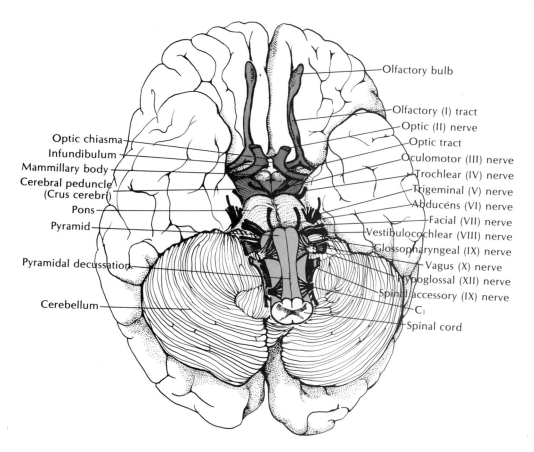

Olfactory bulb

Olfactory (I) tract

Optic (II) nerve

Optic tract

Oculomotor (III) nerve

Trochlear (IV) nerve

Trigeminal (V) nerve

Abducens (VI) nerve

Facial (VII) nerve

Vestibulocochlear (VIII) nerve

Glossopharyngeal (IX) nerve

Vagus (X) nerve

Hypoglossal (XII) nerve

Spinal accessory (IX) nerve

C₁

Spinal cord

Optic chiasma

Infundibulum

Mammillary body

Cerebral peduncle (Crus cerebri)

Pons

Pyramid

Pyramidal decussation

Cerebellum

Fig. 7-23. Inferior view of brain showing major subdivisions and cranial nerve attachments. Midbrain, medulla oblongata, and spinal cord are shaded.

the nerve is about 5 cm long, invested by meninges, and passes posteriorly through the optic foramen to the **optic chiasma,** which rests on the tuberculum sellae. Fibers at the chiasma partially decussate at the optic chiasma to continue posteriorly as the **optic tract,** which winds around the cerebral peduncle and terminates in the lateral geniculate body of the mesencephalon.

LESION

Optic (II): **Depending on the location of the lesion, blindness may be total in both eyes (at the chiasma), in one eye only (along an optic nerve), or affect both visual fields (along the optic tract).**

The **oculomotor nerve (III),** from its emergence at the oculomotor sulcus, passes between the posterior cerebral and superior cerebellar arteries. Lateral to the clinoid processes it pierces the dura to traverse the cavernous sinus and enter the orbital cavity through the supraorbital fissure. The oculomotor nerve is accompanied by the abducens and nasociliary nerves, all of which pass between the heads of the lateral rectus muscle. Its terminal distribution to the extraocular muscles will be described with the section on the orbit.

LESION

***Oculomotor (III)*: Pupil dilatation, inability to focus, and eyeball directed downwards and outwards, and ptosis (drooping of upper eyelid) on the damaged side.**

The **trochlear nerve (IV)** is motor in function. It is the most slender of the cranial nerves, yet has the longest intracranial course. Originating from the dorsum of the brain stem, it winds around the midbrain, enters the edge of the tentorium cerebelli, and passes with the oculomotor nerve between the posterior cerebral and the superior cerebellar arteries. It continues forward around the cerebral peduncle to penetrate the dura and enter the cavernous sinus between the third and sixth nerves. It passes along the lateral wall of the sinus to enter the orbit through the supraorbital fissure above the origin of the ocular muscles. Its orbital course will be described with the orbital cavity.

LESION

***Trochlear (IV)*: No obvious dysfunction occurs if this nerve is damaged. However, if patient looks downward and outward, he has double vision.**

The large **trigeminal nerve (V)** is both motor and sensory. It is formed by a large **sensory root from the trigeminal ganglion** and a smaller **motor root.** The trigeminal ganglion occupies a cavity (cavum trigeminale) in the dura at the sutural area between the petrous portion of the temporal and the greater wing of the sphenoid bones. From the ganglion the **ophthalmic division (V_1)** courses anteriorly to enter the orbit by way of the supraorbital fissure, the **maxillary division (V_2)** passes through the foramen rotundum to the pterygopalatine fossa, and the **mandibular division (V_3)** traverses the foramen ovale to reach the infratemporal fossa. The motor root passes independently through the foramen ovale to join the mandibular nerve in the infratemporal fossa. Attached to branches of the trigeminal nerve by sensory roots are the four small parasympathetic ganglia of the head. The trigeminal nerve, however, has no parasympathetic components in its brain stem nuclei. The **ophthalmic division** will be discussed with the orbital cavity; the **maxillary division** will be described with the pterygopalatine fossa, and the **mandibular division** has been covered with the infratemporal fossa.

LESION

***Trigeminal (V)*: If the lesion involves the entire nerve, extensive anesthesia is present on the affected side over the face, scalp, conjunctiva, gingivae, anterior two-thirds of tongue, mucous membrane of nose, hard and soft palate, cheek, and lips. The four muscles of mastication**

on the injured side are also paralyzed. If only the 1st or 2nd divisions are damaged, only sensation over appropriate distribution is lost. If the 3rd division is divided, there is sensory loss to the lower facial region and motor loss to the muscles of mastication.

The **abducens nerve (VI)** is motor in function and pierces the dura at the dorsum sellae to pass below the posterior clinoid process and enter the cavernous sinus at the lateral side of the internal carotid artery. Traversing the sinus, it passes through the supraorbital fissure to enter the orbital cavity above the ophthalmic artery. Its distribution will be discussed with the orbital cavity.

LESION

Abducens (VI): **With damage to this nerve the eye cannot move laterally beyond the midpoint.**

The **facial nerve (VII)** contains both motor and sensory fibers. The large **motor** and smaller **sensory (nervus intermedius) roots** traverse the internal acoustic meatus (in company with the eighth nerve) to unite at the **geniculate ganglion** located on the sharp posterior bend of the seventh nerve within the facial canal. From the geniculate ganglion the **major petrosal nerve,** transmitting preganglionic parasympathetic fibers, traverses the petrous portion of the temporal bone to enter the middle cranial fossa through the hiatus of the facial canal. This branch then courses forward between the dura and the trigeminal ganglion, passing deep to the latter, to unite with the **deep petrosal nerve.** The latter carries postganglionic sympathetic fibers from the internal carotid plexus with their cell bodies located in the superior cervical sympathetic ganglion. These two nerves unite to form the **nerve of the pterygoid canal** (Vidian nerve), which passes through the pterygoid (Vidian) canal in the sphenoid bone to terminate in the pterygopalatine ganglion. Distal to the geniculate ganglion the facial nerve passes through the facial canal to emerge from the skull at the stylomastoid foramen. Within the facial canal the nerve gives rise to the **branch to the stapedius muscle, communicating twigs** to the auricular branch of the vagus, and the **chorda tympani nerve.** The latter turns upward through a separate small canal to enter the tympanic cavity through the posterior wall and courses forward on the internal surface of the tympanic membrane. Arching across the handle of the malleus, it leaves the tympanic cavity by passing through the anterior wall. It then traverses the petrotympanic fissure and joins the lingual branch of the trigeminal nerve in the infratemporal fossa. The **chorda tympani** carries preganglionic parasympathetic fibers to the submandibular ganglion and special taste fibers to the anterior two-thirds of the tongue. The **terminal branches** of the facial nerve have been described with the face.

LESION

Facial (VII): **Symptoms vary in extent and severity, depending on site of the lesion. However, paralysis of the facial muscles always occurs.**

The taste fibers to the anterior two-thirds of the tongue and the corneal reflex may or may not be affected. Spontaneous paralysis of the facial nerve is often called Bell's palsy.

The **vestibulocochlear nerve (VIII),** entirely sensory in function, consists of two parts, the cochlear and vestibular portions, which differ in peripheral endings, central connections, and function. The eighth nerve courses with the facial nerve through the internal acoustic meatus, where it divides at the termination of the canal into its respective parts. The **cochlear portion,** the nerve of hearing, consists of bipolar neurons associated with the **spiral ganglion** of the cochlea. The peripheral processes pass to the spiral organ **(of Corti),** while the central processes pass from the modiolus to the lateral end of the internal acoustic meatus. The **vestibular portion,** the nerve of equilibrium, consists of the bipolar cells of the **vestibular ganglion** located in the superior part of the lateral end of the internal acoustic meatus. Peripheral processes of these bipolar neurons pass to the utricle, the saccule, and the ampullae of the semicircular ducts. The central processes join with those of the cochlear division to form the eighth nerve.

LESION

The eighth cranial nerve has two functional components, auditory and vestibular. Partial or complete deafness may result from damage of any of its auditory nerve components, such as, organ of Corti, spiral ganglion, or cochlear division of the eighth nerve. Lesions of the vestibular division will cause disturbances of equilibrium—dizziness, loss of balance, or nystagmus. Meniere's syndrome is a fairly common disturbance of the vestibular apparatus causing dizziness (vertigo), nausea, ringing in the ears (tinnitus), and often deafness. Its etiology is obscure, and its cure still eludes us.

The **glossopharyngeal nerve (IX)** is both motor and sensory and exits through the jugular foramen in company with the tenth and eleventh nerves. Two **sensory ganglia,** the **superior** and the **inferior (petrosal),** are associated with the glossopharyngeal nerve as it passes through the foramen. The **tympanic branch** (nerve of Jacobson), transmitting preganglionic parasympathetic fibers to the otic ganglion, passes through a small canal (of Jacobson) within the temporal bone to the tympanic cavity. Here it joins with branches of the facial nerve to form the **tympanic (promontory) plexus,** which supplies the mucous membrane of the tympanic cavity. From this plexus, the **minor petrosal nerve** is reconstituted and courses through the petrous portion of the temporal bone to run forward in the middle cranial fossa, traverses the foramen ovale, and synapses in the **otic ganglion.** At its exit from the jugular foramen, the glossopharyngeal nerve passes deep to the styloid process, gives a **branch to the stylopharyngeus muscle,** then joins with branches of the vagus and sympathetic fibers to form the **pharyngeal plexus.** **Terminal branches** of the glossopharyngeal nerve supply the posterior third of the tongue with general and special (taste) sensation.

LESION

Glossopharyngeal (IX): Severing this nerve results in loss of sensation of posterior one-third of tongue and of the pharynx. The patient may also have some difficulty in swallowing.

The **vagus nerve (X)**, containing both motor and sensory fibers, has the most extensive course and distribution of any of the cranial nerves. It leaves the cranial cavity through the jugular foramen in company with the ninth and eleventh nerves. The superior and inferior sensory ganglia of the vagus are located within the jugular foramen or just below it. Passing to the cervical region in company with the internal jugular vein, the vagus nerve gives a **recurrent branch** to the meninges, **auricular branches** to the ear, **pharyngeal branches** to the pharyngeal plexus and soft palate, the **nerve to the carotid sinus, cervical cardiac branches,** and a **superior** and an **inferior recurrent laryngeal branch** to the larynx. At the root of the neck, the left vagus passes between the common carotid and subclavian arteries to pass anterior to the arch of the aorta and posterior to the root of the lung, where it joins with the right vagus to form the **esophageal plexus.** The right vagus passes anterior to the subclavian artery, then descends along the trachea to the posterior aspect of the root of the lung and joins the left vagus as above. The **inferior laryngeal nerves** loop under the arch of the aorta on the left side and the subclavian artery on the right to reach the tracheoesophageal groove and ascend to supply all the intrinsic musculature of the larynx, except the cricothyroideus. In the thorax, the vagus gives off **cardiac** and **pulmonary branches** and then forms the **esophageal plexus.** After passing through the esophageal hiatus, the right vagus reforms as the **posterior vagal trunk,** the left as the **anterior vagal trunk,** to aid in the formation of the **celiac plexus.** The vagi contribute to the innervation of all abdominal viscera, except those portions of the gastrointestinal tract distal to the splenic flexure of the colon.

LESION

Vagus (X): Trauma to the vagus nerve will result in unilateral paralysis of larynx and palate, with hoarseness due to loss of inferior (recurrent) laryngeal nerve. Unilateral anesthesia of larynx on the same side will also be present. If both vagi are cut in the head or neck, tachycardia and decreased respiration result, and the patient is often unable to breathe or speak because of loss of innervation of both vocal cords.

The **spinal accessory nerve (XI)** is a motor nerve formed from both cranial and spinal components. The smaller **cranial portion** unites with the spinal part to pass through the jugular foramen, then separates to join with the vagus for distribution to the pharynx and larynx. It contributes to the supply of the musculus uvulae, the levator veli palatini, the pharyngeal constrictor muscles, and muscles of the larynx and esophagus. The **spinal portion,** originating in the motor cells of the ventral horns of the first through the fifth cervical nerves, passes superiorly along the side of the spinal cord and through the foramen magnum, where it joins with the cranial portion. It continues a short distance with the latter as it exits through the jugular foramen, then separates to pass inferiorly behind the jugular vein, the stylohyoideus, the digastric, and the upper part of the sternocleidomastoideus. It crosses the posterior triangle of the neck, picking up communicating twigs of the cervical plexus (C_2 and C_3), and passes to the deep surface of the trapezius to supply this muscle and the sternocleidomastoideus.

LESION

Spinal Accessory (XI): **If damaged, the sternomastoid and trapezius muscles are paralyzed.**

The **hypoglossal nerve (XII),** the motor nerve to the tongue musculature, passes through the hypoglossal canal to descend almost vertically to a point opposite the angle of the mandible. It courses deep to the internal carotid artery and internal jugular vein, then lies between the artery and the vein, deep to the stylohyoideus and digastric muscles. At the intermediate tendon of the digastric it loops around the occipital artery and passes anteriorly between the hyoglossus and mylohyoideus muscles to **supply the intrinsic and extrinsic muscles of the tongue.** Communicating twigs from the first loop of the cervical plexus join the hypoglossal nerve and run with it for a short distance before most of the fibers leave the hypoglossal nerve as the superior limb (descendens hypoglossi) of the ansa cervicalis to supply infrahyoid muscles of the neck. Some of the fibers from this communication with the cervical plexus continue with the hypoglossal nerve to branch from the latter as individual twigs to supply the thyrohyoideus and geniohyoideus muscles.

LESION

Hypoglossal (XII): **The affected side of the tongue becomes wrinkled and atrophied. When the tongue is protruded, the tip deviates towards the side of the lesion.**

CRANIAL PARASYMPATHETIC GANGLIA (Table 7-15)

Located within the orbital cavity, the small **ciliary ganglion** receives preganglionic parasympathetic fibers from the **ciliary branch of the oculomotor nerve.** From cell bodies within the ganglion the postganglionic parasympathetic fibers leave through the **short ciliary nerves** to penetrate the sclera at the area cribrosa and supply the ciliary and sphincter pupillae muscles. Sensory fibers from the **nasociliary nerve** traverse the ganglion without synapsing and become components of the short ciliary nerves, which supply general sensation to the eyeball. Sympathetic fibers from the ophthalmic plexus also pass through the ganglion and are distributed with the short ciliary nerves to innervate the dilator pupillae muscle and the smooth muscle of orbital blood vessels.

Within the pterygopalatine fossa, the **pterygopalatine ganglion** is attached to the

Table 7-15. Parasympathetics of Head

Nerve	Nucleus of Preganglionic Cell Bodies	Ganglion of Postganglionic Cell Bodies	Transmission Pathway
Oculomotor (III)	Edinger–Westphal	Ciliary	Oculomotor → motor twigs → ciliar ganglion → short ciliaries → middle of eyeball → ciliary muscle and sph muscle of iris
Facial (VII)	Superior salivatory	Pterygopalatine	Facial → major petrosal → nerve of pterygoid canal → pterygopalatine g glion → maxillary → zygomatic → c municating to lacrimal → lacrimal gl from ganglion other branches (sphe palatine, greater and lesser palatine) pass to midregion of face
		Submandibular	Facial → chorda tympani → lingual submandibular ganglion → subman lar and sublingual glands
Glossopharyngeal (IX)	Inferior salivatory	Otic	Glossopharyngeal → tympanic nerve tympanic plexus → minor petrosal → otic ganglion → parotid gland
Vagus (X)	Motor nucleus of vagus	Diffuse	Vagus → cardiac nerves → cardiac p → heart; pulmonary nerves → pulm plexus → lungs; esophageal plexus → gastric trunks → vascular plexuses (e celiac) → postganglionic cell bodies myenteric and submucosal plexuses digestive organs

maxillary division of the trigeminal nerve by two **short sensory roots.** Preganglionic parasympathetic fibers pass from the **nervus intermedius of the seventh nerve** to the ganglion through the **major petrosal nerve.** Postganglionic parasympathetic fibers are distributed with branches of the maxillary division of the trigeminal nerve to the lacrimal gland. Branches from the ganglion also supply the nasopharynx, nasal cavity, palate, upper lip, and gingiva. Postganglionic sympathetic fibers, transmitted along the internal carotid plexus, pass as the **deep petrosal nerve** to join the **major petrosal nerve** in the middle cranial cavity and form the **nerve of the pterygoid canal.** The sympathetic fibers pass through the ganglion to be distributed with its branches. **Sensory fibers,** by way of the two roots to the ganglion from the maxillary division of the fifth cranial nerve, also pass through the ganglion and are distributed as sensory components of the ganglionic branches.

   The **otic ganglion,** located in the infratemporal fossa, is attached to the mandibular division of the trigeminal immediately distal to the foramen ovale. Its preganglionic parasympathetic fibers are derived from a branch of the glossopharyngeal nerve, **the tympanic nerve,** which passes to the promontory plexus. From the latter, the **minor petrosal nerve** is reconstituted and transmits the preganglionic parasympathetic fibers to the otic ganglion. Postganglionic fibers are distributed to the parotid gland by the auriculotemporal branch of the trigeminal. Sympathetic fibers passing through the ganglion are derived from the plexus surrounding the middle meningeal artery and are distributed to the blood vessels of the parotid, along with sensory fibers derived from the mandibular nerve.

The **submandibular ganglion** is adjacent to the submandibular gland. Its preganglionic parasympathetic fibers are derived from the seventh cranial nerve through the **chorda tympani.** The latter nerve arises from the facial nerve within the facial canal to course through the middle ear cavity and traverses the petrotympanic fissure to reach the infratemporal fossa where it joins the lingual nerve. The preganglionic fibers pass to the submandibular ganglion by way of two short roots that suspend the ganglion from the lingual nerve. Postganglionic parasympathetic fibers are distributed to the submandibular and sublingual glands. Fibers from the sympathetic plexus around the external maxillary artery may pass through the ganglion and continue with its branches. Sensory fibers from the lingual nerve are distributed with the branches of the ganglion to the submandibular and sublingual glands, and to mucous membrane of the oral cavity.

Orbital Cavity

The pyramidal-shaped **orbital cavity** presents a base, an apex, and four walls (Fig. 7-24). The optic foramen is located at the **apex** while the quadrangular **base** opens onto the face and is formed about equally by the frontal, maxillary, and zygomatic bones. Each bone transmits cutaneous nerves: the supraorbital in the frontal, the infraorbital in the maxillary, and the zygomaticofacial and zygomaticotemporal in the zygoma. The **medial walls** of the cavities are parallel and about 2 to 3 cm apart. They are separated by the nasal cavities and are formed by the fragile lacrimal bone and the orbital plates of the ethmoid and palatine bones. The strong **lateral walls,** at right angles to each other, are formed by processes of the zygomatic and the greater wings of the sphenoid bones. The fossa for the lacrimal gland is in the superolateral portion of the lateral wall. The **superior wall,** or roof of the cavity, is formed by the orbital plate of the frontal bone. The orbital plate of the maxillary and a small part of the zygomatic bone forms the **inferior wall,** or floor. In the floor of the orbital cavity, the **infraorbital groove** continues anteriorly as the **infraorbital canal** to open onto the face at the **infraorbital foramen.**

The **orbital periosteum** is a funnel-shaped sheath attached to the bony walls of the orbital cavity, continuous posteriorly with the outer layer of the dura and anteriorly with the periosteum covering the external surface of the skull. It encloses the contents of the orbit, except for the zygomatic nerves and the infraorbital nerves and vessels that lie between the periosteum and the bone. The thin, membranous **fascia bulbi (Tenon's capsule)** encloses the eyeball, forming a socket in which it moves, and separates the eyeball from the orbital fat pad. Its smooth inner surface forms, with the sclera, the **periscleral space.** Expansions from the fascial sheath pass to the lateral and medial recti muscles as the **check ligaments.** The sheath reflects onto and encloses the eye muscles as they attach to the sclera. Posteriorly the bulbar fascia is perforated by ciliary vessels and nerves at the **lamina cribosa sclerae.**

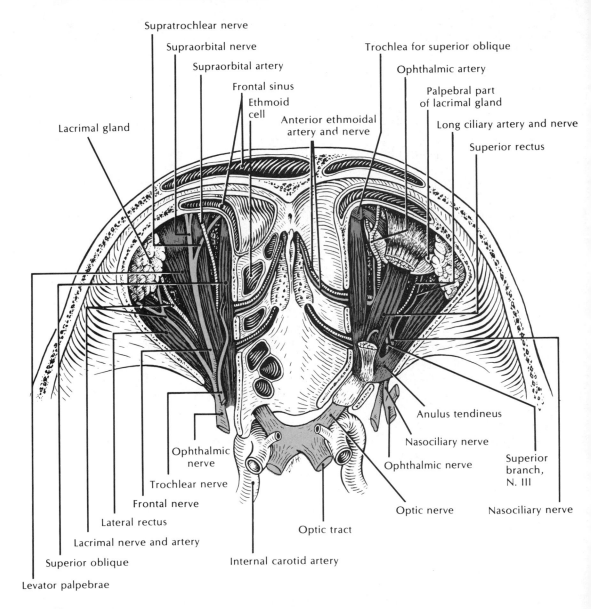

Fig. 7-24. Superior view of the interior of the orbit. (Hollinshead WH: Textbook of Anatomy, 3rd ed. New York, Harper & Row, 1974)

EXOPHTHALMOS

Exophthalmos is an abnormal protrusion of the eyeball. It is associated with hyperthyroidism (toxic goiter), or an aneurysm of one of the arteries in the orbit, which pushes the eye forward (pulsating exophthalmos).

Table 7-16. Extrinsic Muscles of the Eye

Muscle	Origin	Insertion	Action	Nerve
Rectus superior, inferior, medial, and lateral	All originate from fibrous cuff fixed posteriorly to optic foramen and anteriorly to dural sheath of optic nerve, and insert superiorly, inferiorly, medially, and laterally by bandlike aponeuroses into sclera just behind corneoscleral junction		See below	Lateral rectus, abducens; oculomotor, others
Levator palpebrae superioris	Orbital roof anterior to optic foramen	Upper tarsal plate and superior fornix of conjunctivum	Elevates upper lid	Oculomotor
Superior oblique	Roof of orbital cavity between superior and medial recti and anterior to optic foramen	Slender tendon passes through fibrous ring (trochlea), reverses direction to insert deep to superior rectus	See below	Trochlear
Inferior oblique	Floor of orbital cavity lateral to lacrimal canal	To sclera between superior and lateral recti	See below	Oculomotor

Action of Eye Muscles

Action	Muscle
Adduction	**Medial,** superior, and inferior recti
Abduction	Inferior and superior oblique; **lateral rectus**
Elevation	Inferior oblique; **superior rectus**
Depression	Superior oblique; **inferior rectus**
Medial rotation	Superior rectus; superior oblique
Lateral rotation	Inferior oblique; inferior rectus

Muscles

The **extrinsic muscles** of the eye consist of four straight muscles, the superior, inferior, medial, and lateral recti; two oblique muscles, the superior and inferior, and the levator palpebrae superioris (Table 7-16). The **recti muscles** originate from the margin of a **fibrous cuff,** which is fixed posteriorly to the periosteum and the dural sheath of the optic nerve at the optic foramen, and laterally to the margins of the supraorbital fissure. The lateral rectus is split into upper and lower heads at its origin, with vessels and nerves entering the orbital cavity between the two heads. The recti muscles spread out like the staves of a barrel to insert into a bandlike aponeurosis encircling the sclera just behind the corneoscleral junction. Each muscle has a fascial sheath, with adjacent sheaths joining to form a **fibromuscular cone.**

The **levator palpebrae superioris** is separated superiorly from the superior rectus and inserts into the tarsal plate of the upper eyelid and the superior fornix of the conjunctivum. The tendon of the **superior oblique** passes through a fascial pulley, the **trochlea,** attached to the medial wall of the orbit and reverses direction before inserting into the sclera. The **inferior oblique,** located in the anterior part of the orbital cavity, has an origin on the floor of the orbital cavity, apart from the other extrinsic ocular muscles. The lateral rectus is supplied by the **abducens nerve,** the superior oblique by the **trochlear nerve,** and the remaining muscles by the **oculomotor nerve.** The medial, superior, and inferior recti acting together move the eye medially, while the inferior and superior oblique and the lateral rectus muscles acting in concert shift it laterally. The inferior oblique and superior rectus direct the eye upward; the superior oblique and inferior rectus move it downward. The superior rectus and superior oblique medially rotate the eye, and the inferior oblique and inferior rectus are lateral rotators. Additional actions include the recti muscles acting as retractors and the oblique muscles as protractors to keep the eyeball in balance.

Blood Vessels and Nerves

As the internal carotid artery leaves the cavernous sinus, it gives off the **ophthalmic artery** that passes through the optic foramen. Inferior to the optic nerve, it pierces the dural sheath and lies free within the fibromuscular cone of the orbital cavity. It courses above the optic nerve to give several branches to the structures within the cavity (Fig. 7-25). A very important small branch, the **central artery to the retina,** pierces the optic nerve sheath about 1 to 2 cm posterior to the eyeball. It courses in the center of the optic nerve to reach and supply the retina.

CENTRAL ARTERY OF RETINA

The central artery of the retina is an end artery. Therefore, if it is blocked, there are no anastomoses with other arteries to sustain circulation to the retina. The result is sudden, total blindness of the affected eye. An occlusion may result from several causes, such as a tumor, a thrombus, or massive edema.

Several **short posterior ciliary arteries** arise from the ophthalmic artery to pierce the sclera and form a plexus in the choroid. Additional branches, the **long posterior ciliary arteries,** extend anteriorly to anastomose with the anterior ciliary artery at the margin of the iris. The **anterior ciliary arteries** arise from muscular branches to the recti muscles. They pierce the sclera just behind the corneoscleral junction to supply the ciliary body and iris, and give twigs to the deep conjunctival plexus. The relatively large **lacrimal branch** of the ophthalmic artery begins near the optic foramen and courses with the lacrimal nerve along the lateral wall above the lateral rectus muscle. It supplies the latter, the superior oblique muscle, the lacrimal gland, upper eyelid, and the conjunctiva. Terminal branches of the ophthalmic artery include the **supraorbital, supratrochlear, dorsal nasal, anterior ethmoidal,** and **posterior ethmoidal,** which pass from the orbital cavity to anastomose freely with branches of the external carotid in the upper face.

Venous drainage of the orbital cavity is accomplished by the **superior** and **inferior**

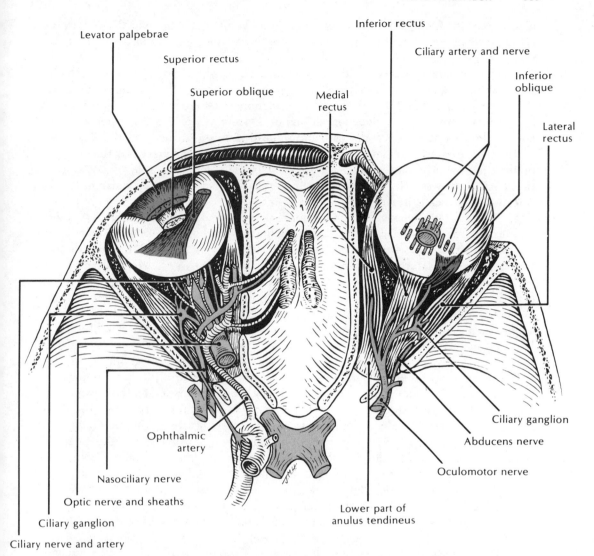

Fig. 7-25. Deep aspect of the orbital cavity. (Hollinshead WH: Textbook of Anatomy, 3rd ed. New York, Harper & Row, 1974)

ophthalmic veins. The former, formed by the junction of the supraorbital and supratrochlear veins, drains into the cavernous sinus. The inferior ophthalmic vein, originating in the floor of the cavity, communicates through the infraorbital fissure with the pterygoid plexus or through the supraorbital fissure with the cavernous sinus.

The **optic nerve** enters the orbital cavity through the **optic foramen,** while all the other nerves to the orbit traverse the supraorbital fissure. The optic nerve passes anterolaterally and slightly inferiorly from the optic foramen to penetrate the posterior aspect of the eyeball, slightly medial to the posterior pole of the eye. Its meningeal coverings fuse with the sclera of the eyeball. The optic nerve is slightly longer than its course and, therefore, does not interfere with movements of the eyeball. Within the retina, the optic nerve fibers spread out as the third-order neurons in the visual pathway.

The **oculomotor nerve** is motor to the superior, medial, and inferior recti, the inferior oblique, and the levator palpebrae superioris muscles. It also carries preganglionic parasympathetic fibers to the ciliary ganglion. Passing between the heads of the lateral rectus muscle, it divides into an **upper division,** supplying the superior rectus and the levator palpebrae superioris, and a **lower division** to the medial and inferior recti and the inferior oblique muscles. The **trochlear nerve** innervates the superior oblique muscle as it passes along the superior border of that muscle. The **abducens nerve** passes between the heads of the lateral rectus to course on the ocular surface of this muscle, which it innervates.

The smallest division of the trigeminal nerve, the **ophthalmic nerve,** is sensory and terminates as frontal, lacrimal, and nasociliary branches, which are distributed to structures within the orbital cavity. The **frontal nerve** passes through the supraorbital fissure above the lateral rectus muscle to course between the levator palpebrae superioris and the orbital plate, and terminates as **supraorbital** and **supratrochlear branches,** which supply the eyelids, forehead, and scalp. The **lacrimal nerve,** passing just below the frontal branch, follows the upper border of the lateral rectus muscle to supply the lacrimal gland and terminates as twigs to the conjunctivum, skin of the eyelids, and the skin over the zygomatic process. Communications between the lacrimal nerve and the zygomatic branch of the maxillary division of the trigeminal form the pathway for postganglionic parasympathetic fibers to pass from the pterygopalatine ganglion to the lacrimal gland. The **nasociliary nerve** courses between the heads of the lateral rectus muscle to cross the medial wall of the orbital cavity above the optic nerve. Its branches include the **long ciliary nerves,** carrying sensory and postganglionic sympathetic fibers to the iris; **posterior ethmoidal nerves,** supplying the ethmoidal and sphenoidal air sinuses; the **infratrochlear nerve** innervating the lacrimal sac, conjunctivum, eyelids, and upper nose; and the **anterior ethmoidal nerve.** The latter passes through the anterior ethmoidal foramen to enter the anterior cranial fossa, courses along the cribriform plate, and enters the nasal cavity by traversing the nasociliary slit. It then terminates by dividing into the **internal** and **external nasal branches** (Table 7-17).

The **ciliary ganglion** located in the posterior third of the orbital cavity is a synaptic station for parasympathetic neurons and has been described on page 333.

Lacrimal Apparatus

The **lacrimal gland** is located in the superolateral aspect of the orbital cavity, partly within the **lacrimal fossa** and partially embedded in the upper eyelid (Fig. 7-26). It drains through six to ten **lacrimal ducts** that pierce the superior fornix of the conjunctivum to empty onto the opposing palpebral and ocular surfaces of the conjunctivum. The action of blinking spreads a uniform layer of lacrimal fluid over the conjunctivum. At the medial canthus, a **punctum lacrimale** opens at the summit of a **lacrimal papilla** on the free margin of each lid as the beginning of the **lacrimal canaliculi.** The latter drain the lacrimal fluid into an expansion of the proximal end of the nasolacrimal

Table 7-17. Nerve Distribution to Orbit

Nerve	Origin	Course	Distribution
Optic (II)—Actually a brain tract. Optic "nerve" formed by nerve cell processes of tertiary neurons in visual pathway, neuronal elements in retina include (1) rods and cones, (2) bipolar neurons, and (3) neurons of ganglion layer.			
	Posterior pole of eyeball conveys nerve cell processes of neurons in ganglion layer of retina	Traverses orbital cavity to pass through optic foramen and joins nerve of other eye at chiasma; extends to brain as optic tract	Nerve of vision
Motor nerves (III, IV, VI)—All arise from brain stem, traverse cavernous sinus, and enter orbit via supraorbital fissure to extrinsic eye muscles			
Oculomotor (III)		Divides into superior and inferior divisions	Superior division to superior rectus and levator palpebrae superioris; inferior division to medial and inferior rectus and inferior oblique, and conveys preganglionic parasympathetic fibers to ciliary ganglion; postganglionics to ciliary muscle and circular (sphincter) muscle of iris
Trochlear (IV)		Along superior border of superior oblique	Superior oblique
Abducens (VI)		Along medial aspect of lateral rectus	Lateral rectus
Ophthalmic (V₁)—From trigeminal ganglion traverses cavernous sinus to enter orbit via supraorbital fissure and trifurcates into lacrimal, frontal, and nasociliary nerves			
Lacrimal		Follows lateral wall of orbit to reach lacrimal gland	Sensory to lacrimal gland
Supraorbital and supratrochlear	Terminal branches of frontal	Run on superior aspect of levator palpebrae superioris to reach foramen or notch of same name	Supply skin of forehead and scalp to lambdoidal suture
Posterior ethmoidal	Nasociliary	Passes through posterior ethmoidal foramen	Sensory to ethmoidal air cells
Anterior ethmoidal	Nasociliary	Passes through anterior ethmoid foramen, reenters cranial cavity, runs along cribriform plate and, via nasal slit, traverses nasal cavity to emerge on dorsum of nose	Internal nasal—to mucosa of anterosuperior portion of nasal cavity; external nasal—to skin on dorsum of nose
Infratrochlear	Nasociliary	Along medial wall of orbit to upper lid	Skin of upper lid
Long ciliary	Nasociliary	Pierces area cribrosa of sclera to enter middle tunic of eyeball	Sensory to eyeball

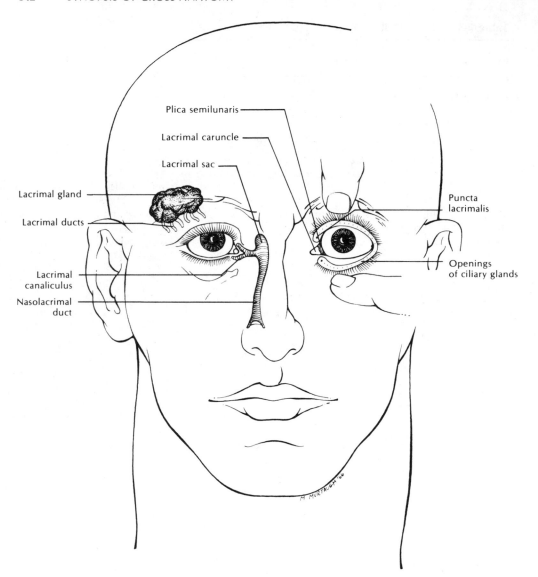

Plica semilunaris

Lacrimal caruncle

Lacrimal sac

Lacrimal gland

Lacrimal ducts

Lacrimal canaliculus

Nasolacrimal duct

Puncta lacrimalis

Openings of ciliary glands

Fig. 7-26. Lacrimal apparatus.

duct, the **lacrimal sac.** The **nasolacrimal duct** passes through the nasolacrimal canal of the maxilla and lacrimal bones to open into the inferior meatus of the nasal cavity.

The lacrimal gland is innervated by parasympathetic fibers originating from the facial nerve. Postganglionic fibers from the pterygopalatine ganglion travel with a communicating branch from the zygomatic nerve to the lacrimal branch of the ophthalmic division of the trigeminal nerve. It receives its blood supply by way of the lacrimal branch of the ophthalmic artery.

Conjunctivum

The **conjunctivum** is the mucous membrane of the eye that lines the inner surface of the eyelids, and covers the outer surfaces of the sclera and cornea. It forms a partial sac where it reflects from the deep surface of the lid on to the eye as the **conjunctival fornix.** There is a fornix for each lid, that is, a superior and an inferior fornix.

The conjunctivum is divided into a vascular **palpebral part** that lines the back of the eyelids, and a thin, transparent, bulbar portion covering the sclera and cornea. A vertical fold of bulbar conjunctivum at the medial angle of the eye forms the **plica semilunaris.** The **lacrimal caruncle** is a small mound of modified skin at the inner canthus. The conjunctivum is innervated by a rich plexus from **ciliary, lacrimal,** and **infratrochlear nerves,** whose fibers largely terminate as free nerve endings.

Eyeball

MYOPIA

Myopia is a condition in which the eyeball is too long and the focus of objects lies anterior to the retina, rather than on it.
Hyperopia is the opposite condition; the focus lies posterior to the retina.
Obviously either condition causes faulty vision, which is corrected by lenses that will focus the image on the retina.

The **eyeball,** approximately 2 to 3 cm in diameter, is spherical with a slight anterior bulge (Fig. 7-27). Its wall consists of three concentric coats: an outer fibrous, a middle vascular, and an inner nervous tunic. The sclera and cornea compose the outer **fibrous tunic,** with the **sclera** forming a firm cup covering the posterior five-sixths of the eyeball. Its outer surface is smooth, separated from the bulbar fascia by loose connective tissue, and perforated posteriorly by the optic nerve, and the central artery and vein at the lamina cribrosa. At the corneoscleral junction, it is continuous with the transparent **cornea,** which bulges slightly over the anterior one-sixth of the eyeball. The cornea is avascular and receives its nutrients by diffusion from a capillary network at its margin. The cornea is richly supplied with free sensory nerve endings derived from the ciliary nerves.

The **vascular tunic** lies internal to the sclera and consists of the choroid, the ciliary body, and the iris. The **choroid** is a thin, highly vascular membrane, and is brown in color from pigmented cells. It covers the posterior two-thirds of the eyeball and extends anteriorly to the ora serrata. The choroid consists of a dense capillary network of small arteries and veins held together with connective tissue. It is loosely connected to the sclera, except at the entrance of the optic nerve, where it is firmly fixed. It is intimately attached to the inner pigmented layer of the retina. The **ciliary body,** consisting of a thickening of the vascular tunic as the ciliary ring, ciliary processes, and ciliary muscle, connects the choroid at the **ora serrata** to the peripheral circumference of the iris. The **ciliary ring** extends from the ora serrata to the **ciliary processes,** while the latter, sixty to eighty small projections, are continuous peripherally with the ciliary ring

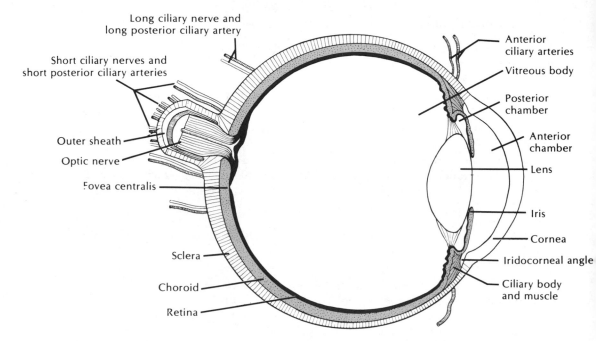

Fig. 7-27. Cross section of the eyeball. (Hollinshead WH: Textbook of Anatomy, 3rd ed. New York, Harper & Row, 1974)

and give attachment anteriorly to the **suspensory ligament of the lens.** The nonstriated **ciliary muscle** originates at the posterior margin of the scleral spur and inserts into the ciliary ring and ciliary processes. It contracts in accommodation to draw the ciliary processes forward, which relaxes the suspensory ligament permitting the natural elasticity of the lens to become more rounded and so effects focus. The **iris** is a thin, pigmented, contractile diaphragm with a central aperture, the **pupil.** The opening varies in size. It constricts by the action of the circularly arranged **sphincteric muscle** that is innervated by parasympathetic fibers from the ciliary ganglion and third cranial nerve, or by dilating through the action of the radially arranged **dilator muscle**—innervated by sympathetic fibers from the superior cervical ganglion. The iris, the conspicuous colored portion of the eye, separates the **anterior chamber** (posterior to the cornea and anterior to the iris) from the **posterior chamber** (posterior to the iris and anterior to the suspensory ligament and ciliary processes). Both chambers are continuous with each other through the pupil and are filled with a clear, refractile fluid, the **aqueous humor.**

GLAUCOMA

Glaucoma is a condition of the eye characterized by increased intraocular pressure. It occurs when the drainage of the aqueous humor fails to keep up with the production of the fluid by the ciliary processes. The buildup of pressure may cause severe pain and damage to

the nerve fibers in the retina, especially at the optic disc. If the pressure continues, blindness results. Drugs that reduce the rate of aqueous humor production and thus lower the intraocular pressure, will prevent permanent retinal and optic nerve injury.

The **nervous tunic** or **retina** is developmentally an evagination of the brain. It has an outer pigmented layer and an inner light-receptor portion. Three regions of the retina are differentiated: the optic, ciliary, and iridial portions.

DETACHMENT OF THE RETINA

Detachment of the retina may occur in trauma, such as a blow to the head. The tear in the retina causes blindness in the corresponding field of vision. The actual detachment occurs between the sensory part of the retina and the underlying pigmented layer. Fluid accumulates between these layers, forcing the thin, friable retina to billow out toward the vitreous humor. The retina may be reattached by various surgical procedures, such as photocoagulation by laser beam, cryosurgery, or scleral resection.

The optic part is light-sensitive and contains a three-neuron pathway. The first neuron constitutes the light-receptor **rods** and **cones;** the second, **bipolar cells;** and the third, the **ganglion cells,** whose axons form the fibers of the optic nerve. The optic portion occupies the posterior part of the bulb and ends at the ora serrata. It is firmly attached at the ora serrata anteriorly and at the entrance of the optic nerve posteriorly. The **ciliary portion** of the retina begins at the ora serrata and continues anteriorly to line the internal surface of the ciliary body. The pigmented **iridial (iris) portion** of the retina covers the posterior aspect of the iris.

COLOR BLINDNESS

Color blindness is the inability to distinguish colors correctly. To have normal color vision, one must have three types of cones in the retina. Each type responds only to a particular color, for example, red, green, or violet. When all three cones are equally stimulated, sensation of white results. Very few people are truly color blind, but many are color weak since they have difficulty in distinguishing hues. Color blindness is controlled by sex-linked recessive genes and manifests itself much more frequently in the male than female.

The **macula lutea** is a yellowish, oval area at the posterior pole of the eye that presents a slight central depression, the **fovea centralis,** the area of greatest visual acuity. Light rays are focused at the fovea if the eyes are correctly accommodating. The **optic disc,** the site of emergence of the optic nerve, is about 3 mm medial to the macula lutea. Axons of the third-order neuron of the retina converge at this point in an area devoid of light-receptor cells, known as the **blind spot** of the eye. At the center of the optic disc, the central artery of the retina emerges and distributes to the retina.

CHOKED DISC OR PAPILLEDEMA

A choked disc or papilledema is the swelling of the optic disc, so that it protrudes into the vitreous body. Changes in the disc, as viewed through an ophthalmoscope, reflect evidence of the condition of the brain proper since the optic nerve and disc are an extension of the brain. Edema of the papilla suggests increased intracranial pressure that compresses the thin veins from the optic nerve causing congestion and edema of the retina's optic disc. Several pathologic conditions may cause the increase in intracranial pressure, for example, brain abscess, aneurysm, brain tumor, or hydrocephalus.

The refractile **lens,** a transparent biconvex body more flattened anteriorly than posteriorly, is composed of laminated transparent lens fibers. It is enclosed by a transparent **capsule** and held in place by the **suspensory ligament.** Its shape is modified in focusing by the action of the ciliary muscle. Posterior to the lens, the refractile **vitreous humor,** or **body,** occupies the central portion of the eyeball.

CATARACT

A cataract is a condition in which the lens becomes opaque or milk-white in appearance. When the lens has a cataract, little light is transmitted to the retina and blurred images and poor vision results. To remove this obstruction of light, the lens with the cataract is removed and light can freely pass through the eye to the retina. Loss of the lens requires the patient to wear thick convex lenses that focus a clear sharp image on the retina.

Mouth and Pharynx

ORAL CAVITY

The **oral cavity** is subdivided for descriptive purposes into the **vestibule** and the **mouth cavity proper** (Fig. 7-28). The former is the cleft separating the lips and cheeks from the teeth and gingivae, or gums. With the mouth closed the vestibule communicates with the mouth cavity through the interval between the last molar teeth and the rami of the mandible. At the lips the skin of the face is continuous with the mucous membrane of the oral cavity. The bulk of the **lips** is formed by the orbicularis oris mus-

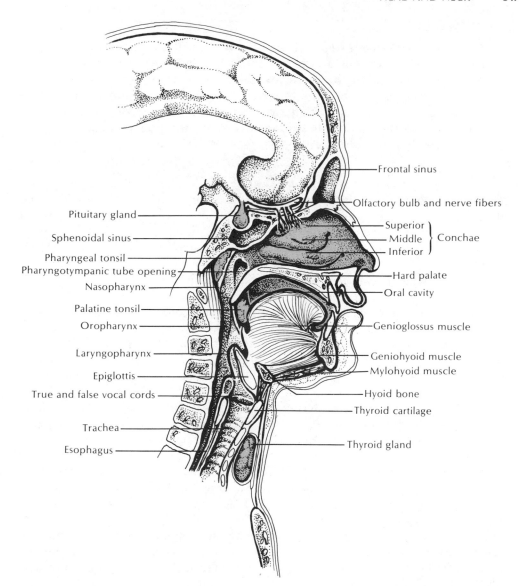

Frontal sinus

Olfactory bulb and nerve fibers

Superior
Middle } Conchae
Inferior

Hard palate

Oral cavity

Genioglossus muscle

Geniohyoid muscle

Mylohyoid muscle

Hyoid bone

Thyroid cartilage

Thyroid gland

Pituitary gland

Sphenoidal sinus

Pharyngeal tonsil

Pharyngotympanic tube opening

Nasopharynx

Palatine tonsil

Oropharynx

Laryngopharynx

Epiglottis

True and false vocal cords

Trachea

Esophagus

Fig. 7-28. Hemisection of head showing nasal and oral cavities, and pharynx.

cle and contains a vascular arch arising from labial branches of the facial artery. The **cheeks,** consisting for the most part of the buccinator muscle and the buccal fat pads, are pierced by the parotid duct, which opens into the vestibule opposite the upper second molar tooth. The **gingiva** is composed of dense fibrous tissue covered by a smooth vascular mucosa. It attaches to the alveolar margins of the jaws where it embraces the necks of the teeth as the **periodontal membrane.** Thirty-two **teeth** are normally present in the adult: two incisors, one canine, two premolars, and three molars are found in each half of the upper and lower jaws.

PYORRHEA

Pyorrhea is a discharge of pus around the teeth with progressive necrosis of the alveolar bone and looseness of the teeth. Recession of the gums continues until the teeth lose their bony and gingival attachments and fall out. When the teeth are lost, the inflammatory symptoms subside. Good oral hygiene with semi-annual professional cleaning of the teeth will usually prevent pyorrhea, the principal cause of loss of teeth in the adult.

Posteriorly the mouth cavity communicates with the pharynx through the **fauces.** The **hard** and **soft palates** make up the roof of the mouth, and the floor is formed by the tongue and mucous membrane. Anteriorly the tongue lies more or less free in the mouth, with a median fold of mucous membrane, the **frenulum linguae,** passing from the floor of the mouth to the under surface of the tongue. Two transverse **sublingual folds** overly the sublingual glands and the minute orifices of these glands open along the summit of the fold. Near the midline on each sublingual fold, the sublingual papillae surround the openings of the submandibular ducts.

The tongue, a mobile mass of muscle and mucous membrane, functions in taste, chewing, swallowing, and speech. It is shaped like an upside-down high-topped shoe, with the sole of the shoe being the dorsum of the tongue and the upper portion of the shoe the root of the tongue. At the back of the tongue a V-shaped groove, the **sulcus terminalis,** is flanked anteriorly by a ridge of large **circumvallate papillae,** and divides the dorsum of the tongue into two parts, an **anterior** two-thirds, **horizontal** or **palatine portion** and a **posterior** one-third, **vertical or pharyngeal portion.** A small pit, the **foramen cecum,** is present at the vertex of sulcus terminalis. **Fungiform** and **filiform papillae** are distributed over the palatine, or horizontal portion of the tongue. With the mouth open only the palatine portion of the tongue is visible; the less apparent vertical pharyngeal portion forms the anterior wall of the oropharynx and is related inferiorly to the epiglottis. On the ventral surface of the tongue, the midline **frenulum linguae,** flanked by the **deep lingual veins,** attaches the tongue to the floor of the mouth. A delicate fringed ridge of mucous membrane, the **fimbriated fold,** is present on the lateral aspect of the ventral surface of the tongue. Numerous lymph follicles constituting the **lingual tonsil** are present on the pharyngeal portion of the tongue. As the mucous membrane reflects onto the epiglottis, it forms **median** and **lateral glossoepiglottic folds.**

TONGUE TIE

Tongue tie (ankyloglossia) results from a congenital shortening of the frenulum, which attaches the tongue to the floor of the mouth. Such a condition often causes a severe speech impediment. Recognizing the condition early and simply cutting the frenulum frees the tongue and cures the problem.

Muscles of the Tongue

The muscles of the tongue consist of four pairs of intrinsic and four pairs of extrinsic muscles (Table 7-18). The muscles of either side are separated by a midline fibrous septum. The **superior longitudinal muscle** extends from the tip to the root of the tongue, and the **inferior longitudinal muscle** is present in the interval between the extrinsic genioglossus and the hyoglossus muscles. The **transversus linguae** extend from the septum to the sides of the tongue, while the **verticalis linguae** originate from the dorsum of the tongue and sweep inferiorly and laterally to interdigitate with the other tongue musculature.

Extrinsic muscles include the fan-shaped **genioglossus,** radiating from the genial tubercle of the mandible into the tongue, with its lowermost fibers inserting into the hyoid bone. The quadrilateral **hyoglossus,** under cover of the mylohoideus, passes from the body and greater horn of the hyoid to the sides of the tongue. The sliplike **styloglossus** sweeps forward from the tip of the styloid process and stylomandibular ligament to blend with, and insert into, the hyoglossus and palatoglossus with some of its fibers extending along the side of the tongue as far as the tip. The **palatoglossus** originates from the palate, passes to the side of the tongue, and forms the anterior pillar of the palatine fossa.

All the muscles of the tongue, except the palatoglossus (pharyngeal plexus), are innervated by the hypoglossal nerve. The intrinsic muscles act to alter the shape of the tongue, and the extrinsic group functions to change the position and, to a limited extent, the shape of the tongue. The anterior two-thirds of the tongue is supplied with special (taste) sensation by the chorda tympani branch of the facial nerve and with general sensation by the lingual branch of the mandibular division of the trigeminal nerve. The glossopharyngeal nerve supplies both general and special sensation to the posterior one-third of the tongue.

Table 7-18. Muscles of the Tongue

Muscle	Origin	Insertion	Action	Nerve
Genioglossus	Genial tubercle of mandible	Ventral surface of tongue and body of hyoid bone	Aids in protrusion, retraction, and depression of tongue	Hypoglossal
Hyoglossus	Body and greater cornu of hyoid bone	Sides of tongue	Depresses and draws tongue laterally	Hypoglossal
Styloglossus	Styloid process	Sides of tongue	Aids in retraction and elevation of tongue	Hypoglossal
Palatoglossus	Soft palate	Dorsum and sides of tongue	Elevates tongue and narrows fauces	Pharyngeal plexus
Longitudinalis linguae (superior and inferior); transversus and verticalis linguae	Form intrinsic musculature of tongue; named according to their relationship		Alter shape of tongue	Hypoglossal

INJURY TO HYPOGLOSSAL NERVE

If cranial nerve XII is damaged or divided, the half of the tongue on the same side of the lesion becomes atrophic and wrinkled. When the tongue is protruded, the muscles of the intact side push the tip of the tongue towards the paralyzed side. Asking a patient to stick out his tongue tests the function of the hypoglossal nerve.

The **roof of the mouth** is a vaulted dome formed by the hard and soft palates. The **hard palate** is composed of the palatine processes of the maxillary bone anteriorly, and the horizontal laminae of the palatine bones posteriorly. The **soft palate** consists of muscles, glands, and the palatine aponeurosis. The soft palate is attached anteriorly to the posterior margin of the hard palate and laterally to the wall of the pharynx. Posteriorly in the midpoint of its free margin, it forms the conical **uvula** directed inferiorly into the fauces. During deglutition the soft palate elevates to help close the nasopharynx.

CLEFT PALATE AND HARE LIP

Cleft palate and hare lip may occur separately or together, and in varying degrees or extent. Developmentally during the 8th to 9th week of gestation, the palate is formed by fusion of the two lateral palatine processes and the single central premaxilla. Failure of union of these structures leaves a gap or cleft at the fusion site. A hare (cleft) lip is the failure of the medial nasal processes to merge with the maxillary processes. Although they sometimes are bilateral, single cleft palate and cleft lip are the most common anomalies.

Muscles of the Palate

Five pairs of muscles are associated with the soft palate (Table 7-19). The **palatoglossus** and **palatopharyngeus** originate from the palatine aponeurosis and the posterior part of the hard palate and are contained within the palatoglossal and palatopharyngeal folds, respectively. The palatoglossus inserts into the dorsum and side of the tongue. The palatopharyngeus splits to pass to either side of the levator veli palatini and musculus uvulae, reunites, and then blends with the salpingopharyngeus before inserting into the thyroid cartilage and pharynx. Both **musculi uvulae** arise from the posterior nasal spine and unite as they pass backward to insert into the mucous membrane of the uvula. The rounded **levator veli palatini** originates from the temporal bone adjacent to the opening of the carotid canal. It passes obliquely downward and inserts into the palatine aponeurosis, interdigitating with fibers of the muscle from the opposite side. The flat, triangular **tensor veli palatini muscle** arises from the scaphoid fossa, spine of the sphenoid, and cartilaginous portion of the pharyngotympanic tube. It tapers to form a rounded tendon that hooks around the pterygoid hamulus, turns medially, and then spreads out to form the palatine aponeurosis attached to the posterior border of the hard palate.

Table 7-19. Muscles of the Palate

Muscle	Origin	Insertion	Action	Nerve
Tensor veli palatini	Scaphoid fossa, spine of sphenoid, and cartilaginous portion of pharyngotympanic tube	Tendon passes around hamulus of pterygoid to insert into soft palate	Tenses soft palate	Mandibular division of trigeminal
Levator veli palatini	Petrous portion of temporal bone and cartilaginous portion of pharyngotympanic tube	Midline of soft palate	Elevates soft palate	Pharyngeal plexus
Palatoglossus	Soft palate	Dorsum and sides of tongue	Narrows fauces and elevates tongue	Pharyngeal plexus
Palatopharyngeus	Soft palate	Posterior border of thyroid cartilage and musculature of pharynx	Elevates pharynx and helps to close nasopharynx	Pharyngeal plexus
Uvulus	Palatine aponeurosis	Mucous membrane of uvula	Elevates uvula	Pharyngeal plexus

The region of communication between the oral cavity and the pharynx, the **fauces,** is bounded superiorly by the soft palate, inferiorly by the dorsum of the tongue, and laterally on either side by the palatoglossal and palatopharyngeal arches, which enclose the palatine tonsils.

PHARYNX

The **pharynx** is a wide muscular tube, about 12 to 14 cm long, and lined with mucous membrane. It extends from the base of the skull to the level of the sixth cervical vertebra, where it becomes the esophagus. The attachments of the pharynx, from superior to inferior, are: the pharyngeal tubercle at the base of the skull, medial pterygoid lamina, pterygomandibular raphe, inner aspect of the ramus of the mandible, hyoid bone, and thyroid and cricoid cartilages. The muscular wall of the pharynx consists of the three overlapping constrictor muscles (superior, middle, and inferior), the stylopharyngeus and palatopharyngeus muscles. Various fascial layers and a mucous membrane also contribute to the wall of the pharynx.

The pharynx is subdivided into the nasal, oral, and laryngeal portions. Anteriorly the wall is interrupted by, and related to, structures associated with openings into these portions. The **nasal pharynx** is situated above the level of the soft palate, posterior to the nasal cavity, and is related superiorly to the sphenoid bone and basilar portions of the occipital bone. It is the widest part of the pharynx and normally remains patent for the passage of air. Anteriorly it is bounded by the **choanae** (internal nares), which open into the nasal cavity. Laterally it receives the opening of the **auditory** or **pharyngotympanic tube,** around which the mucous membrane is raised as the **torus tubarius.** A mucous membrane covering the salpingopharyngeal muscle, the **salpingopharyngeal fold,** descends vertically from the torus tubarius. The roof and the posterior wall of the nasopharynx form a continuous curve, and contain an aggregate of lymphoid tissue between the roof and pharyngeal recesses, the **pharyngeal tonsils.** When enlarged these are called **adenoids.**

ADENOIDS

Adenoids is a pathologic condition of the pharyngeal tonsil characterized by infection and hypertrophy of its lymphoid tissue. The marked enlargement of the gland blocks the internal nares (choana), which necessitates the person to breathe through the mouth. If the condition persists, the child develops a characteristic facial expression called adenoid facies defined as a "dull expression, with open mouth seen in children with adenoid growth." Infection may spread to the lymphoid tissue surrounding the pharyngotympanic tube (tubal tonsil), causing swelling and closure of the tube. Recurring attacks of middle ear infection usually follow, which may result in temporary or permanent hearing loss.

The pharyngeal isthmus, located between the nasal and oral parts of the pharynx, is bounded laterally by the palatopharyngeal arch and the mucous membrane covering the palatopharyngeal muscle. It is closed during swallowing by the elevation of the soft palate and contraction of the superior constrictor muscle of the pharynx.

PHARYNGOTYMPANIC TUBE INFECTIONS

The pharyngotympanic (eustachian) tube connects the middle ear cavity with the nasopharynx. Such a communication provides a pathway for nasal and oral infections to spread to the middle ear and in time to the mastoid antrum and mastoid air cells. Prior to the advent of antibiotics, mastoid abscesses were very common sequelae of respiratory infections. Now they are rare because penicillin and other antibiotics help eliminate the offending pathogenic organisms, and cure the middle ear infections. Since the mastoid is separated only by a thin plate of bone from the temporal lobe of the brain and the cerebellum, a mastoid abscess or mastoiditis may cause a brain abscess and meningitis, a very serious complication of upper respiratory infections.

The **oropharynx** is located posterior to the oral cavity. The lower portion of the anterior wall is formed by the root of the tongue and by the epiglottic cartilage. Three mucous membrane folds, a median glossoepiglottic fold between the tongue and the epiglottic cartilage, and two lateral glossoepiglottic folds between the epiglottis and the junction of the tongue and pharynx, bound depressions, the **epiglottic valleculae.** Each lateral wall houses a mass of lymphoid tissue, the **palatine,** or **true tonsil,** located between the **palatine arches.** The latter are formed anteriorly by the palatoglossus muscle and posteriorly by the palatopharyngeus muscle. The lingual tonsil is a diffuse collection of lymphoid tissue at the root of the tongue.

TONSILLECTOMY

Tonsillectomy is a common operation, especially in children. Frequent bouts of tonsillitis originate in the many deep crypts that are embedded in the tonsillar mucosal surface. Because the palatine tonsils are very close to the common carotid artery, severe hemorrhage may follow a careless operation. The highly vascular tonsillar tissue may also bleed profusely after the operation.

The **laryngopharynx** lies behind the larynx, extending from the inlet of the larynx to the cricoid cartilage, where it becomes the esophagus. The posterior and lateral walls of the laryngopharynx have no characteristic features. Superiorly, the anterior wall presents the **inlet** of the larynx with the epiglottis anteriorly, the aryepiglottic folds laterally, and the **piriform recesses** to either side of the folds. These recesses lie between the aryepiglottic membrane medially, and the thyroid cartilage and thyrohyoid membrane laterally. Inferiorly, the muscles and mucous membrane on the posterior aspect of the arytenoid and cricoid cartilages form the anterior wall of the laryngopharynx.

Muscles of the Pharynx

Most of the wall of the pharynx is formed by the three paired superior, middle, and inferior constrictor muscles, which overlap or telescope into one another (Table 7-20, Fig. 7-29). All the constrictors insert posteriorly into the median raphe, with the **inferior constrictor** originating from the oblique line of the thyroid and arch of the cricoid cartilages; the **middle constrictor** from the greater cornu of the hyoid bone and the stylohyoid ligament, and the **superior constrictor** from the medial pterygoid plate and hamulus, pterygomandibular raphe, and mylohyoid line of the mandible. Fibers from the inferior constrictor ascend obliquely toward the medial raphe to overlap the middle constrictor. Fibers from the middle fan out to descend internal to the inferior, and ascend to overlap the superior constrictor. Fibers of the superior constrictor form a gap inferiorly through which the stylopharyngeus muscle passes, and are deficient superiorly at the pharyngeal recess. The latter is filled in by the levator and tensor veli palatini.

The thin conical **stylopharyngeus** passes from the tip of the styloid process anteroinferiorly to interdigitate and insert between the superior and middle constrictors. The **salpingopharyngeus** descends vertically from the pharyngotympanic tube to insert with the **palatopharyngeus,** which passes from the palatine aponeurosis into the wall of the pharynx internal to the constrictors. All the muscles are innervated by the vagus and glossopharyngeal nerves through the pharyngeal plexus, except the stylopharyngeus, which is supplied solely by the glossopharyngeal nerve (Table 7-21).

Table 7-20. Innervation to Mouth and Pharynx

Nerve	Origin	Course	Distribution
Hypoglossal (XII)	Brain stem	Exits cranial cavity via hypoglossal canal, deep to posterior belly of digastric picks up and conveys fibers of ansa cervicalis; in neck, lies on hyoglossus and passes deep to intermediate tendon of digastric and mylohyoid to reach floor of mouth	Supplies all muscles o tongue except palato-glossus
Lingual (V)	Posterior division of mandibular (V$_3$)	Passes on lateral surface of medial pterygoid to reach tongue; receives and conveys fibers of chorda tympani	General sensation to terior ⅔ of tongue; chorda tympani supp taste to anterior ⅔ of tongue, conveys pre-ganglionic fibers to su mandibular ganglion; postganglionics to su mandibular and subli gual glands
Lingual (IX)	Glossopharyngeal (IX) near styloid process	Passes anteriorly to tongue	General and taste ser tion to posterior ⅓ of tongue
Lingual (X)	Vagus (X) near thyroid cartilage	Passes directly to tongue	General and taste ser tion to root of tongu
Greater and lesser palatine	Maxillary (V$_2$)	From V$_2$ (pterygopalatine ganglion) pass through palatine canal to reach hard palate	Sensory to hard (grea and soft (lesser) pala convey postganglioni to glands of palate
Pharyngeal plexus	Glossopharyngeal (IX) and vagus (X)	Twiglets form plexus as the nerves traverse the neck	Vagus motor, glosso-pharyngeal–sensory; sopharyngeal also su plies motor innervati to stylopharyngeus

Nasal Cavity

Situated above the hard palate and divided by the nasal septum, the **nasal cavity** opens anteriorly at the external nares (nostrils) and posteriorly into the nasopharynx at the internal nares (choanae) (Fig. 7-30). The **external nares** are kept patent by the presence of U-shaped greater alar cartilages. The oblong **internal nares** are rigid, being bounded by bone.

The horizontal **floor** of the nasal cavity is formed by the superior surface of the hard palate, the palatine process of the maxilla, and the horizontal plate of the palatine

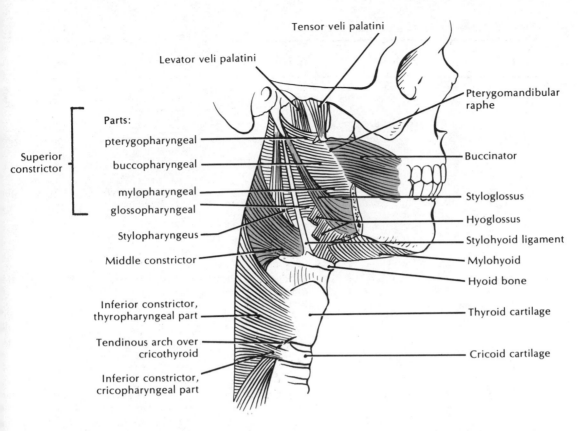

Fig. 7-29. Musculature of the pharynx. (Hollinshead WH: Textbook of Anatomy, 3rd ed. New York, Harper & Row, 1974)

bone. It is approximately 7 to 8 cm long and 1 to 2 cm wide. The long, very narrow **roof** is formed anteriorly by the upper nasal cartilage and nasal bones, and posteriorly by the cribriform plate of the ethmoid. The latter is pierced by twelve to twenty filaments of the olfactory nerve. The osseous portion of the **medial wall** of the nasal cavity, or nasal septum, is formed by the thin vertical (perpendicular) plate of the ethmoid superiorly and the vomer inferiorly, with the septal cartilage present anteriorly and between the bones noted above.

DEVIATED SEPTUM

A deviated septum is deflected laterally from the midline of the nose. The deviation usually occurs at the junction of the osseous with the cartilaginous portion of the septum. If the deformity is severe, it may entirely block the nasal passageway. Even though the blockage may not be complete, infection and inflammation develop, causing nasal congestion, occlusion of the paranasal orifices, and chronic sinusitis.

Table 7-21. Muscles of the Pharynx

Muscle	Origin	Insertion	Action	Nerve
Inferior constrictor	Side of cricoid and oblique line of thyroid cartilages	Median raphe of pharynx	Constricts pharynx in swallowing	Pharyngeal ple
Middle constrictor	Greater and lesser cornua of hyoid and stylohyoid ligament	Median raphe	Constricts pharynx in swallowing	Pharyngeal ple
Superior constrictor	Continuous line from medial pterygoid plate, pterygoid hamulus, pterygomandibular ligament, and side of tongue	Median raphe; superiormost fibers reach pharyngeal tubercle of skull	Constricts pharynx in swallowing	Pharyngeal ple
Stylopharyngeus	Styloid process	Superior and posterior borders of thyroid cartilage and musculature of pharynx	Raises pharynx	Pharyngeal ple
Palatopharyngeus	Soft palate	Posterior border of thyroid cartilage and musculature of pharynx	Elevates pharynx and helps to close nasopharynx	Pharyngeal ple
Salpingopharyngeus	Cartilaginous portion of pharyngotympanic tube	Musculature of pharynx	Opens pharyngotympanic tube during swallowing	Pharyngeal ple

The **lateral wall** of the nasal cavity presents bony projections, the **conchae,** which shelter a number of openings. The conchae (turbinate bones) are three curled bony plates projecting from the lateral wall into the nasal cavity that are covered by thick mucous membrane. The **superior concha,** a process of the ethmoid, is very short; the **middle concha,** also a process of the ethmoid, is larger; the **inferior concha,** longer than the middle, is an individual bone located midway between the middle concha and the floor of the nasal cavity. The **meatuses** are air-flow tracts lying deep to, or under cover of, their respective conchae. The area above the superior concha, into which the sphenoidal air sinuses open, is designated the **sphenoethmoidal recess.** The short, narrow **superior meatus** between the superior and middle conchae receives drainage from the posterior and middle ethmoidal air sinuses. Between the middle and inferior conchae, the rather extensive **middle meatus** receives the **infundibulum,** an anterosuperior funnel-shaped opening of the frontal air sinus. Located posterior to the infundibulum a deep, curved groove, the **hiatus semilunaris,** drains the anterior and middle ethmoidal and the maxillary air sinuses. The **bulla ethmoidalis** is a prominent bulging of the ethmoidal air sinuses forming the upper margin of the hiatus semilunaris. The **inferior meatus** is the horizontal passage deep to the inferior concha, into which the **nasolacrimal duct** opens.

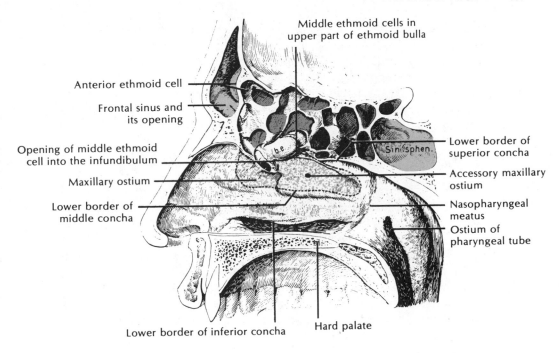

Middle ethmoid cells in
upper part of ethmoid bulla

Anterior ethmoid cell

Frontal sinus and
its opening

Opening of middle ethmoid
cell into the infundibulum

Maxillary ostium

Lower border of
middle concha

Lower border of inferior concha

Hard palate

Lower border of
superior concha

Accessory maxillary
ostium

Nasopharyngeal
meatus

Ostium of
pharyngeal tube

Sin. sphen.

Fig. 7-30. Lateral wall of the nasal cavity and sinuses. (Hollinshead WH: Textbook of Anatomy, 3rd ed. New York, Harper & Row, 1974)

NASAL POLYPS

Nasal polyps are protruding growths of the mucous membrane that usually hang down from the posterior wall of the nasal septum. In the rhinoscopic mirror, the polyps appear as bluish white tumors, which may fill the nasopharynx. If untreated nasal polyps usually undergo cystic degeneration but they are easily removed with a nasal snare and cautery.

Mucoperiosteum, consisting of mucous membrane closely adherent to periosteum, lines the nasal cavity, except for the area of the vestibule. The latter is lined with skin, while part of the roof, the superior concha, and adjacent septum are lined with olfactory epithelium. The mucoperiosteum is continuous through the nasolacrimal duct with the conjunctivum, through various apertures with the mucous membrane lining the several air sinuses, and through the choanae with the mucous membrane of the pharynx. It is thick and spongy due to the presence of rich sinusoidal venous plexuses and numerous cells. It functions to moisten and warm the incoming air.

Arteries and Nerves

All the vessels to the nasal cavity form a rich irregular anastomosis deep to the mucous membrane, with the **sphenopalatine artery,** a branch of the maxillary, as the principal supply. Branches of the sphenopalatine artery include the **posterior lateral nasal branch,** supplying the conchae and the ethmoidal, frontal, and maxillary air sinuses; a **posterior septal branch** to the upper part of the septum, and the **nasopalatine artery** that continues anteriorly on the septum to pass through the incisive foramen and anastomose with the greater palatine artery supplying the hard palate. The ophthalmic artery gives **anterior** and **posterior ethmoidal branches,** which supply the anterior portion of the superior and middle conchae, adjacent septal areas, and give twigs to the frontal and ethmoidal air sinuses.

NOSEBLEED

Nosebleed, or epistaxis, is frequent because of the exposure of the nose to trauma, the rich venous plexus deep to the conchal mucosa, and the general extensive vascularity of the nose. Bleeding, either arterial or venous, usually occurs on the anterior part of the septum and can be arrested by firm packing of the anterior nares. If the point of bleeding is in the posterior region, plugging of both the anterior and poterior nares may be necessary. In extreme emergency, ligation of the external carotid artery may be needed to control the hemorrhage.

Nerves to the nasal cavity include bipolar neurons of the olfactory nerve, nerves of general sensation from branches of both the ophthalmic and the maxillary divisions of the trigeminal nerve, and the autonomic nerves from the pterygopalatine ganglion.

The cells of origin of the **olfactory nerve** are limited to the small area of **olfactory epithelium** lining a portion of the roof and adjacent surfaces of the septum and superior nasal concha. Peripheral processes begin as osmoreceptors, and central processes pass through the **cribriform plate** to synapse in the **olfactory bulb,** which gives rise to the **olfactory tract** leading to the brain. Anterior and posterior ethmoidal nerves are branches of the nasociliary nerve from the ophthalmic division of the trigeminal. The **posterior ethmoidal nerve** passes through the posterior ethmoidal foramen to the posterior ethmoidal and sphenoidal air sinuses, while the **anterior ethmoidal nerve** passes through the anterior ethmoidal foramen to reenter the anterior cranial fossa. It then crosses the cribriform plate and enters the nasal cavity by way of the nasal slit (fissure) at the side of the crista galli. In the nasal cavity this nerve divides into an **external nasal branch** to pass down the nasal bone to supply the skin on the dorsum of the nose, and an **internal nasal branch,** which sends a medial branch to the superoanterior part of the nasal septum and a lateral branch to the anterior portion of the superior and middle conchae. From the maxillary division of the trigeminal nerve, **sensory twigs** pass to the **pterygopalatine ganglion** and distribute with its branches. These include the **lateral posterior superior nasal branch** to the posterior part of the superior and middle conchae; the **nasopalatine branch,** which crosses the roof of the nasal cavity to supply the septum and then follow the nasopalatine artery through the incisive foramen to supply the anterior portion of the hard palate. The **greater palatine nerve** passes through the palatine canal to give branches to the inferior concha and terminate in the mucous membrane of the hard and soft palates.

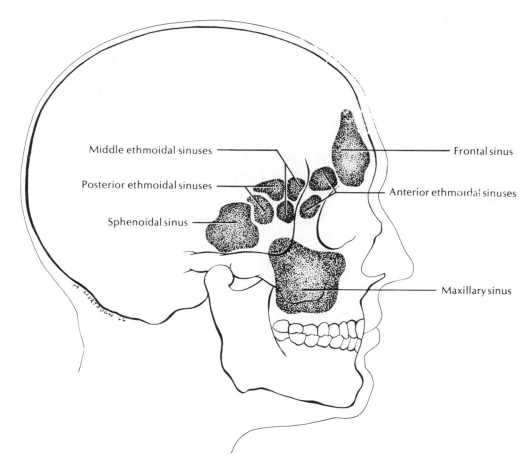

Middle ethmoidal sinuses ———

Posterior ethmoidal sinuses ———

Sphenoidal sinus ———

————— Frontal sinus

————— Anterior ethmoidal sinuses

————— Maxillary sinus

Fig. 7-31. Paranasal air sinuses.

Paranasal Air Sinuses

The bilateral **paranasal air sinuses** are located in bones adjacent to the nasal cavity (Fig. 7-31). The **sphenoidal air cavities,** occupying the body of the sphenoid, are rarely symmetrical. Their ostia usually open into the middle or superior part of the sphenoethmoidal recesses. The large **maxillary sinuses** are four-sided, hollow pyramids between the lateral walls of the nasal cavities and the infratemporal fossae. The slitlike ostium of each sinus opens into the posterior part of the hiatus semilunaris. The sockets of the upper molar or premolar teeth may project into these sinuses. The numerous **ethmoidal air cells** are usually limited to the lamina papyracea and the orbital portion of the ethmoid bone. The posterior air cells drain into the superior meatus, and the middle and the anterior into the hiatus semilunaris. **Frontal sinuses,** forming the brow ridges, are developmentally anterior ethmoidal air cells that migrate or extend into the frontal bone. Each drains inferiorly into a middle meatus by way of an infundibulum.

SINUSITIS

Sinusitis is a frequent complaint resulting from infection of the paranasal sinuses. This condition occurs most commonly in the frontal and maxillary sinuses. In health the sinuses contain air but when infected they collect fluid, which can be demonstrated radiologically. Since all paranasal sinuses drain into the nasal cavity, they are easily infected by nasal secretions. Maxillary sinusitis may be confused with toothache since only a thin layer of bone separates the roots of the teeth from the sinus cavity. Ethmoidal and sphenoidal sinusitis may infect the cranial meninges because only a thin bony plate intervenes between these sinuses and the subarachnoid space.

HEADACHES

Headaches have many causes. Sustained, tense, involuntary contraction of the epicranius muscles causes the common tension headache. A migraine or sick headache is periodic, usually affecting only one side of the head and is accompanied by nausea, vomiting, and visual disturbances. It is probably due to vascular agitation, such as vigorous vasoconstriction followed by vasodilation of the cerebral vessels.

Sinus headache involves severe pain over the sinuses, especially in the frontal and maxillary regions and is caused by pressure in the paranasal sinuses when the drainage is blocked by inflammatory or allergic edema. If the sinusitis is chronic, it is often relieved by surgically enlarging the openings of the sinus into the nasal cavity.

Pterygopalatine Fossa

The **pterygopalatine fossa** is an elongated, triangular area between the posterior aspect of the maxillary bone and the pterygoid processes of the sphenoid bone (Fig. 7-32). The **medial wall** opens into the nasal cavity through the sphenopalatine foramen. The **roof** is formed by the greater wings of the sphenoid. The **lateral wall** is relatively open as the pterygomaxillary fissure. Openings into the pterygopalatine fossa include the **sphenopalatine foramen** at the junction of the roof and the medial wall, for the passage of vessels and nerves to the nasal cavity; the **greater** and **lesser palatine canals** inferiorly, for the passage of the greater and lesser palatine nerves and arteries; posteriorly the **foramen rotundum,** for the maxillary division of the trigeminal nerve, and the **pterygoid** canal, for the passage of its nerve and artery. The fossa communicates with the orbital cavity by way of the **inferior orbital fissure** and with the infratemporal fossa by way of the **pterygomaxillary fissure.**

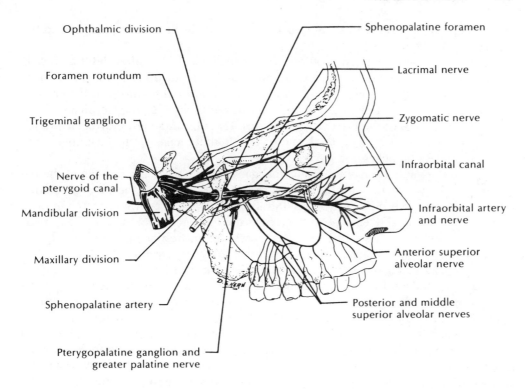

Ophthalmic division

Foramen rotundum

Trigeminal ganglion

Nerve of the pterygoid canal

Mandibular division

Maxillary division

Sphenopalatine artery

Pterygopalatine ganglion and greater palatine nerve

Sphenopalatine foramen

Lacrimal nerve

Zygomatic nerve

Infraorbital canal

Infraorbital artery and nerve

Anterior superior alveolar nerve

Posterior and middle superior alveolar nerves

Fig. 7-32. Pterygopalatine fossa.

Arteries and Nerves

Contents of the pterygopalatine fossa include the third portion of the **maxillary artery** and its companion **veins,** the **maxillary division of the trigeminal nerve,** and the **pterygopalatine ganglion.** The pterygopalatine, or third portion of the maxillary artery, lies in the fossa lateral to the pterygopalatine ganglion and gives branches to the nasal and orbital cavities, palate, upper teeth, and face. As the artery enters the fossa, its **posterior superior alveolar branch** descends on the maxillary tuberosity to enter the superior alveolar canal and supply the gingivae and upper molar and premolar teeth. The **greater palatine artery** passes through the palatine canal to emerge at the greater palatine foramen. It passes forward on the hard palate supplying glands and mucous membrane of the palate and gingivae, and anastomoses with the nasopalatine branch through the incisive foramen. Within the palatine canal the greater palatine artery gives rise to the **lesser palatine branch,** which emerges at the lesser palatine foramen to supply the soft palate and palatine tonsil. The **artery to the pterygoid canal** passes posteriorly to supply twigs to the pharynx, pharyngotympanic tube, and tympanic cavity. **Pharyngeal branches** of the maxillary are distributed to the upper pharynx and pharyngotympanic tube by way of the pharyngeal canal. The major blood supply of the nasal cavity is the **sphenopalatine artery,** which passes through the sphenopalatine foramen to give posterior lateral nasal branches to the conchae, meatuses, and sinuses.

It then descends on the nasal septum to anastomose with the greater palatine and the superior labial arteries as it passes through the incisive foramen.

The terminal direct continuation of the maxillary is the **infraorbital artery.** It enters the orbital cavity through the infraorbital fissure to traverse the infraorbital groove and canal and emerges on the face at the infraorbital foramen. In the canal it gives branches to the inferior oblique and inferior rectus muscles, the lacrimal sac, and mucous membranes of the maxillary air sinuses. Its **anterior superior alveolar branch** supplies the incisor and canine teeth and its terminal branches on the face ramify and anastomose with branches of the facial artery.

From the midportion of the trigeminal ganglion the maxillary division of the trigeminal nerve passes through the foramen rotundum, traverses the pterygopalatine fossa and enters the orbit by way of the infraorbital fissure. It continues in the infraorbital groove and canal to emerge on the face at the infraorbital foramen as the infraorbital nerve. The branches of the maxillary nerve are entirely sensory and supply the skin and the mucous membrane of the lower eyelid and upper lip.

Within the cranial cavity the **maxillary division** of the fifth cranial nerve gives branches that follow the middle meningeal artery to supply the meninges. The zygomatic, sphenopalatine, and posterior superior alveolar branches arise from the pterygopalatine portion of the maxillary nerve. The **zygomatic branch** enters the orbital cavity through the infraorbital fissure and sends a **zygomaticotemporal branch** through the lateral wall of the orbit to the temporal fossa, a **communicating branch** to the lacrimal nerve that carries postganglionic parasympathetic fibers from the pterygopalatine ganglion to the lacrimal gland, and a **zygomaticofacial branch** passing through the inferolateral angle of the orbit to supply the skin over the prominence of the cheek. The **maxillary nerve** sends two short sensory roots to the pterygopalatine ganglion to be distributed with branches from the ganglion; an orbital branch to the periosteum of the orbit and the posterior ethmoidal sinuses; a **greater palatine branch,** that passes through the palatine canal and out the greater palatine foramen to supply the hard and soft palates, the middle and inferior meatuses, and the inferior concha, and a **lesser palatine branch,** which follows the same course as the greater palatine but emerges through the lesser palatine foramen to supply the soft palate, uvula, and palatine tonsil. Additional branches of the sphenopalatine nerve include the **posterior superior nasal branch** that traverses the sphenopalatine foramen and distributes to the superior and middle conchae and posterior ethmoidal sinus; direct branches to the superior and middle conchae, posterior ethmoidal sinus, and posterior part of the septum; a **pharyngeal branch** passes through the pharyngeal canal to the nasopharynx, and the **nasopalatine (long sphenopalatine) branch** supplies the roof of the nasal cavity and the nasal septum. The latter nerve courses downward on the septum to pass through the incisive canal and communicate with the greater palatine branch. The **posterior superior alveolar nerve** passes through the pterygomaxillary fissure to reach minute foramina on the tuberosity of the maxilla and enters the posterior alveolar canals to supply upper molar teeth, gingivae, and mucous membrane of the cheek.

Within the infraorbital canal, the infraorbital nerve gives off a **middle superior alveolar branch** that supplies the maxillary sinus and the premolar teeth, an **anterior superior alveolar branch** to the maxillary sinus and the canine and incisor teeth, and a **nasal branch** that innervates the anterior part of the inferior meatus and floor of the nasal cavity (Table 7-22).

Table 7-22. Distribution of Olfactory (I), Maxillary (V₂) Nerves and Nerve of Pterygoid Canal

Nerve	Origin	Course	Distribution
Olfactory (I)	Bipolar neuronal cell bodies in olfactory mucosa; peripheral processes have osmoreceptors; central processes form filaments of olfactory nerve	Filaments pass through cribriform plate; processes synapse in olfactory bulb	Olfactory mucosa on superior concha and adjacent lateral wall of nasal cavity for sense of smell
Maxillary (V₂)	Trigeminal ganglion	Exits cranial cavity via foramen rotundum, traverses pterygopalatine fossa, and leaves through infraorbital fissure to reach floor of orbital cavity	Sensation to midportion of face; postganglionic autonomic fibers conveyed by most of its branches

Within pterygopalatine fossa the pterygopalatine ganglion is suspended from V₂; most distributing branches are branches of the maxillary nerve or pterygopalatine ganglion. They convey both sensory and postganglionic autonomic fibers (see nerve of pterygoid canal).

Nerve	Origin	Course	Distribution
Nerve of pterygoid canal	Formed by junction of major petrosal transmitting preganglionic parasympathetic fibers (from facial nerve), and deep petrosal from carotid plexus transmitting postganglionic sympathetic fibers (cell bodies in superior cervical ganglion)	Traverses pterygoid canal to terminate in pterygopalatine ganglion	Parasympathetics synapse in ganglion; distributing branches convey both sympathetic and parasympathetic postganglionic processes and sensory fibers to midportion of face
Nasopalatine	Pterygopalatine ganglion	Nasal septum	Sensory and autonomics to nasal mucosa
Sphenopalatine (Lateral nasal branches)	Pterygopalatine ganglion	Lateral wall of nasal cavity	Sensory and autonomics to nasal mucosa
Greater and lesser palatines	Pterygopalatine ganglion	Palatine canal	Mucosa of hard and soft palate
Posterior superior alveolar	Maxillary (V₂)	Through pterygomaxillary fissure to reach small foramina on tuberosity of maxilla	Upper molar teeth
Zygomatic	Maxillary (V₂)	Arises from maxillary in infraorbital groove, courses along lateral aspect of orbit to reach foramina in zygoma	Terminal zygomaticofacial and zygomaticotemporal branches supply skin of face; communicating branch to lacrimal nerve conveys postganglionic parasympathetics to lacrimal gland
Infraorbital	Terminal branch of maxillary (V₂)	Traverses infraorbital groove, canal, and foramen	Skin of cheek, lower eyelid, side of nose, upper lip; in infraorbital canal, gives middle and anterior superior alveolar branches to upper teeth

The terminal (facial) portion of the maxillary nerve emerges at the infraorbital foramen as the **infraorbital nerve.** It divides into the **inferior palpebral branch** to the lower lid, the **lateral nasal branch** to the skin at the side of the nose, and the **superior labial branch** to the skin, mucous membrane, and glands of the upper lip.

Ear

External Ear

For descriptive purposes the ear is divided into external, middle, and internal portions. The **external ear** comprises the pinna (auricula) and the external auditory meatus. (Fig. 7-33). The **pinna** collects sound waves and directs them into the external auditory canal where they strike the tympanic membrane. Parts of the pinna include the **concha,** the well of the ear leading into the external auditory meatus; the **helix,** the outer rim of the external ear beginning at the concha and ending at the **lobule;** the **antihelix,** rimming the concha opposite the helix; the **tragus,** the small lip overlapping the concha; the **fossa triangularis,** the triangular depression above the concha, and the **scapha,** which forms a depressed groove in front of the helix.

EAR WAX

Ear wax deposits may block the external auditory meatus, causing temporary deafness. Forceful efforts to remove the wax can result in impaction and infection. The wax can usually be dislodged by vigorous irrigation of the canal with warm water.

The lateral third of the **external auditory meatus** is cartilaginous; the remainder is formed by the tympanic part of the temporal bone.

Fig. 7-33. External ear.

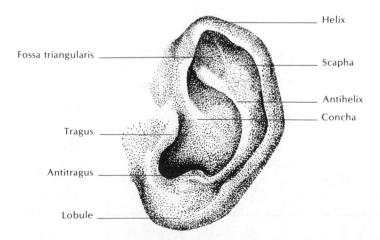

INFECTIONS OF THE EXTERNAL AUDITORY CANAL

Infections of the external auditory canal often result from infected hair follicles. These small abscesses or boils are extremely painful because of the tightness of the skin lining the canal. Such severe pain may cause nausea and vomiting because the nervous impulses of pain reflexly stimulate the vagus nerve, which supplies parasympathetic innervation to the upper gastrointestinal tract.

The **tympanic membrane** slopes obliquely inferomedially with the lateral surface slightly concave; the maximal point of the concavity is designated as the **umbo.** The handle of the malleus can be seen through the membrane extending inferiorly to the umbo. A flaccid, less tense portion of the tympanic membrane, the **pars flaccida,** lies above the lateral process of the malleus. The whole of the peripheral margin, except for the flaccid part, is lodged in the **tympanic groove.**

Middle Ear

The **middle ear,** or **tympanic cavity,** is filled with air and communicates anteromedially with the pharynx through the **pharyngotympanic (auditory) tube** and posterosuperiorly with the tympanic antrum and mastoid air cells through the aditus (Fig. 7-34). The tympanic cavity contains the three auditory ossicles, stapedius and tensor tympani muscles, tympanic plexus of nerves, and chorda tympani nerve. The tympanic cavity has a roof, floor, and four walls. The **roof** of the cavity, the **tegmen tympani,** is a thin plate of bone separating the cavity from the middle cranial fossa. The **floor** of the cavity, or **jugulum,** is a thin plate of bone separating the cavity from the jugular fossa and the bulb of the jugular vein. The **posterior (mastoid) wall** contains an opening, the **aditus,** leading from the **epitympanic recess** into the **tympanic antrum.** In the lower part of the posterior wall a small conical projection, the **pyramid,** lodges the **stapedius muscle,** and lateral to the pyramid an opening admits the chorda tympani nerve. Owing to the convergence of the medial and lateral walls, the **anterior (carotid) wall** is narrow, and an opening in the upper portion leads to the canal that houses the tensor tympani muscle. An opening in the midportion leads into the pharyngotympanic tube. Between the canal for the tensor tympani and the pharyngotympanic tube, a septum is prolonged posteriorly on the medial wall as the **processus cochleariformis,** which affords a pulley around which the tensor tympani tendon turns laterally to its insertion. In the lower portion of the anterior wall a thin lamina of bone separates the cavity from the carotid canal. The **medial (labyrinthine) wall** forms the boundary between the middle and internal parts of the ear, where anteriorly the rounded **promontory** is formed by the underlying basal turn of the cochlea. Above the posterior part of the promontory is a depression in which the **fenestra vestibuli (oval window)** is closed by the foot plate of the stapes. Above the oval window, an anteroposterior ridge demarcates the position of the canal containing the facial nerve. Immediately above the ridge of the facial canal a horizontal projection indicates the site of the lateral semicircular canal. Below and behind the promontory is a fossa that contains the **fenestra cochleae (round window),** an opening into the scala tympani, which is closed by the **secondary tympanic membrane.**

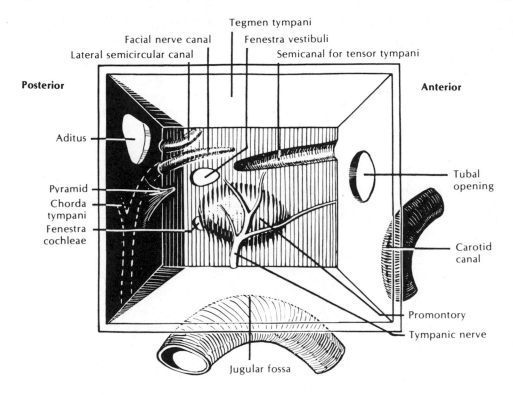

Tegmen tympani
Facial nerve canal
Lateral semicircular canal
Fenestra vestibuli
Semicanal for tensor tympani

Posterior

Anterior

Aditus

Tubal opening

Pyramid
Chorda tympani
Fenestra cochleae

Carotid canal

Promontory

Tympanic nerve

Jugular fossa

Fig. 7-34. Middle ear cavity. (Gardner E, Gray DJ, O'Rahilly R: Anatomy: Philadelphia, Saunders, 1975)

The auditory ossicles, the malleus (hammer), incus (anvil), and stapes (stirrup), extend in a chain from the lateral to the medial wall of the middle ear cavity. The **malleus** is described as having a head, neck, handle, and anterior and lateral processes. The **head** of the malleus articulates with the body of the incus. The **handle, or manubrium,** of the malleus is attached along its length to the tympanic membrane. It extends to the umbo and receives into its medial surface the insertion of the tendon of the tensor tympani muscle. Just above this insertion the chorda tympani nerve passes between the manubrium and the long crus of the incus. The **incus** consists of a body, and short (horizontal) and long (vertical) crura. The **lenticular process** is a small knob on the end of the long crus, which articulates with the stapes. The **short crus** is attached to the posterior aspect of the epitympanic recess by way of the ligament of the incus and the **body** receives the articulating head of the malleus. The **stapes** presents a head, neck, foot plate, and anterior and posterior limbs. The incus articulates at the concave socket on the head of the stapes, and the **foot plate** is attached by the **annular ligament** to the margin of the fenestra vestibuli. The stapedius muscle inserts into the posterior surface of the **neck** of the stapes.

Located immediately behind the epitympanic recess, the **tympanic antrum** communicates through the **aditus** with the mastoid air cells. The **tegmen tympani** separates the epitympanic recess and the tympanic cavity from the middle cranial fossa. Immediately above and behind the external auditory meatus, the lateral wall of the recess is formed by the squamosal portion of the temporal bone.

OTITIS MEDIA

Otitis media is an acute infection of the middle ear cavity with a reddening and outward bulging of the eardrum, which may rupture. The infection may spread to the mastoid air cells, causing mastoiditis, a very serious complication prior to the advent of antibiotics. Acute otitis media is usually caused by the spread of infection from the nasopharynx along the pharyngotympanic (eustachian) tube, and into the middle ear.

Internal Ear

The **internal ear** is located within the petrous portion of the temporal bone and consists of the osseous labyrinth, which contains the membranous labyrinth. The **osseous labyrinth** is composed of bony cavities: the vestibule, three semicircular canals, and the cochlea, all filled with perilymph, in which the membranous labyrinth is suspended. The **vestibule** is centrally located. The superior, posterior, and lateral semicircular canals open into it posteriorly by way of five openings, with the adjoining ends of the superior and posterior canals forming a common terminal canal. Laterally, the **fenestra vestibuli** is closed by the foot plate of the stapes, while the medial wall of the vestibule presents depressions with small openings for the emergence of filaments of the eighth cranial nerve.

OTOSCLEROSIS

Otosclerosis is a pathologic process that deposits new bone around the oval window, which may immobilize the stapes. The resulting deafness may be cured by making a new opening into the internal ear (fenestration). The ankylosed stapes is then freed and attached to the new membranous window. Many attachment techniques are available, but they all have the same objective—to reestablish the functional movement of otic ossicles.

The **semicircular canals** are each approximately 0.8 mm in diameter, and each presents a dilatation, the ampulla, at one end. The vertical **superior canal** is at right angles to the similarly vertical **posterior canal,** which parallels the posterior surface of the petrous portion of the temporal bone. The horizontal **lateral canal** is in the angle between the superior and posterior canals and bulges into the medial wall of the tympanic cavity.

Located anterior to the vestibule, the bony **cochlea** resembles a snail's shell. It is a tapering tube that spirals about 2½ turns around a central core, the **modiolus.** The osseous tube opens into the tympanic cavity through the **fenestra cochlea** (round window) and is closed in the fresh state by the secondary tympanic membrane. The basal, or first, turn around the modiolus bulges as the promontory on the medial wall of the tympanic cavity. The modiolus is thick at the base and tapers rapidly to the apex, with a thin narrow shelf of bone, the **spiral lamina,** turning around the modiolus like the threads of a screw.

Two membranous sacs, the utricle and the saccule, the three semicircular ducts, and the cochlear duct compose the **membranous labyrinth,** which lies within the bony labyrinth but does not completely fill it. The membranous labyrinth contains the fluid endolymph, while the fluid perilymph occupies the space between the membranous and osseous labyrinths. Lying in the posterosuperior part of the vestibule, the **utricle** receives the openings of the semicircular ducts. The smaller **saccule** lies in the anteroinferior part of the vestibule. A short canal, the **ductus reuniens,** extends from the lower part of the saccule to the cochlear duct. The blind **endolymphatic duct** leaves the posterior part of the saccule, is joined by the short **utriculosaccular duct** from the utricle, then traverses the aqueduct of the vestibule. It ends under the dura mater on the posterior surface of the petrous portion of the temporal bone as a dilatation, the **endolymphatic sac.** The **semicircular ducts** and their terminal **ampullae** are attached to the convex sides of the semicircular canals and open into the utricle.

LESIONS OF THE VESTIBULAR NERVE

Lesions of the vestibular portion of the vestibulocochlear nerve causes vertigo (dizziness) and nystagmus. The latter is an involuntary rapid movement of the eyeball, and is recognized by a rhythmic oscillation, a slow movement in one direction followed by a rapid jerk back, which is repeated in rapid succession in both eyes.

Within the bony canal of the cochlea the **membranous labyrinth** consists of a closed spiral tube, the **cochlear duct,** which is separated from the internally located **scala vestibuli** by the **vestibular membrane,** and from the externally placed **scala tympani** by the **basilar membrane.** The basilar membrane supports the **spiral organ (of Corti),** which contains the peripheral nerve endings associated with sound reception. The scala vestibuli and scala tympani are continuous with each other at the apex of the cochlea through a small opening, the **helicotrema** (Table 7-23).

NERVE DEAFNESS

Nerve deafness may result from a lesion of the cochlear division of the eighth nerve, its ganglion, the organ of Corti, or the cochlear duct. Tone deafness is a dysfunction of the hair cells in the organ of Corti, so that musical sounds cannot be perceived. Conduction deafness is a defect in the sound conducting apparatus, such as the external auditory meatus, eardrum, or otic ossicles. Word deafness is a lesion of the auditory center of the brain (superior temporal gyrus), where the sounds are heard but convey no meaning to the individual.

Table 7-23. Nerves in Ear

Nerve	Origin	Course	Distribution
Vestibulocochlear	Cochlear division—bipolar neuronal cell bodies in spiral ganglion of modiolus; peripheral processes pass to organ of corti central processes and join vestibulocochlear nerve	Traverses internal auditory meatus with facial nerve to reach brain stem	Hearing
	Vestibular division—bipolar neuronal cell bodies in vestibular ganglia; peripheral processes in equilibratory receptors; central processes join vestibulocochlear nerve	Traverses internal auditory meatus with facial nerve to reach brain stem	Equilibrium
Nerve to stapedius	Facial (VII) in vertical portion of facial canal	Passes directly to stapedius	Stapedius
Chorda tympani	Facial (VII) in vertical portion of facial canal	From iter chorda posterioris crosses tympanic cavity to pass through iter chorda anterioris to reach infratemporal fossa and join lingual nerve	Taste to anterior ⅔ tongue; preganglionic fibers to submandibular ganglion; postganglionics to submandibular and sublingual glands
Tympanic plexus	From tympanic branch of glossopharyngeal (IX) and twigs of facial (VII); gives rise to minor petrosal	Tympanic branch arises in jugular foramen, passes through tympanic canaliculus, crosses floor of middle ear cavity to reach promontory; minor petrosal traverses petrosphenoidal fissure to reach otic ganglion in infratemporal fossa	Sensory to middle ear; postganglionics from otic ganglion to parotid gland

Larynx

The **larynx** is specially modified for vocalizations. Situated anteriorly in the neck below the hyoid bone and the tongue, it has a marked anterior projection, the laryngeal prominence (Adam's apple). The larynx is related anteriorly to skin and fascia, and laterally to the thin strap muscles of the neck, the thyroid gland, the great vessels of the neck, and the vagus nerves. Posteriorly it is separated from the vertebral column and prevertebral muscles by the laryngopharynx.

The **skeleton of the larynx** is formed by three single cartilages: the thyroid, cricoid, and epiglottic; and three paired cartilages, the arytenoids, corniculates, and cuneiforms (Fig. 7-35). The thin, leaf-shaped, **epiglottic cartilage** forms the anterior boundary of the inlet (aditus) and the vestibule of the larynx. The superior end of the

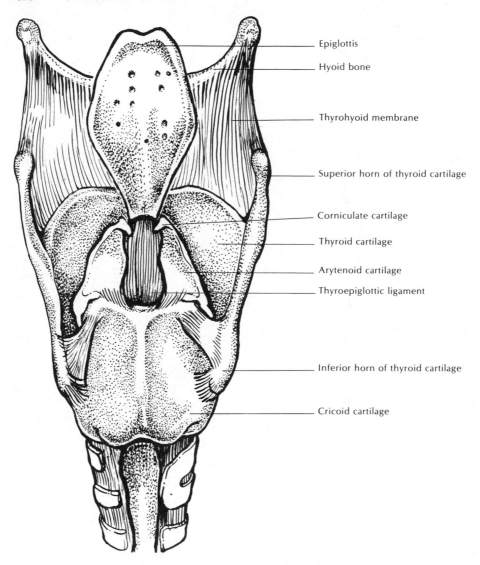

Epiglottis

Hyoid bone

Thyrohyoid membrane

Superior horn of thyroid cartilage

Corniculate cartilage

Thyroid cartilage

Arytenoid cartilage

Thyroepiglottic ligament

Inferior horn of thyroid cartilage

Cricoid cartilage

Fig. 7-35. Posterior aspect of larynx. (After Romanes GJ, ed: Cunningham's Textbook of Anatomy, 11th ed. New York, Oxford University Press, 1972)

cartilage is broad and free, with the lateral margins enclosed in the aryepiglottic folds, while the lower end is pointed and connected to the thyroid cartilage by the thyroepiglottic ligament. The large **thyroid cartilage** has two broad quadrilateral **laminae,** which are fused anteriorly, open posteriorly, and separated anterosuperiorly by the V-shaped **thyroid notch.** The junction of the notch and the fused laminae forms the **laryngeal prominence.** The posterior border of the thyroid cartilage is thick and rounded, and is prolonged upward as the **superior cornua** and downward as the **inferior cornua.** The superior border gives attachment to the **thyrohyoid membrane,** which is pierced by the internal laryngeal nerve and the superior laryngeal vessels. The thyrohyoid membrane, between the greater horn and the hyoid bone, is free posteriorly and en-

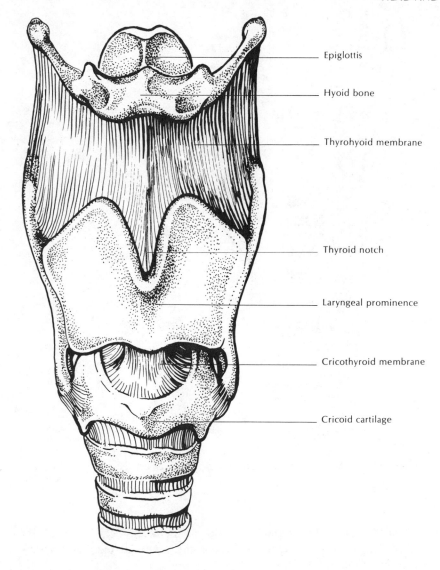

Epiglottis

Hyoid bone

Thyrohyoid membrane

Thyroid notch

Laryngeal prominence

Cricothyroid membrane

Cricoid cartilage

Fig. 7-36. Anterior view of larynx. (After Romanes GJ, ed: Cunningham's Textbook of Anatomy, 11th ed. New York, Oxford University Press, 1972)

closes the small cartilago triticea. The relatively flat lateral surface of the thyroid cartilage gives attachment to the sternothyroideus, thyrohyoideus, and inferior pharyngeal constrictor muscles. The medial (internal) surface is smooth and gives attachment to the thyroepiglottic, vestibular, and vocal ligaments, and to the thyroarytenoideus and vocalis muscles. The short, thick, inferior horn articulates with the cricoid cartilage (Figs. 7-36 and 7-37).

The signet ring-shaped **cricoid cartilage** presents posteriorly a broad, quadrilateral **lamina** with two convex facets on the upper border that articulate with the base of the arytenoid cartilages. The anterior arch gives attachment to the cricothyroid mus-

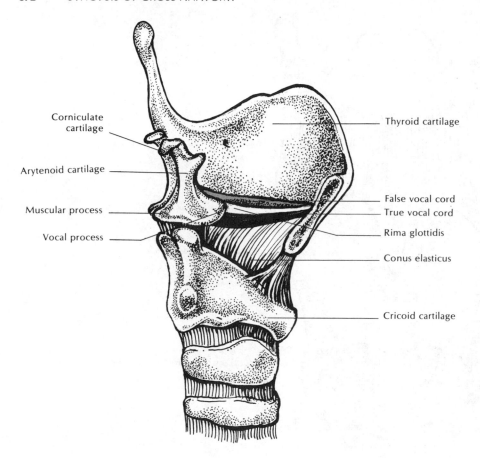

Fig. 7-37. Right half of thyroid cartilage removal to expose vocal cords. (After Romanes GJ, ed: Cunningham's Textbook of Anatomy, 11th ed. New York, Oxford University Press, 1972)

cles and the cricovocal ligament. The superior border of the cricovocal ligament attaches to the vocal process of the arytenoid cartilage and presents a free edge, which attaches to the vocal ligament.

The two **arytenoid cartilages** are three-sided pyramidal structures with their bases articulating on the upper border of the cricoid lamina and their apices curving posteromedially. The posterolateral angle of the base presents the thick, projecting **muscular process** that gives attachment to laryngeal muscles, and a spinelike **anterior (vocal) process** for attachment of the vocal ligament (Fig. 7-37). The **corniculate cartilages** are small conical bodies at the apex of the arytenoid cartilages located in the posterior edge of the aryepiglottic folds. The small, rod-shaped **cuneiform cartilages** are embedded in the aryepiglottic folds above the level of the corniculate cartilages.

Interior of the Larynx

The interior of the larynx is smaller than might be expected. It is subdivided into three portions by the vocal (true vocal cords) and vestibular (false vocal cords) folds, which extend anteroposteriorly and project inwardly from the sides of the cavity. The upper subdivision, the **vestibule,** extends from the aditus to the vestibular folds. It diminishes in width from superior to inferior, and its anterior wall is longer than the posterior. The anterior wall is formed, in part, by the epiglottic cartilage, the thyroid lamina, and the thyroepiglottic ligament. The lateral walls are formed by the **aryepiglottic folds,** which cover the aryepiglottic muscle, and the posterior wall is the interarytenoid membrane.

The middle portion, the **ventricle,** is the smallest of the three regions and is bounded by the vestibular folds above and the vocal folds below. The soft, flaccid **vestibular folds** stretch anteroposteriorly across the side of the cavity, and the interval between them, the **rima vestibuli,** is wider than the interval between the **vocal folds.** The latter are sharp, prominent bands that enclose the **vocal ligament.** They are prismatic in cross section and appear pearly white in the fresh state. The elongated interval between the vocal folds, the **rima glottidis,** forms the narrowest part of the laryngeal cavity, and the shape of the opening varies with respiration and vocalization. The vocal folds and the interval between them form the glottis (Fig. 7-37).

The remainder of the laryngeal cavity, the **infraglottic portion,** extends from the rima glottidis to the trachea. Superiorly it is narrow and compressed from side to side, then it gradually widens to become circular as it becomes the trachea.

FUNCTIONS OF THE LARYNX

The larynx has three functions. The inlet acts as a sphincter in swallowing to prevent food or drink from entering the trachea. The rima glottidis acts also as a sphincter to close the larynx during coughing, sneezing, or whenever an increased intrathoracic pressure is needed, for example, micturition, defecation, or parturition. The third function is vocalization. In the latter, short bursts of expired air cause vibration of the closely approximated vocal cords that produce the sound in vocalization. Direct visualization of the larynx, including the vocal cords, is possible through a laryngoscope.

Muscles

The suprahyoid and infrahyoid musculature of the neck aids in phonation (Table 7-24). The former muscles elevate the larynx in the production of high notes, while the latter depress the larynx in the formation of low notes. The intrinsic muscles control the airway through the larynx. The cricothyroideus, lateral cricoarytenoideus, transverse arytenoideus, and thyroarytenoideus **adduct,** or close, the vocal folds, while the important posterior cricoarytenoideus, "the safety muscle of the larynx," **abducts** the vocal folds and opens the glottis. In phonation the thyroarytenoideus tenses the vocal folds, while the vocalis muscle acts to relax them (Fig. 7-38).

The **cricothyroideus** bridges the lateral portion of the interval between the cricoid and thyroid cartilages. The **posterior cricoarytenoideus** passes from the lamina of the cricoid cartilge to the muscular process of the arytenoid cartilage. The only unpaired muscle in the larynx, the **transverse arytenoideus,** extends from the posterior aspect of one arytenoid cartilage to the other. The **oblique arytenoidei** lie on the poste-

Table 7-24. Muscles of the Larynx

Muscle	Origin	Insertion	Action	Nerve
Cricothyroideus	Arch of cricoid	Inferior horn and lower border of thyroid cartilage	Chief tensor of vocal ligament	External larynge
Posterior cricoary-tenoideus	Posterior surface of lamina of cricoid cartilage	Muscular process of arytenoid carti-lage	Abductor of vocal fold	Inferior larynge
Transverse artyten-oideus (only un-paired muscle of larynx)	Passes from posterior aspect of one ary-tenoid cartilage to other		Closes rima glot-tidis	Inferior larynge
Oblique arytenoi-deus	Muscular process of arytenoid carti-lage	Some fibers into apex of arytenoid, most prolonged as aryepiglottic mus-cle	Closes rima glot-tidis	Inferior larynge
Lateral cricoaryten-oideus	Upper border of cricoid arch	Muscular process of arytenoid carti-lage	Adducts vocal folds	Inferior larynge
Thyroarytenoideus	Inner surface of la-mina of thyroid	Anterolateral sur-face of arytenoid cartilage	Slackens vocal folds and closes rima glottidis	Inferior larynge
Thyroepiglottis	Anteromedial sur-face of lamina of thyroid cartilage	Lateral margin of epiglottic cartilage	Aids in closure of laryngeal inlet	Inferior larynge
Vocalis	Anteromedial sur-face of lamina of thyroid cartilage	Vocal process	Adjusts tension of vocal ligament	Inferior larynge

rior aspect of the arytenoid cartilage, where they cross, like the limbs of an X, superficial to the transverse arytenoideus. Some of the fibers of the oblique arytenoideus insert into the apex of the arytenoid cartilage, but most of the fibers are prolonged anteriorly as the **aryepiglottic muscle,** which inserts into the margin of the epiglottis. The **lateral cricoarytenoideus,** applied to the upper border and side of the cricoid arch, passes posterosuperiorly to insert into the muscular process of the arytenoid cartilage. The **thyroarytenoideus muscle** extends as a sheet between the thyroid and arytenoid cartilages, where the uppermost fibers continue superiorly as the **thyroepiglottic muscle,** and the deepest fibers stretch from the thyroid cartilage to the lateral side of the vocal process as the **vocalis muscle.** All the intrinsic muscles of the larynx are supplied by the inferior (recurrent) laryngeal branch of the vagus nerve, except the cricothyroideus, which is innervated by the external laryngeal branch of the superior laryngeal nerve of the vagus.

Arteries and Nerves

The innervation of the larynx is from the **vagus nerve.** The **superior laryngeal branch,** originating high in the cervical region, passes inferiorly to divide opposite the

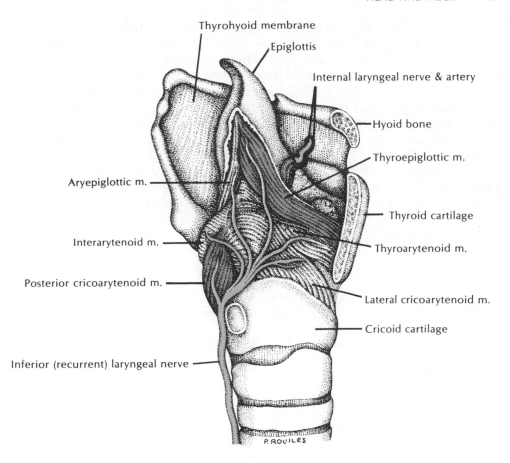

Thyrohyoid membrane

Epiglottis

Internal laryngeal nerve & artery

Hyoid bone

Thyroepiglottic m.

Aryepiglottic m.

Thyroid cartilage

Interarytenoid m.

Thyroarytenoid m.

Posterior cricoarytenoid m.

Lateral cricoarytenoid m.

Cricoid cartilage

Inferior (recurrent) laryngeal nerve

P. ROUILES

Fig. 7-38. Muscles and nerve supply of the larynx.

Table 7-25. Innervation of Larynx

Nerve	Origin	Course	Distribution
Internal laryngeal	Superior laryngeal branch of vagus (X)	Pierces thyrohyoid membrane to reach interior of larynx	Sensory to larynx above level of vocal folds
External laryngeal	Superior laryngeal branch of vagus (X)	Follows course of superior thyroid artery	Cricothyroid
Inferior (recurrent) laryngeal	Vagus (X)	Right hooks around subclavian artery, left hooks around arch of aorta; both ascend in tracheoesophageal groove to reach larynx	Sensory to larynx below level of vocal fold; motor to all intrinsic muscles of larynx except cricothyroid

hyoid bone into **internal** and **external laryngeal nerves.** The former pierces the thyro-hyoid membrane to supply sensation to the mucous membrane above the vocal folds, while the **external laryngeal nerve** passes along the external surface of the larynx to supply the cricothyroideus muscle. The **right recurrent (inferior) laryngeal branch** of the vagus nerve loops around the right subclavian artery, and the **left recurrent (inferior) branch** loops around the arch of the aorta. Both nerves then ascend in the tracheoesophageal groove to supply all the intrinsic muscles of the larynx except the cricothyroideus, and to supply sensation to the interior of the larynx below the vocal folds (Table 7-25).

The **superior laryngeal branch of the superior thyroid artery** accompanies the internal laryngeal nerve. It passes through the thyrohoid membrane to ramify and supply the internal surface of the larynx. The **inferior laryngeal branch of the inferior thyroid artery** accompanies the recurrent laryngeal nerve in the tracheoesophageal groove to supply the larynx from its inferior aspect.

Index

Page numbers followed by the letter f represent figures; the letter t indicates tabular material.

hemiazygos veins, 129
hemorrhage
    of cerebral arteries, 325
    extradural, 313
    subarachnoid, 314
    subdural, 313
hepatic artery, 164
hepatic ducts, 163
hepatic flexure, 159
hepatic portal system, 28
hepatic veins, 165
hepatoduodenal ligament, 155, 161
hepatopancreatic ampulla, 155, 163
hernia
    esophageal, 146
    femoral, 223
    inguinal, 142
herniated disc, 234
herpes zoster, 38–39
Hesselbach's triangle, 142
hiatus of diaphragm, 146
hiatus semilunaris, 356
hiccup, 273
hilus of lung, 108
hinge joints, 12
hip joint, 248–250, 250f
    dislocation of, 250
Hirschsprung's disease, 159
horizontal fissure of lungs, 109
Horner's syndrome, 271
humeral circumflex arteries, 53, 70
humerus, fractures of, 74
hyaline cartilage, 5–6, 12
hydrocele, 184
hydrocephalus, 322–323
hydronephrosis, 170–171
hymen, 188
hyoglossus muscle, 349
hyoid bone, 260
hyperopia, 343
hypertension, renal, 172–173
hypogastric artery, 199
hypogastric autonomic plexus, 199
hypoglossal canal, 311
hypoglossal nerve, 312, 349
    injury to, 350
hypophysis, 316, 319
hypothalamus, 318–319
hypothenar compartment, 92
hypothenar eminence, 60, 82

ileocecal orifice, 157
ileocecal valve, 156–157
ileocolic artery, 161
ileum, 156
ilia, 46
iliac arteries, 161, 199, 201
iliac circumflex artery, 142, 219, 226

iliac circumflex vein, 219
iliac crest, 136, 212
iliac spine, 136
    anterior superior, 139, 212
iliococcygeus muscle, 192
iliocostalis cervicis muscle, 283t
iliocostalis lumborum muscle, 283t
iliocostalis thoracis muscle, 283t
iliofemoral ligament, 249
iliohypogastric nerve, 143, 215
ilioinguinal nerve, 143
iliolumbar artery, 201
iliopectinal line, 139
iliopsoas muscle, 225
iliotibial tract, 213
incus, 366
infections
    of ear, 365, 367
    of pharyngotympanic tube, 352
    of scalp, 304, 306
inferior alveolar nerve, 294–295, 304
inferior colliculi, 319
inferior extremity, 212–257
    arteries of, 247f
    compartments of, 214f
    fasciae of, 212–213
    gluteal region of, 220–221, 222f, 223
    joints of, 248–257. *See also specific joints*
    leg, 234, 236–242. *See also* leg
    nerves of, 215, 216f, 217f, 218f, 219
    surface anatomy of, 212
    thigh, 221, 223–234. *See also* thigh
    veins of, 217f, 218f, 219
inferior longitudinal muscle, 349
inferior meatus, 356
inferior mediastinum, 126
inferior oblique muscle, 338
inferior ramus of ansa cervicalis, 273
infraanal fascia, 194
infrahyoid muscles, 265
infrahyoid portion of external investing layer of
    fascia, 261
infraorbital artery, 362
infraorbital canal, 335
infraorbital foramen, 335
infraorbital groove, 335
infraorbital nerve, 294, 362, 364
infrapatellar fat pad, 251
infraspinatus muscle, 51–52
infratemporal fossa of face, 301–304
infratrochlear nerve, 340
infundibulopelvic ligaments, 205–206
infundibulum, 123
inguinal arteries, 142
inguinal canal, 140, 142
inguinal ligaments, 139–140, 212
inguinal region, 139–142, 140f
inguinal rings, 140
inguinal triangle, 142
inion, 306
injections, in gluteal region, 196, 221

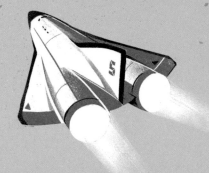
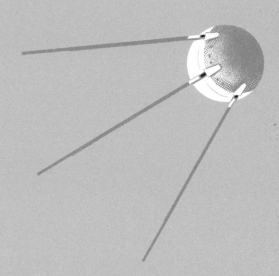

SPACE
EXPLORERS

WRITTEN BY
LIBBY JACKSON

ILLUSTRATED BY
LÉONARD DUPOND

ALADDIN
New York London Toronto Sydney New Delhi

BEYOND WORDS
Hillsboro, Oregon

To Lillie, Jacob, Rufus, and Florence
—Libby Jackson

R. Buckminster Fuller said, "We are all astronauts on a little spaceship called Earth" and I believe that it's time we take care of our small and fragile craft.
—Léonard Dupond

ALADDIN
An imprint of Simon & Schuster
Children's Publishing Division
1230 Avenue of the Americas
New York, NY 10020

BEYOND WORDS
8427 N.E. Cornell Road, Suite 500
Hillsboro, Oregon 97124-9808
503-531-8700 / 503-531-8773 fax
www.beyondword.com

This Beyond Words/Aladdin edition August 2021
Text copyright © 2020 by Libby Jackson
Illustrations copyright © 2020 by Léonard Dupond

Originally published in Great Britain in 2020 by Wren & Rook
Hardcover ISBN: 978-1-5827-07648
Paperback ISBN: 978-1-5263-62124

For information about special discounts for bulk purchases, please contact
Simon & Schuster Special Sales at 1-866-506-1949 or business@simonandschuster.com.

The Simon & Schuster Speakers Bureau can bring authors to your live event.
For more information or to book an event contact the Simon & Schuster Speakers Bureau
at 1-866-248-3049 or visit our website at www.simonspeakers.com.

Art Director: Laura Hambleton
Designer: Pete Clayman
The text of this book was set in Intro Book

Manufactured in China 0521 SCP

10 9 8 7 6 5 4 3 2 1

Library of Congress Control Number: 2020935833

THESE STORIES ARE NOT CREATIONS OF IMAGINATION,
NO MATTER HOW UNBELIEVABLE THEY MAY AT FIRST SEEM.

THEY ARE ALL TRUE, TOLD JUST
AS THEY REALLY HAPPENED ...

CONTENTS

WE HAVE LIFT-OFF!

1957–1958

Beep … Beep … Beep …

On October 4, 1957, people all around the world tuned in their radios and heard a mysterious beeping noise. But they weren't listening to a radio station broadcasting from Earth, they were listening to a spacecraft, called Sputnik, broadcasting from space! Sputnik, which means "fellow traveler" in Russian, was the first human-made object to circle around the Earth in orbit. It was put there by the Soviet Union, kicking off the new era of space travel.

This mysterious object was put into space by an even more mysterious man, known only as the Chief Designer. His name—Sergei Korolev—and identity were kept a secret from the world for his entire life, but he was the great mastermind behind the Soviet Union's early space program.

The Soviet Union was a huge country that used to span across eastern Europe and northern Asia, before it was broken up into several smaller countries. It was in conflict with America, and these two big countries were constantly trying to prove that they were better than each other at absolutely everything. When Sputnik went into orbit, and flew around the world in just 96 minutes, there was cheering and celebrations in the Soviet Union, but the rest of the world was shocked. If the Soviets could launch a satellite broadcasting a radio signal, then what would be next? Spy cameras? Nuclear bombs? Lots of people in the US were really worried.

But there was one man living in America who was more upset than most: a German immigrant called Wernher von Braun. He was very cross that the Chief Designer had beaten him and his American team into space. While Sergei and Wernher never met, they had lots in common. They were both clever scientists who devoted their lives to designing rockets and dreamed of sending them to the Moon, or even Mars. But during the Second World War, they were on opposite sides, and designing rockets that were used for much darker reasons—to launch bombs deep into enemy territory.

As the end of the war approached, Wernher was still living in Germany when he received instructions from the German Army to destroy all the plans and drawings of the rockets he had designed. This was meant to stop the information from falling into enemy hands. But Wernher wasn't about to wipe out his life's work, so he disobeyed the orders. He then hatched a daring plan, to surrender to the Americans—Germany's enemy during the Second World War—and take as much of his work with him as he could. The US Army welcomed him with open arms and gladly put him to work making new rockets for them. He was once again able to dream of the Moon and Mars.

But after Wernher had left for America, some Soviet engineers raided the German sites where he used to work. They discovered that he had left some things behind. They rounded up the useful information and sent it all back to their Chief Designer, who gladly added the know-how they gleaned to his own designs for rockets.

After the Chief Designer's spectacular success with Sputnik, Sputnik 2 was launched just one month later, but this time a dog named Laika was inside! This maddened the American government, who were desperate to show the world that they could send rockets and satellites into space too. So, the US Navy was ordered to launch the rocket and satellite that they had been developing, called Vanguard, and they invited television crews and newspapers to watch.

The cameras were focused on the launch pad as the clock counted down:

5 ... 4 ... 3 ... 2 ... 1 ... LIFT-OFF!

The Vanguard rocket engines roared into life and the rocket was on its way! Up it went ... a whole yard off the ground, when suddenly the rocket stopped and started going backwards! Everyone watched in horror as the rocket slumped back to the ground and turned into a giant

EVERYONE WATCHED IN HORROR AS THE ROCKET SLUMPED BACK TO THE GROUND AND TURNED INTO A GIANT FIREBALL

fireball. America's first attempt to get into space was a disaster, and "Flopnik" was soon on the front pages of newspapers the world over.

But Wernher had a plan. He had a different rocket ready and waiting, and had access to another satellite as well. There was just one problem—the rocket wasn't powerful enough to send the satellite into orbit. There was no time to redesign anything, but if the rocket could be filled with a more powerful fuel, then it would work. This had never been done before though, and it might be impossible. If there was one person who could crunch the numbers and find the solution, saving the Americans from more embarrassment, it was Mary Sherman Morgan.

Mary was a brilliant chemist and engineer. She had grown up on a farm in North Dakota, but had to run away from home to go to university. Sadly, she never got to finish her degree as the Second World War broke out and she stopped studying to help the war effort. But, even without a degree, she soon became an expert in the chemistry of explosives. She may have been a woman working as an engineer, which was very unusual for the time, but Mary was the perfect person for the job.

Mary wasn't put off by the challenge; she trusted that her brain would find the answer. She needed to make a more powerful fuel that would work inside Wernher's rocket. She got to work and tried all sorts of different ideas, but one by one, she and her team ruled out each option. Then, after months of searching, she came up with the solution—a fuel cocktail! None of the existing fuels were powerful enough for the rocket, but Mary found that if she mixed two fuels together, in just the right way, then bingo, she could make the fuel Wernher needed. After months of calculations, Mary had done it!

Mary's new fuel was named Hydyne and on January 31, 1958, it was loaded into Wernher von Braun's Juno I rocket. Everyone held their breath as the US made their second attempt at joining the space race.

The clock counted down to launch:

5 ... 4 ... 3 ... 2 ... 1 ... LIFT-OFF!

Mary's fuel ignited and the rocket headed upwards! This time there was no fireball and the rocket soared into the sky, taking the Explorer 1 satellite all the way to space, putting it safely into orbit, circling around Earth. The satellite even had scientific instruments on board that sent information about space back to scientists on the ground. The Americans had done it! The mission was a success and Wernher's brilliant achievements were celebrated far and wide. But back in the Soviet Union, it was the Chief Designer's turn to be upset. When he saw the images of Wernher smiling and proudly holding his model of the Explorer 1 high, he was jealous. As the secret Chief Designer for the Soviet Union, he would never be allowed to have his photograph taken and celebrated like that. His rocket may have made it into space first, but it was Wernher who was being recognized for his great achievements. The Chief Designer wasn't ready to quit yet, though. He had plans for bigger and better things. The Americans might have at long last joined their rivals in space, but Sergei would do all he could to make sure they wouldn't catch up with the Soviets. The space race was on!

THE FIRST SPACE TRAVELER
APRIL 12, 1961

Yuri Gagarin sat and thought about what was about to happen as the engineers strapped him into the tiny Vostok spacecraft. He knew the inside of this small, shiny, spherical capsule back to front—all the dials and switches, the windows by his feet—he had memorized it all. In the months leading up to this flight, he had spent many long hours in a mock-up of the spacecraft at the training facilities in Moscow, learning everything he needed to know about how it worked. He had passed all his exams and even sat inside this, the real thing, a few days earlier. But this time it was different.

Yuri was a Soviet space traveler known as a cosmonaut. If everything went according to plan, in just a couple of hours' time, the five rocket engines beneath him would fire up and blast him into the sky. Ten minutes later, the last engines would stop and Yuri would be floating in orbit, circling around the Earth at about 17,500 miles an hour—the first human being ever to travel into space!

It was going to be a dangerous journey, one that no person had ever tried before. And no one knew for certain how the human body would react to floating in space. Some people worried that Yuri's lungs might explode, or that his eyes would bulge out of the sockets in his head. Dogs, rabbits, monkeys, and even fruit flies had successfully made the journey to space and back. But some of the animals had become confused when they found themselves hovering in the spacecraft, unable to feel the pull of gravity. The same might happen to Yuri!

Just in case, all of the controls inside the spacecraft would be operated by engineers back at Mission Control on Earth. Yuri had a code, in case of emergencies, that he could use to override Mission Control, but if there was a problem and he wasn't able to activate the code, his life was in the hands of those back home. Every engineer and scientist who had built the Vostok had taken extra care when tightening the bolts and readying the rockets, knowing that the life of a fellow human was at stake.

Yuri sat in his seat inside his bright orange spacesuit and waited. It had been an early morning start and it would be a couple of hours before the launch so he listened to music to pass the time. Yuri was calm, but the people in Mission Control were pacing nervously, checking their plans.

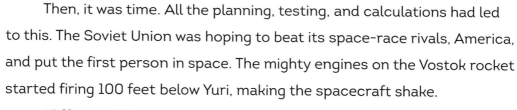

Then, it was time. All the planning, testing, and calculations had led to this. The Soviet Union was hoping to beat its space-race rivals, America, and put the first person in space. The mighty engines on the Vostok rocket started firing 100 feet below Yuri, making the spacecraft shake.

"Off we go!" Yuri shouted excitedly over the radio channels.

"Goodbye my friends, see you soon!" he continued.

"See you soon!" replied Mission Control.

The rocket lifted off the ground, slowly at first, but quickly picking up speed. Yuri could feel himself being pressed into his seat as the force of the engines built. After a couple of minutes, the protective cover that had been in front of Yuri's window dropped away and he got his first breathtaking view of Earth. "I can see clouds, the landing site ... It's beautiful!" he yelled over the radio, but he was being drowned out by the huge noise of the metal tower that was taking him into space.

As the rocket engines took Yuri higher and higher, the Earth got farther and farther away. He described to Mission Control the clouds, mountains, and forests he could see getting smaller and smaller. Then, just as he thought the view couldn't get any more beautiful, Yuri saw the horizon out of the window and the incredible curve of the Earth, bright against the darkness of space.

JUST AS HE THOUGHT THE VIEW COULDN'T GET ANY MORE BEAUTIFUL, YURI SAW THE HORIZON

After the deafening roar of the blast off, suddenly the rocket engines shut down and everything fell silent. Yuri felt his stomach lift and the forces from the engines that were pressing him into his seat vanished. He was floating, held in place only by his seatbelts. He had

14

made it! He was in space! Over 200 miles above the Earth's surface where he had been just minutes earlier. As it sped around the world, the Vostok spacecraft was rotating slowly, which meant Yuri saw the blue Earth drift by his window, and then the vast blackness of space, speckled with stars. Yuri was mesmerized. He was the first person ever to see the world like this and he couldn't take his eyes off the view. He watched in wonder as the Sun got closer to the horizon, and then he became the first person to sit in space and watch the Sun set. It was lighting up the world with a beautiful halo that then transformed into a vivid rainbow of color around the edge of Earth.

But it wasn't only the view that was fascinating Yuri. The feeling of weightlessness, of floating in space, was incredible. He reported back to Mission Control that far from his eyes popping out, or brain exploding, he felt brilliant. "My health is excellent, excellent, excellent!" he exclaimed. He then tried eating and drinking from some food tubes, and was pleased to find that there were no problems at all. It was only when Yuri took out his notepad and pencil that he found that floating could become an issue. After he had taken some notes, he placed the notebook and pencil in front of him, hovering in the air, ready for his next observation. A little while later, when he went back to use them again, Yuri found that the pencil had drifted off, out of reach!

After just over an hour, Yuri and his Vostok spacecraft had nearly made a complete journey around the world and it was time to head home. "It's too soon!" thought Yuri sadly. The silence that had surrounded him ended and the rocket engines fired once more, slowing down the spacecraft and sending it falling back to Earth. As soon as the engines stopped, the spacecraft started to tumble, end over end, spinning everything around, out of control. Yuri kept calm and watched his instrument panel. The spacecraft should have separated into two halves by now, leaving Yuri in the only part that was able to return to Earth. But with the two parts still joined together, it couldn't fly like it should.

The spacecraft continued spiraling, but Yuri stayed calm. He knew that the Soviet Union was very, very big and even if his spacecraft didn't land where it should, he was sure that he would still land in his own country. So instead of worrying Mission Control, he decided to tell them everything was going well!

The spacecraft plunged back into the Earth's atmosphere, still twisting and turning over and over. After ten long and scary minutes, the straps holding the two parts of the spacecraft together finally gave way and the Vostok capsule slowed down. The air became thicker, and the spacecraft slowed down even more. The outside of the capsule got very hot as it slammed into the air particles. Yuri could see a purple light out of his window, hear crackling sounds, and feel the spacecraft scorching with fiery heat. The deceleration forces pressed down on Yuri, making him feel ten times heavier than usual. His vision became blurry and gray as the blood drained out of his head, and Yuri had to fight to clear his vision by clenching every muscle in his body to force the blood back into his head.

Four miles above the surface, the hatch of Yuri's capsule blasted off, and seconds later Yuri followed in his ejector seat, flung out into the air. This was the only way down because the Vostok capsule couldn't land softly or safely back on the ground. His parachute opened and as he looked around he saw a river and bridge he recognized from training. He was landing in Saratov!

Yuri floated back down to Earth, landing in a field, less than two hours after he had blasted off—having traveled all the way around the world!

Yuri was safe and had achieved the incredible, but there was no time to celebrate just yet. Everyone waiting in Mission Control had no idea how Yuri was doing and waited desperately for news. Did the parachute work? Was he alive? Where was he? Yuri had to find a telephone and let them know the mission was a success. He climbed up the nearest hill and saw a farmer named Anna, with a young girl. They had seen Yuri fall from the sky. Seeing a strange figure dressed in orange, they wondered what was happening. As Yuri approached them, waving, he asked to use a telephone, but they were scared. The little girl turned and started running away. "Don't be afraid, I am a friend!" called Yuri. Relieved that the strange man could speak their language and seemed friendly, Anna let Yuri use their phone. When they heard the news that Yuri was alive and well, everyone in Mission Control whooped and cheered as champagne corks popped! They had done it! A human had been to space, circled the Earth, and made it home safely. This was another triumph for the Soviet Union, and put them a leap further ahead of their rivals, America, in the great space race.

THE HUMAN COMPUTER

Separate schools

Separate parks

Separate theaters

Separate swimming pools

Separate seats on the bus

All because of the color of your skin. It might be hard to imagine but this was daily life for black Americans up until the 1960s. America, and many other places around the world, used to divide people based on the color of their skin, with white people always taking priority over anyone else. It was a racist, prejudiced, and tension-filled place to live. But this is the country that a boundary-breaking black girl was born into in 1918, and she refused to let her intellect be held back by the ignorant world around her.

Katherine Johnson grew up on a farm in West Virginia. It was clear from a young age that her mind was special. She was an extremely curious child, who used to follow her brother to school before she was old enough to attend. When she finally started school, Katherine jumped ahead a class because she was so clever. But education was not always available to everyone, especially black Americans. Despite this, Katherine's father was determined that all his children would go to high school and college.

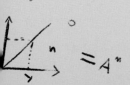

So when there were no schools near their house for African Americans, he moved his family almost 100 miles away and worked an extra job to make sure his children could get a full education.

Katherine's thirst for knowledge continued throughout school, and after jumping ahead several more classes, she finished high school at just 14 years old. Katherine then went to university to study French, but knowing how gifted she was, the math professor at the university threatened that if Katherine didn't turn up for math lessons, they would come and find her and put her in their math class. Katherine ended up taking every math class the university had to offer, and graduated from university with the highest degree possible in math and French. One of her math professors suggested she find a job as a research mathematician, and when Katherine admitted she didn't know what this was, he told her to go and find out.

But the rest of America wasn't ready for an inquisitive black woman with Katherine's brain. Jobs for women of all races were very limited at the time. Women were expected to stay at home and raise children, while men went to work. It was even harder for black women. The few jobs available for those with an education included being a teacher or nurse. So Katherine started teaching at a primary school when she left university, passing her curiosity on to children. But she never gave up on her ambitions, and when a friend let her know that a company was making a small step forwards in equality, and hiring black women who were good at math, Katherine moved her family and even turned down another job to make her dream come true.

It wasn't just opportunities for women and racial prejudice that made the 1950s different from the world we know today. The technology that we use in our everyday lives—the Internet, mobile phones, and computers—did not exist. They were all waiting in the wings of the future, ready to be invented. Televisions were starting to appear in people's homes, but not everyone had one.

THE INTERNET, MOBILE PHONES, AND COMPUTERS DID NOT EXIST

This meant that the teams of people who designed satellites, airplanes, and rockets in the 1950s had no electronic devices to help them. If engineers and scientists wanted to know how fast a rocket could go or work out complicated sums, they had to make these calculations by hand, without an electronic computer. Every mistake had to be rubbed out and corrected, and copies painstakingly made. Each complicated mathematical formula would have to be carefully worked out one step at a time, and sometimes the calculations would take days or even weeks to complete.

So when Katherine joined the National Advisory Committee for Aeronautics (NACA) in June 1953, she was hired to be a computer, a human computer, and was finally going to find out what a research mathematician did. Human computers calculated and solved the complex sums that scientists and engineers gave them. They used mechanical calculators that were about the size of an old desktop computer, but rather than being full of electronics, they used gears and cogs that would add numbers together, a bit like an abacus. Everything was written down at large desks with ink and pencil. While it was pretty much only white men who were hired to be the scientists and engineers, it was women who were hired to be human computers. They were thought to have more patience and attention to detail than men, and so be better at the calculations. Despite their intelligence,

WOMEN WERE MASTERS OF THE NUMBERS

these women were considered lowly, earning a lot less than men. But the women were masters of the numbers, using their mathematical knowledge to work out all the problems that the engineers gave them. The work could be tedious and long-winded, but not for Katherine. She loved math, where there is always a right answer, you just have to work it out. In spite of the segregation, and the racial and gender discrimination around her (which meant black women had to use different bathrooms, offices, and cafeterias from white women), Katherine soon showed that race and gender have nothing to do with the brilliance of a person's mind.

During the 1950s, America became obsessed with space. The US and Soviet Union were in a hugely competitive race to claim the prize of putting the first person in space. NACA became NASA—the National Aeronautics and Space Administration—and everyone, including Katherine, focused their attention on putting humans in space. Getting to space was basically a huge math problem, and Katherine wanted to understand the reasons behind the math problems she was being asked to calculate. So with her typical confidence, she asked to go along to the space program briefings with the male engineers and scientists. When she was told women didn't attend these meetings, Katherine challenged this answer, asking "Is there a law against it?" There wasn't, so Katherine attended the meetings.

These knowing minds got to work, trying to figure out how a rocket could fly a person into space. The problem of sending anything into space is a bit like when a soccer player is taking a free kick. Aiming the ball at the goal, the soccer player wants to get the kick just right, to put it in the top

John Glenn, the astronaut who would be inside of the Friendship 7 capsule on the Mercury 6 mission, wanted to make sure the calculations were correct, as his life relied on them. He asked for a second opinion and he knew who he wanted to double-check the numbers—Katherine. "If she says they're good, then I'm ready to go," declared Glenn. It took Katherine a day and a half, carefully working through each minute of the mission and calculating the capsule's orbit. While some people might get overwhelmed by having the fate of someone's life rely on their calculations, Katherine kept a cool head. Every single number she calculated matched the computer's answers, and with everyone's minds at rest that they could trust these new-fangled computers, John Glenn's mission was cleared to proceed.

On February 20, 1962, John Glenn was hurled into space in the Friendship 7 spacecraft. After three orbits of the Earth, he splashed down in the Atlantic Ocean. The calculations made by the computer and confirmed by Katherine were correct. Finally, nearly a year after their rivals in the Soviet Union, the Americans had completed their first crewed orbit of the Earth.

Katherine's work on the Mercury missions weren't her only successes at NASA. Her proudest moment came when she helped calculate the flight path of the Apollo 11 mission that saw humans land on the Moon. She continued to work at NASA for over three decades, retiring in 1986. Her brilliant mind and pioneering role as a black woman working in mathematics was recognized in 2015, when she received the highest honor given to American citizens—the Presidential Medal of Freedom.

A NICE DAY FOR A SHORT SPACEWALK

MARCH 18, 1965

As soon as their Voskhod spacecraft reached space, Soviet cosmonauts Alexei Leonov and Pavel Belyayev started to work through their list of jobs. They checked that Alexei's spacesuit, his air supply, and the cameras were all working. Everything needed to be perfect—there was no room for mistakes because Alexei was getting ready to do something that no one had ever tried before, something very dangerous ...

ALEXEI WAS GOING TO BE THE FIRST PERSON EVER TO WALK IN SPACE

Alexei was going to be the first person ever to walk in space. He would be alone out there, attached to the spacecraft by just a safety tether. And because there is no air in space, the only thing keeping Alexei alive would be the air in his backpack and his precious spacesuit.

Spacewalking was the next big step in the space race between the two space rivals, America and the Soviet Union. The Soviets had put the first satellite and the first person, Yuri Gagarin, in space. Now the competition was on for the first steps. Spacewalking would be absolutely necessary on future missions, as it would allow astronauts and cosmonauts to access the outside of a spacecraft so they could conduct experiments or fix broken parts. Getting this right would also mean that people could walk on the Moon one day, which was the ultimate goal. If everything went well, and Alexei took those first space steps, then the Soviet Union would score another victory against their archrivals ...

Once Alexei and Pavel were happy that everything was working, Alexei headed into a long, cylindrical tube attached to the side of the Voskhod 2 capsule. Pavel closed the hatch behind him. This was the airlock: an inflatable contraption that would create a room between the spacecraft, where Pavel would stay with air around him, and space, where there was no air at all. With the hatch between them safely and firmly closed, Pavel opened up the vents in the airlock and the air around Alexei rushed out into the vacuum of space. The door to space swung open and Alexei caught a glimpse of Earth. It was deep blue and beautiful next to the star-speckled darkness around it.

28

Slowly, carefully, Alexei poked his head out, squinting at first because the Sun's rays were blinding him. As he blinked, everything became clear and the magnificent view gripped him. It was like nothing he, or anyone, had ever seen before. The whole world was beneath him, over 60 miles straight down, moving nearly 28,000 miles an hour.

Gradually, Alexei slid out of the airlock and into the waiting expanse. He felt like a bird with wings outstretched. He was flying in space, soaring above Earth. The feeling was extraordinary and he was the first person to ever do this! Everything was silent except for his heart pounding in his chest. And the views were awesome—he was surrounded by the stars on one side and the beauty of the Earth on the other.

As Alexei kicked the side of the hatch to move himself away from the capsule, the force of his push started him tumbling uncontrollably. The only thing to stop him from floating off forever was a single piece of rope anchoring him to the capsule. Pavel was watching Alexei on the television screen from inside of the spacecraft and panicked when Alexei disappeared. "Where are you? Are you OK?" he called out over the radio. Alexei kept spinning out into the darkness of space, into the great unknown. And then he felt a huge tug and he was suddenly yanked back towards the spacecraft by the safety rope. Pavel saw Alexi come in to view again and warned him "Be careful!" Alexei smiled. He knew that he needed to stay calm and not be afraid—any decision out here really was a matter of life and death.

Alexei tried to take in every bit of the amazing view, but after only ten minutes, Pavel radioed that it was time to get back inside of the airlock. Alexei felt like he was a young child again, and had the same sense of crushing disappointment that came when his mother would call for him to come in from playing. He desperately wanted to stay out longer, but knew that he must do as he was told.

Alexei was to get back by pulling himself along the safety rope to the airlock, but when he tried to do this, he realized that something had gone horribly wrong. He seemed to have shrunk! His hands and feet couldn't reach the fingertips or toes of his spacesuit anymore, so he couldn't grip the rope. Alexei forced himself to stay calm. He knew that he couldn't have actually shrunk because that was impossible. But what was going on?

Alexei realised that if he hadn't changed shape, then his spacesuit must have. The difference in pressure between the air inside his suit and the airless vacuum of space had caused his suit to balloon in size. And now his gloves were far too big! Alexei tried again to move his fingers to grip the rope, but it was impossible. It felt like he had giant mittens on. If he couldn't hold the rope, then how was he going to get back inside the spacecraft?

IF HE COULDN'T HOLD THE ROPE, THEN HOW WAS HE GOING TO GET BACK INSIDE THE SPACECRAFT?

Alexei didn't lose his nerve though. He was a cosmonaut, and cosmonauts are trained to keep cool under pressure. Instead he thought about how he could solve the problem. Alexei realised that the only way to shrink his suit back to the right size was to let some air out of it. This was very dangerous though. If he let too much out he could suffocate and die! But he had no choice—there was no other way to haul himself back to safety, and he'd soon run out of air. Without telling Mission Control, or even

Pavel, Alexei daringly opened the pressure valve on his suit and let out the air little by little until he could start to feel his gloves tighten around his hands again.

Alexei let out a huge sigh of relief—it had worked! He quickly heaved himself back along the rope to the airlock and safety. He was supposed to enter it feet first, so he could close the outer hatch behind him, but he was so tired and hot that he went in head first. Alexei was drenched in sweat and it was sticking to his skin and getting in his eyes. He could barely see but he had to close the hatch. With one final burst of energy, he managed to do a somersault, turn around and close the door to space. Phew! Pavel let the air back into the airlock and Alexei climbed back inside the spacecraft and took off his visor. Finally, he was safe again, or so he thought ...

Alexei and Pavel settled down in their seats and prepared for the rest of their mission, which still had almost a day left. Alexei grabbed his notebook and colored pencils and made some sketches of the amazing sights he'd just seen. He couldn't wait to share his adventure with his wife Svetlana and his daughter Vika when he got home.

But with just five minutes to go before they were due to start their journey back to Earth, Alexei and Pavel discovered that one of the computers wasn't working properly. To get back to Earth, a spacecraft has to slow down by firing its rocket engine for just the right amount of time. It is also important that this happens at just the right moment, so that the spacecraft lands in the right place. The spacecraft was meant to do this automatically, but the problems with the computer meant that the cosmonauts would have to control the rocket engines themselves. Alexei and Pavel were going to have to take on the role of the spaceship's computer! This included pointing the spacecraft in the right direction, firing the rocket engines at exactly the right time and for the right length of time, and even choosing the landing site. It was vital that they got everything correct. If they made any mistakes their return journey could go horribly wrong. They could land in the wrong country or even in the sea. If the spacecraft fell back to Earth too quickly, the cosmonauts might pass out from too much force on them. Or the

THEY HAD ONLY ONE SHOT TO GET THIS RIGHT

spacecraft could even break up completely on its journey home! No one had ever tried to fire the rocket engines manually before, but this is exactly what Alexei and Pavel were going to have to do. To make things even worse, the spacecraft was low on fuel and they had only one shot to get this right.

The cosmonauts spent the next orbit planning their new way home. Alexei chose a landing site that would be far from any towns or cities, and Pavel pointed the spacecraft in the right direction. They tried to tell Mission Control what they were planning, but they didn't get a reply. Mission Control hadn't heard and had no idea what the cosmonauts were going to do. At the moment Alexei calculated, Pavel fired the rocket engine to slow down the Voskhod 2 capsule and they started heading back home. As they re-entered Earth's atmosphere, they felt ten times heavier than usual, and the blood vessels in their eyes started to burst. Suddenly, with a large jolt, the parachute of the Voskhod 2 capsule opened and everything became calm as they swayed gently, drifting down to Earth.

Even though they were out of immediate danger, they weren't out of trouble yet. Back at Mission Control no one knew where the crew had landed or even if they were alive. The only thing they knew for certain was that the spacecraft was far from where it should have landed. Alexei and Pavel also had no idea where they were, but they did know that the recovery forces were a long way off. Bracing themselves, they opened the hatch of their capsule into the unknown once more. There was a huge bang as the explosive bolts detonated to release the hatch. But when they opened their eyes, they were surprised to see that it hadn't moved. Looking out of the window they saw that it was stuck in place by a huge tree. They realised they were going to have to move the capsule otherwise they'd be trapped forever! So, with all their might, the cosmonauts rocked their capsule backwards and forwards, wiggling it free. Eventually they moved it enough and the hatch gave way, letting fresh air rush into the cabin. Alexei and Pavel took a deep breath, and their lungs filled with ice-cold air.

They had landed in the middle of a dense forest, covered in deep snow. The capsule's parachute was caught on a tree and fluttering in the wind, high above them. Alexei and Pavel looked at each other and realized that they might be here for some time.

Alexei hurriedly started sending out signals on their radio transmitter, tapping the messages out in Morse code, letting everyone know they were OK. As the two men huddled in the cold, waiting to be rescued, Alexei made sure he had his pistol close at hand—there would be bears and wolves in these woods and two cosmonauts would make a tasty dinner for them.

There was nothing they could do but wait. The hours passed and the

light through the thick clouds got dimmer, but no one came. The cosmonauts were all alone and were going to have to sleep out in the cold for the night. They huddled in their Voskhod 2 capsule and waited. This was not the return to Earth they had imagined.

The next morning a low-flying aircraft woke them from their fitful slumber and soon they saw people heading towards them on skis—they were finally going to be rescued! The rescuers brought warm clothes and an axe so they could start chopping down trees to make a fire and a clearing in the thick trees. It would be another night before Alexei and Pavel could ski to a waiting helicopter. Several days later than planned, they headed home to see their families, where they told them about their amazing and unexpected adventures, and the world's first spacewalk. The Soviets had beaten their rivals, America, with another first in the great space race.

A VERY IMPORTANT PACKED LUNCH

Every living thing on Earth needs food and water to say alive. Astronauts, even if they may seem superhuman living and working up in space, are no different. But unlike most Earthlings, they can't pop out to the supermarket. And, while you might be used to the odd pea flying off your plate, imagine if your whole meal did! So how do astronauts get their food? What can you serve for dinner when everything, including peas, floats away? Can you even swallow food in space?

These questions have challenged scientists since the beginning of the space age. Some even worried that it might not be possible to eat food in space, and that astronauts might choke if they tried. So when Yuri Gagarin became the first person to orbit Earth in 1961, he was sent to space with a very important packed lunch. He was very happy to report that there were no problems swallowing or digesting the first-ever space meal—a tube of puréed meat followed by chocolate sauce!

Yuri's menu may sound delicious, but early space food was anything but yummy. The people who made it were far more worried about stopping the food from floating away than it tasting good. Their answer was to put mushed up food in tubes, just like tubes of toothpaste, so that astronauts and cosmonauts could squeeze it straight into their mouths. They also made weird bite-sized cubes of dried food squished together and covered in jelly. Whether you ate it from tubes or in cubes, everyone agreed on one thing: space food was bland and disgusting.

It was so bad that in the spring of 1965, one astronaut decided to take his own lunch to space. The Gemini III mission would be the first time that the US launched two people into space at the same time and the crew, John Young and Gus Grissom, would also be trying out a new menu. But John thought he would surprise Gus with something tastier than the horrible space food they would be eating. As they were making the final preparations to get into the capsule, John secretly slipped a corned-beef sandwich into his pocket. Once their mission was underway, he fished it out and handed it to Gus. But John's idea for a secret

JOHN SECRETLY SLIPPED A CORNED-BEEF SANDWICH INTO HIS POCKET

space feast didn't go to plan. When he took the sandwich out of his pocket, it had broken up in to crumbs, which started floating around the cabin. The crumbs threatened to jam the instrument panels and even get in the astronauts' eyes, which would be bad news. John and Gus quickly put the misbehaving snack away and gathered up the crumbs before the sandwich could do any damage, but that was the last time anyone tried to smuggle one into space!

John's messy sandwich showed that terrible food wasn't making the astronauts very happy. When Rita Rapp joined the team at NASA that was responsible for space food, dining among the stars started to get a whole lot better. Rita got to know the astronauts and quickly realized that they would be much happier, and work better, if they were eating tasty food in space. So, Rita got rid of the weird mush and asked the crew what their favorite meals were. She added things such as scrambled eggs, tuna salad, steak, and even corned beef to the menu. Rita would try out lots of different recipes for the astronauts, cooking them over and over again until she found exactly the right combination that made space food taste just like Earth food. Rita and her team didn't just improve the taste of food, they invented new ways of storing it in space and keeping it warm too.

Rita's concepts for packaging food are still being used on the International Space Station (ISS) today. Much of the food is dehydrated, just like instant mashed potatoes. After it has been cooked on Earth, it's put in pouches and freeze-dried, to remove the water. This means that the food will keep for years, and is lighter so it is easier to transport. When astronauts are hungry, they simply add water and enjoy straight out of the wrapper.

As with all things in space, careful planning ensures the pantry always has plenty of food in it, because astronauts can't just pop out to the shops when the cupboard is empty. Instead, every few months cargo ships deliver dried and tinned food, as well as some fresh fruit such as oranges, apples, or tomatoes. Very occasionally, there is a frozen treat as well! One of the cargo ships sometimes carries a small freezer to bring frozen scientific samples back to Earth. Every so often there is room in the freezer on the trip to space and it gets filled with real ice cream. This is the only ice cream that gets eaten in space. The dried "astronaut" ice cream that you can buy isn't real space food. It has never flown in space as it's too crumbly and would make a mess.

Astronauts even get to choose some of their favorite foods to go into space with them. One of the brilliant things about the ISS is that the crew come from all around the world, so there are often exciting things in the pantry, such as Tim Peake's British bacon sandwich and Chris Hadfield's Canadian maple-syrup cookies. Italian astronaut Samantha Cristoforetti decided to use her choices to focus on healthy food, choosing pouches of food such as leeks, mushrooms, and sun-dried tomatoes. She enjoyed putting them together in different combinations, which was as close as she could get to cooking in space. The videos of her making meals show how difficult it is to keep your ingredients from floating away!

Even with all these tasty improvements, things can still sometimes go wrong. One day, Italian astronaut Luca Parmitano was working on the ISS and fancied a snack. He had spied some wasabi peas in the pantry, which a Japanese crew member had left behind. There was no special space packaging for these, they were just in a tin can, straight from the shop.

Excited about his well-earned treat, Luca peeled back the metal lid, but it gave way more quickly than he expected and suddenly all the peas escaped! Before he knew what was happening Luca was in a cloud of green peas. He tried to stuff them back in the tin, but they kept floating away. There was only one thing to do—Luca had to eat them all! And fast—he didn't want any of his crewmates to see what a silly error he had made! Gobble, gobble, gobble. Like a fish feeding, Luca swam around catching the super-spicy wasabi peas in his mouth. He nearly had them all when his tongue started to tingle and suddenly it felt like his mouth was on fire! But there was nothing to do but suffer. Luca wondered if perhaps that was why they had been left in the pantry.

THE CHALLENGES OF FOOD IN SPACE AREN'T OVER YET

The challenges of food in space aren't over yet. Scientists, including Rita, have done incredible things to make it possible to eat tasty food in space, but as plans grow for space travel to venture further into the Solar System, the food needed will have to improve. If humans are going to visit Mars they will need enough food with them for a two-year journey. There won't be any resupply ships to bring fresh oranges from Earth, so the world's space agencies are working out how astronauts can also be farmers in space. Astronauts on the ISS have already grown small amounts of lettuce, cabbage, and even edible flowers. There is still lots of work to do, though, to be able to feed astronauts all the way to Mars and make sure the astronauts are as well fed as Rita would want them to be.

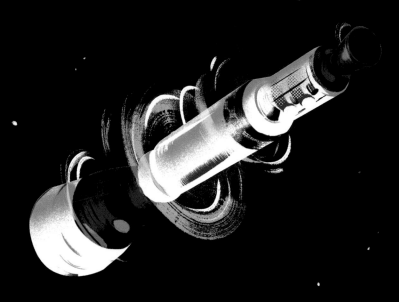

THE SPINNING SPACECRAFT

MARCH 16, 1966

It was a beautiful, warm spring morning in Cape Canaveral, and there was barely a cloud in the crystal-blue sky—the perfect day for the rare sight of a double rocket launch. American astronauts Neil Armstrong and Dave Scott sat in their Gemini VIII spacecraft on one launch pad, with the Agena rocket they were going to chase down in space on another.

THE RACE TO GET HUMANS TO THE MOON WAS IN FULL SWING

The race to get humans to the Moon was in full swing. The Soviet Union had managed to score all the big firsts so far: the first satellite launched, the first person in space, and the first spacewalk. But their rivals were finally making progress. The US had first taken the record for the longest spaceflight, and then smashed it! The Gemini V crew, Gordon Cooper and Pete Conrad, had spent eight days in orbit. Then, Frank Borman and Jim Lovell, in Gemini VII, had proved humans could spend two weeks in space in something smaller than a small two-seater car!

But neither America nor the Soviet Union had yet achieved the ultimate goal to go to the Moon. NASA's plan for getting the Americans there included using two spacecraft, one to get to the Moon and another one to land on it. So, they needed to learn how to get two spacecraft to meet up and join together in space, in maneuvers called rendezvous and docking. This was the mission set for Dave Scott and Neil Armstrong. The Agena rocket would launch first, and then Dave and Neil would follow in their Gemini VIII spacecraft, chasing the Agena, like a high-speed car race. Once they'd caught up with it, Neil would carefully guide Gemini VIII towards it and connect them. Everyone was keeping their fingers crossed that the Agena would behave properly as the last time NASA had tried to launch the rocket, it had exploded on its way into space. But with the race to the Moon against the Soviets in full swing, there was no time to change the plans.

As the Agena blasted off, everyone was nervous, but thankfully it safely reached orbit around the Earth, and just over 90 minutes later Neil and Dave did too. The astronauts spent the next few hours firing the engines to change their speed and direction, closing in on the Agena. Everyone was excited when the astronauts caught sight of the rocket and proved that all the engineers' calculations had been correct.

As commander, it was Neil's job to fly Gemini VIII and control the small rocket engines that moved it. He was really pleased to find that the controls were just like the simulators back on Earth. "Man, this is easy!" he told Mission Control. "There's nothing to it!"

Catching the spacecraft was one thing, but docking was the big prize. As they approached the Agena, Neil carefully lined up Gemini VIII and inched the two spacecraft towards each other. The two docking adapters connected and with a clunk, the two spacecraft became one. "Flight, we are docked!" announced Neil proudly, thrilled with the job well done. There were no computers for the final moments, the success was all down to his skill. Mission Control and the crew allowed themselves just a moment to celebrate as they claimed the first docking in space, but everyone quickly got back to work.

As the Gemini VIII crew flew around the world, they talked to Mission Control through radio signals, but at some points there was no connection.

THE RADIO WENT QUIET. NEIL AND DAVE WERE IN A BLACKOUT

During these "blackouts" the crew were alone. As they flew over the Indian Ocean and towards the Pacific Ocean, the radio went quiet. Neil and Dave were in a blackout, but they kept working through their list of jobs. As Dave looked up from his checklist, he noticed that the cockpit instruments said they were turning when they should be still. "That's rather strange," he thought and showed Neil.

46

The movement was so small neither of them felt it, but one glance out of the window showed that they were indeed moving. Neil took the controls and fired the Gemini's small rockets, called thrusters, to stop the unexpected motion and set them pointing back in the right direction. "That's really odd," Neil said to Dave. They started checking their spacecraft, trying to work out what had happened, and then, minutes later, the movement started again. Once more they were spinning, slowly at first but gradually getting faster.

"Dave, switch off the Agena," said Neil. They both thought that the problem must be with the unreliable rocket. They hoped that if they turned it off, the problem would be solved. Dave followed his commander's instructions, but worryingly it didn't help. Instead, the movement got worse. Neil, always cool and calm, kept trying to stop the motion with Gemini's thrusters, but it was hopeless. He felt like a rodeo rider on a bucking horse: no matter what he did, he couldn't get the spacecraft under control.

"We'd better separate from the Agena," said Dave. Neil agreed—both of them were convinced that it was the source of their problems. Neil pressed the buttons to undock the two spacecraft, expecting that once they were free of the troublesome Agena they would regain control of Gemini VIII. But when they separated, the spinning suddenly got much, much worse!

Over and over they went, tumbling and turning more and more quickly as if on some terrible never-ending fairground ride. They were spinning so fast now that Neil and Dave started to feel dizzy and the sunlight flashed through the windows like a strobe light. Anything that was loose inside the capsule

OVER AND OVER THEY WENT, TUMBLING AND TURNING ... AS IF ON SOME TERRIBLE NEVER-ENDING FAIRGROUND RIDE

had been thrown against the walls. Quickly the astronauts realized that the problem must be with their spacecraft.

Things were getting desperate, and because they were still in a blackout zone, Mission Control couldn't help them. Neil knew that the only hope left was to turn off the Gemini's thrusters and use their precious re-entry rockets to take control. But to do that he needed to move some switches and he was having trouble seeing them because the spacecraft was rotating so fast. His vision was getting narrow as the blood drained out of his head. It became a race against time until Neil and Dave would pass out! If that happened, there would be no way for Mission Control to save them and they would eventually die in an out-of-control spacecraft! Incredibly, they stayed calm and didn't panic. They had trained hard for this sort of problem and all their focus was on fixing it.

Neil reached over his head, fighting tunnel vision and dizziness, and flicked the switch that swapped control to the re-entry thrusters. He fired the jets. It worked! They slowed down and at last Neil managed to regain control. "You've got it!" shouted Dave. They both let out a huge sigh of relief as the blood rushed back into their heads and the crazy fairground ride slowly stopped at last. When Mission Control came back over the airwaves, they ordered Gemini VIII to come back to Earth as soon as possible. The most important thing was Neil and Dave's safety.

The astronauts headed back to Earth, but they were off course and landed a long way from home. They splashed down in the western Pacific Ocean, 6,000 miles away from where they should have landed in the Atlantic Ocean, and several days too early. They were back on Earth, but not safe yet.

The astronauts were trapped inside a ship that was designed for space, not the sea. They had no choice but to wait for the rescue teams to reach them. As the water tossed the little spacecraft around and around, Dave and Neil turned green. They felt like they were on a rollercoaster for a second time—but this time it was a much longer ride! They were bobbing around for three hours before a US Navy ship was able to reach them. Swimmers from the ship had a hard time fighting the waves to get an inflatable ring around the spacecraft to stabilize it, and then at last Neil and Dave were able to climb out and board the ship. They were both very glad that with all the excitement and drama, they hadn't eaten much since they had blasted off less than 12 hours earlier, as they only had one sick bag between them!

Once the astronauts were safely rescued, NASA traced the problem to a faulty thruster on the Gemini VIII spacecraft, probably due to a short circuit. This small rocket was firing all the time and the constant push caused the spacecraft to spin ever faster. The Agena rocket that everyone was so convinced was going to cause problems had behaved perfectly! Without Neil's calm head and quick thinking, the crew would have been doomed. The problem was fixed on all the future Gemini spacecraft, and while the mission hadn't done everything it set out to do, Neil and Dave did successfully rendezvous and dock—the march to the Moon continued!

GOING TO THE MOON FOR CHRISTMAS
December 1968

1968 had not been a good year. Across the globe, people weren't happy. Images of violence, war, and starvation regularly flashed on to television screens and huge protests took place. In America, tensions were running high after the assassination of the civil rights activist Martin Luther King, Jr. Students were demonstrating against the Vietnam War. And many people were grieving after the assassination of the presidential candidate, Robert F. Kennedy.

NASA had been having a bad time too. It was still recovering from the tragic Apollo 1 fire in January 1967 that had claimed the lives of Gus Grissom, Ed White, and Roger Chaffee. No Americans had flown in space while the problems with the Apollo spacecraft were fixed, and precious time was disappearing. The clock was ticking towards President John F. Kennedy's deadline: to land humans on the Moon by the end of the 1960s. To make things worse, the Soviets, America's rivals, were reportedly getting ready to send their Zond spacecraft on a flight to the Moon, possibly with humans on board, before the end of the year. The race to the Moon was heating up and America was in danger of being beaten.

NASA had been developing the Lunar Module—a delicate, spider-like spacecraft that two astronauts would hopefully fly to the Moon's surface. It was the job of the next mission, Apollo 8, to test it in space for the first time, but despite the engineering team's best efforts, the spacecraft just wasn't ready. It looked like Apollo 8 would be delayed and the US would slip behind the Soviet Union in the space race once again.

No one at NASA wanted to admit defeat, though, and so through the long, hot summer, brains started to whirr. A daring thought started to form in the mind of George Low, the manager of the spacecraft program. He knew that if they could pull off his plan, the Americans would be back in the race.

George got his team to check the numbers. Slowly but surely, everyone who asked the question "Can it be done?" had the same reaction. The idea sounded ridiculous, but when they looked closely at the data they all reached the same conclusion, "Yes, it can." Finally, George asked Deke Slayton, the head of the astronaut office, to tell the crew, Frank Borman, Jim Lovell, and Bill Anders.

Deke summoned Frank to his office. "Frank," he began. "We want you, Jim, and Bill to fly a couple of months early, on Apollo 8."

"OK ..." said Frank, surprised by the late change to their mission plan.

Deke carried on "And we want you to go to the Moon for Christmas!" Frank grinned from ear to ear, this was the best Christmas present he'd ever received. Instead of losing time and waiting for the Lunar Module to be ready, they were going to push forwards and send the first humans to orbit the Moon.

On December 21, 1968, Frank, Jim, and Bill put any thoughts about the awful year behind them, as they sat at the very top of the colossal 360-foot-tall Saturn V rocket in their Apollo 8 spacecraft, waiting for launch. They were the first humans to take a ride on this—the biggest and most powerful rocket ever to blast off into space.

THIS WAS THE BEST CHRISTMAS PRESENT HE'D EVER RECEIVED

At 7:51 a.m., not long after the Sun had risen, the mighty Saturn V rocket roared away from the launch pad and took its precious passengers safely into Earth's orbit. After two laps of the Earth to double-check everything was working properly, Mission Control told the crew "You are GO for Trans-Lunar Injection." This meant the crew could fire their rocket engines once more and become the first humans ever to speed towards the Moon at 25,000 miles an hour—ten times faster than a fighter jet!

It took the astronauts three days to reach the Moon. They watched the Earth shrink out of their windows until they could see it hanging in space, the same way we see the Moon in the night sky, and were mesmerized by the view. They could even see nearly all of North and South America together at the same time. But because of the way that the spacecraft was turned, they couldn't see the Moon out of their windows at all.

It was Christmas Eve, and while lots of people on Earth were wrapping presents, filled with excitement for the next day, Apollo 8 was racing towards the Moon. As it got closer, the spacecraft needed to slow down, otherwise the crew would simply slingshot around the Moon with a passing glimpse and head back to Earth. The only way to do this was to use Apollo 8's single rocket engine, and it had to work perfectly, for just the right amount of time. If it took too long, the crew would be doomed and plunge into the Moon's surface!

To make the mission even more difficult, the crew had to fire the engine on the far side of the Moon from Earth. During this time they would be unable to talk to Mission Control because the radio signals they used to communicate couldn't pass through the Moon, leaving the crew in a "blackout" zone. The team in Houston did one last check, and then wished the crew the best of luck as they slipped behind the Moon and the radios went silent. Frank, Bill, and Jim were all alone in the vast Solar System, the most remote place that human beings had ever been. They couldn't see Earth anymore, and they still couldn't see the Moon below them either. They were flying in the complete darkness of the lunar night. The team in Mission Control was helpless. There was nothing they could do except wait for the crew to reappear from the far side of the Moon.

Frank, Bill, and Jim readied their spacecraft to slow down and drop them into lunar orbit. With just two minutes to go before the critical point, they started to fly into the daylight side of the Moon and the Sun rose over the gray lands beneath them. Finally, the mysterious object that the astronauts had traveled 238,000 miles to see came in to view through their windows.

"I got the Moon!" exclaimed Jim as he caught sight of it.

"Do you?" asked Bill, excitedly.

"Right below us!" replied Jim.

The view of the Moon, up close with their own eyes, took their breath away. But, much as they wanted to stop and take in the amazing landscapes, they knew they had to focus on the task at hand. If everything went well, they would have many hours to gaze at this alien world. But if it didn't, they would have many other things to think about. It was back to work.

The astronauts strapped themselves into their seats and waited for the single rocket engine to fire as programmed. Bang on schedule, they felt the rumble of the engine that would slow them down coming to life. For four long minutes, the engine fired and the crew kept a careful watch on the computers as the seconds ticked away. Then, right on cue, everything went silent. Apollo 8 was in orbit, circling endlessly around the Moon.

Back in Mission Control, all of the flight controllers were glued to their seats, staring at the clock and waiting for Apollo 8 to emerge from the blackout. Thirty-two slow minutes passed before their screens started to fill with numbers. Everyone cheered and clapped as they knew they had done it! All around the world, news bulletins interrupted radio and television programs to share NASA's news: "We've got it! Apollo 8 is now in lunar orbit." That evening, people looked up at the night sky with added wonder, realizing that there were three humans, just like them, circling the Moon.

THERE WERE THREE HUMANS ... CIRCLING THE MOON

Safely in lunar orbit, the crew got on with the job they had traveled all this way to do—to study the Moon up close for the first time with human eyes. It was incredible. They flew over the huge craters that had been formed when rocks slammed into the Moon billions of years ago, naming them after friends and colleagues, including the Apollo 1 astronauts who had died in the fire. They took photograph after photograph, and spent each orbit focusing on different parts of the gray scenery beneath them. The photographs would bring important information back to the scientists on Earth who were working out where the Moon landings might be attempted.

As the crew flew behind the Moon for the fourth time, they changed their direction so that rather than pointing the windows straight down towards the Moon's surface, they were looking out along it. This meant that, for the first time since they entered lunar orbit, they could see the Moon's horizon, where the gray landscape met the blackness of space.

At first the astronauts were so busy studying the deep craters and jagged surface they didn't notice the sliver of blue, green, and white that appeared above the horizon. But when Bill caught sight of it he cried out "Oh, my God! Look at that picture over there! Wow, is that pretty!" He immediately started taking photographs of the astounding sight of the delicate Earth rising above the bleak Moon.

Before Apollo 8 could head back to Earth, there was one more important task the crew had to do. At points throughout their mission, the crew became a television show, sending back live transmissions to hundreds of millions of people all over the world. On their last orbit around the Moon, the astronauts crackled on to television screens. They pointed their cameras out of the window, sharing fuzzy images of what they could see and talking about their great adventure. Each astronaut saw the Moon in a different way. Frank thought that it was vast and lonely. Bill loved the beauty of the lunar

Earth is, a grand oasis in the vastness of space. Finally, the crew closed their broadcast. "Good night, good luck, a Merry Christmas, and God bless all of you—all of you on the good Earth."

With that, the crew's work in lunar orbit was almost done. Happy they had done a good job, Frank, Bill, and Jim could focus on coming home. But unless the single rocket engine worked perfectly once more, they would never get to see the good Earth again themselves. The crew and Mission Control readied the computers and checked and double-checked the numbers. Just as when they arrived at the Moon, they would be in a blackout zone and Mission Control would not be able to help them.

The crew passed behind the Moon and once again Mission Control were glued to their seats, waiting helplessly. As the countdown reached zero, the radios crackled to life: "Houston, Apollo 8," called Jim. "Please be advised, there is a Santa Claus!"

"That's affirmative." replied Mission Control. "You are the best ones to know."

Frank, Jim, and Bill splashed down in the Pacific Ocean three days later, on December 27. The photographs of the Moon's surface and the data they'd gathered would help the Americans overcome the next huge hurdle of landing on the Moon. The photograph, Earthrise, that Bill Anders took was shared around the world. It was, and still is, an iconic image that reminds all of humanity just how precious our planet is. In the days and weeks that followed, the crew received many accolades and messages of praise. Among all of them, they remember one in particular ... "You saved 1968."

THE EAGLE'S LANDING

JULY 20, 1969

Deep in concentration, Neil and Buzz had just entered the most critical phase of their mission. They were only 6 miles above the Moon's surface with eight minutes to go until they landed. If they made it, they would become the first humans to step foot on the Moon. As Neil was looking out of the window of their tiny spacecraft, a Lunar Module named Eagle, Buzz was focusing on the dials and displays in the cockpit, when a distracting and unexpected beep in

their headphones caught their attention. This noise was a warning alarm from the computer. Buzz glanced down to the small display to find out what the problem was, but it simply read: 1202. That was all the computer could tell them, and Neil and Buzz had absolutely no idea what it meant. Neil radioed Mission Control: "It's a 1202." He didn't receive the usual chatter in response, though, only silence. As the seconds ticked by, tensions in the cramped cabin rose. It was not the time for things to go wrong. Neil Armstrong and Buzz Aldrin had run simulations of the Moon landing over and over. This mission had been years in the making and they'd prepared for so many faults and errors, but a 1202 was not one of them!

While Neil was called the Commander and Buzz was called the Lunar Module Pilot, at this moment neither of them was actually flying their spaceship. Instead, the Lunar Module's main computer was in charge. This computer, similar in size to a big shoebox and weighing 70 pounds, had just 38K of memory. It was state of the art in the 1960s, but today's smart phones are many thousands of times more powerful. The computer's code was working out where the spacecraft was, how much the rocket engine should be firing, and which way it was pointing, all so that it could automatically guide the Lunar Module to the planned landing point.

Neil was studying the surface of the Moon through the small triangle-shaped windows, watching out for the different craters he had memorized to make sure that they were on course. But the landmarks were going by too early. For some reason they were flying too fast and that meant that the computer was going to miss their carefully chosen landing point.

And then the computer had suddenly sounded its warning: 1202.

They had already made it so far, flying from Earth to the Moon for three days in the Command Module spacecraft with their fellow astronaut Michael Collins. They'd then left Michael in the Command Module, where he would orbit the Moon, and used a smaller spaceship, called the Lunar Module, to descend to the Moon's surface. They were so close to making history and this was not the time to abort their mission. But what was a 1202 alarm and what was the fault causing it? Growing impatient, Neil broke the silence and pushed Mission Control for an answer: "Give us a reading on the 1202 program alarm," he said, an edge of tension creeping in to the always calm voice, and yet silence still followed.

THEY WERE SO CLOSE TO MAKING HISTORY AND THIS WAS NOT THE TIME TO ABORT THEIR MISSION

Twenty-five thousand miles away in Houston, the Flight Control Team were monitoring every aspect of the Lunar Module's descent. The person in charge of the team in Mission Control is called the Flight Director, and that day it was Gene Kranz who was in the hot seat. Gene and his team had also practiced simulation after simulation, to make sure that they were ready for any problem that might happen.

As soon as Gene heard Neil's call, he turned to the person who was responsible for knowing all about the computers. This was Steve Bales, and he didn't know what the alarm meant either. But Steve knew that his colleague Jack Garman, who was supporting him, did. Jack had a handwritten list of all the alarm codes and what they meant, just in case anything happened. He quickly realized that this wasn't a major problem that would jeopardise the mission. Instead, 1202 was a warning that the computer was overloaded and it was prioritizing the most important tasks. Jack and Steve knew that the computer would be OK.

Once the team told Gene that the alarm wasn't anything to worry about, Mission Control let Neil and Buzz know that everything was OK and they should continue. The silence over the radio was just the time it was taking for the experts at Mission Control to do their jobs. There was so much information to know about the spacecraft, and so many different things that could go wrong, that it was impossible for one person to remember everything, but each person knew what they needed to know inside out. Together they worked as a team.

"You're GO on that alarm," Mission Control told Neil and Buzz. "Roger," replied Buzz. They took their eyes off the abort handle and focused again on their descent to the Moon. There was no time to relax, though, they were just minutes from landing.

As Neil looked out of the window again, he could see that they were still off course. With all the distractions of the program alarms, he hadn't been able to concentrate fully on where they were flying, but he could see now that the computer was a bit confused about where the spacecraft was. Worse, though, was that the computer was guiding the Lunar Module to land in a huge crater with steep sides and filled with rocks the size of cars. This was no place to attempt the first lunar landing.

Flying just 500 feet above the Moon's surface, but moving forwards at 30 miles an hour, Neil decided to take over control of the landing from the computer, and maneuver over the crater. He flew past the rocky area and started looking for a better landing site. Buzz kept a watch over the numbers, reporting to Neil and back to Mission Control on their speed and distance from the Moon's surface. At Gene's command, everyone in Mission Control fell silent, knowing that the final moments were in the hands of the astronauts. Neil was a highly trained and very experienced pilot and astronaut, and they trusted him, but they had no idea what he was doing. There was no time for Neil and Buzz to explain what they could see out their windows, or why they weren't landing yet. All everyone in Mission Control could do was watch the fuel gauge, which was starting to look dangerously low. Everyone was holding their breath back on Earth, and up in lunar orbit, where Michael was listening to every word.

THE FUEL GAUGE ... WAS ... DANGEROUSLY LOW. EVERYONE WAS HOLDING THEIR BREATH

Mission Control started counting them down—60 seconds to go ... Neil was still holding the spacecraft so it hovered 65 feet over the surface, flying forwards looking for the perfect landing site.

With just 40 feet until touchdown, the thrusters started kicking up dust from the Moon's surface, blocking Neil's vision. But he focused on the rocks ahead and continued his vertical descent.

30 seconds ...

Everyone was silent.

15 feet to go ...

SHUTDOWN. ENGINE STOP.

With less than 20 seconds of fuel left, Neil and Buzz turned off the rocket engine and Neil radioed four magical words: "The Eagle has landed."

Mission Control radioed back: "We copy you on the ground. You got a bunch of guys about to turn blue. We're breathing again. Thanks a lot."

And with that, just seven years since the American president, John F. Kennedy, declared that America would land on the Moon before the end of the 1960s, Neil Armstrong and Buzz Aldrin had made it! Success hadn't been easy. The Soviets had beaten them to nearly all the big firsts—the first satellite in space, the first person in space, and the first spacewalk. And there'd been the years of planning, the brilliant and creative minds that had worked tirelessly to write the computer programs, build the spacecraft, and plan the missions. Over 400,000 people had been involved in the Apollo missions to get to the Moon. And then there were the failed missions, the explosions, and the mistakes. There had even been the greatest sacrifice of all, the lives lost on both sides of the space race—Americans and Soviets—all in this push to take humankind farther than ever before.

OVER 400,000 PEOPLE HAD BEEN INVOLVED IN THE APOLLO MISSIONS TO GET TO THE MOON

In Mission Control, everyone took a deep breath as excitement started to ripple around the room. There were grins all around, but any cheers and whooping were quickly muffled. Apollo 11 may have landed, but there was still a job to do. This was no time to lose focus or celebrate. "Keep the chatter down," cautioned Gene, as the team worked out whether it was safe to stay on the Moon or not. Around the world people clapped and hugged each other as they heard the news that humans had landed on the Moon.

But for the astronauts on the Moon, there was no time for celebrations. With just a brief handshake and pat on the shoulders, they got to work, checking all the systems in their spacecraft and making sure that everything was ready for them to leave. They didn't plan to leave the Moon immediately—there was exploring to be done!—but just in case anything should go wrong, the Lunar Module needed to be ready to whisk them back to Michael and the relative safety of the Command Module.

Next, the mission plan said that Neil and Buzz should have a meal, and then a four-hour sleep before they ventured out on to the Moon. The mission planners had wanted to make sure that the crew could get some rest if they were tired, or have extra time in case there were any difficulties. But the astronauts were far too excited to go to bed, so Neil recommended to Mission Control that they just get on with it. Without hesitation, they agreed.

Getting ready to do a moonwalk inside such a small, delicate spacecraft was very difficult. Neil and Buzz had to put on their spacesuits and check everything was sealed correctly. If there were any mistakes, their lives could be in danger. The spacesuits were so bulky, with a big backpack containing all the air and cooling systems needed to keep them alive jutting out behind them, that it was hard to get ready without bumping into the sides of the Lunar Module. But just seven hours after they had landed, Neil and Buzz were ready. As the commander, but also more practically, as the astronaut nearest to the door, it would be Neil who would venture out first.

Around the world, people were gathered around televisions, many bought specially for the moment. In Europe, where it was the middle of the night, parents woke up their children, and, dressed in pajamas, they tuned in to watch and listen. In Mission Control, people packed in the rows of consoles to watch the historic moment.

Everyone's eyes were fixed to their screens as a grainy black and white image appeared ... upside down! Someone flicked a switch and the picture was

"THAT'S ONE SMALL STEP FOR MAN, ONE GIANT LEAP FOR MANKIND."

turned the right way up and suddenly, in the dim, hazy images, everyone could make out Neil coming down the ladder, live from the surface of the Moon.

"That's one small step for man, one giant leap for mankind." Neil's now famous words seemed to sum up perfectly the meaning of that first footprint on the Moon. Buzz followed Neil onto the surface about 20 minutes later. "Magnificent desolation," he said, taking in the stark beauty of the lunar landscape they had landed on.

But there wasn't much time for enjoyment or reflection, Neil and Buzz had lots of work to do. The space race might have been driven by the rivalry between the US and the Soviet Union, but now they had come all this way, there was science to be done. The astronauts collected rocks, soil, and dust samples, took photographs of the surface and the view from the Moon, and installed science experiments on the surface that would keep working long after they left. They also took time to remember those from both countries who had paid the ultimate price in the race to the Moon. Neil and Buzz left a fabric badge honouring the Apollo 1 mission, where Roger Chaffee, Gus Grissom, and Edward White lost their lives in a preflight test, and a medal belonging to Vladimir Komarov, who died when his Soyuz spacecraft crashed at the end of his mission.

Finally, after approximately two hours and 20 minutes of bouncing on the Moon's surface, collecting rocks and dust, and a phone call with the president, Richard Nixon, it was time for Neil and Buzz to head back inside the Lunar Module and have their overdue sleep. They had been awake for 22 hours and were extremely tired, but barely got any sleep in their cramped, cold, and noisy Lunar Module.

The next morning, tired and groggy, it was time to start their long journey home, first heading back to join Michael in the Command Module. But before they could, there was one more problem to solve. In all the squeezing and bumping and shuffling to get ready for the moonwalk, one of the switches in the Lunar Module had broken away. Without the switch working, the rocket engine that they needed to blast them away from the Moon's surface simply wouldn't work. If they couldn't fix it, they'd be stranded on the Moon forever. After rummaging around, they discovered that a felt-tip pen would fit in the hole perfectly. When the time came to flick the switch, they simply pushed the felt-tip pen in the hole, jamming the switch in place. Neil and Buzz blasted off from the surface of the Moon without any more hitches, and a few hours later, they had docked and were hugging Michael as the members of the Apollo 11 crew were reunited.

Four days later, on July 24, Neil, Buzz, and Michael splashed down in the Pacific Ocean. With the crew safely back on Earth the mission was complete and only then did Gene Kranz and all the teams in Mission Control celebrate. The success of the Apollo 11 mission was met with admiration and jubilation all around the world. Ten other people walked on the Moon in the three years that followed. The rocks that the Apollo missions brought back helped scientists unravel some of the secrets of our Solar System, and the footsteps that they left changed the way humanity saw itself. Today, over 50 years later, those footsteps are still there, waiting for the next generation to come back and explore the Moon once more.

THE CHOSEN ONES

The first astronauts and cosmonauts ventured into the complete unknown. They sat in spacecraft on top of experimental rockets, which, if everything went well, would jet them into space faster than a speeding bullet. During this terrifying ride, their bodies would feel like an elephant was sitting on their chest because of the forces from the rocket. Finally, when they found themselves floating above the Earth, unsure if the human body could even survive without gravity, there were just the metal walls of their spacecraft between them and the airless vacuum of space. Space travel was not for the faint-hearted. So how did space agencies find people who were willing and excited to go into space?

Both the Soviet Union and America turned to military jet pilots, people who are trained to fly planes at great speeds. Military jet pilots are used to dealing with low oxygen levels and lots of G-force—the forces that push on them when accelerating quickly—and they know how to make split-second, life-or-death decisions. These are just some of the skills needed to conquer space and so pilots make great candidates.

But even with all of these qualities, NASA wanted only the fittest pilots in all of America. So they pulled together a shortlist and put the chosen pilots through one of the most extreme medical tests ever created. Doctors and nurses stuck probes, electrodes, and other devices in the candidates to test their physical limits. They put cold water in their ears to confuse their balance, which made their eyeballs flutter uncontrollably. They were laid down on a table that could tilt, to see how their blood circulation would cope without the effects of gravity. They were locked in a sensory deprivation chamber, with no light or sound, and not told how long they would be in there for. They had to balance on a chair blindfolded, and had to describe what they could see in smudges of ink, known as inkblots.

After all this testing, there were still people who were excited about going into space, and NASA selected seven of them to train as their first astronauts. They were announced to the world's media in a blaze of publicity, and named the Mercury Seven. The Soviets were much more secretive about their selection, but they also announced a group of 20 people who would train for their space program. Both groups were solely made up of white men, a reflection of the racial and gender prejudices at the time.

But one person didn't agree with this. Randy Lovelace, the doctor NASA had appointed to put the men through the horrendous medical tests, believed women were just as capable. After NASA had selected its astronauts, Randy set up another experiment. He recruited a group of accomplished female pilots and put them through the same painful and confusing tests. He found that the women were just as good as the men, in some cases scoring better and in general the women seemed to have better pain thresholds and complained

THE WOMEN WERE JUST AS GOOD AS THE MEN

less! Thirteen women passed the medical testing, and were nicknamed the Mercury 13, but despite appeals, NASA refused to consider recruiting women. The Soviet Union, always wanting to beat the Americans, recruited a group

of women to train as cosmonauts in 1962, including Valentina Tereshkova, who became the first woman in space, in 1963. It wasn't until 1978 that NASA eventually caught up. The space agency recruited 35 astronauts to fly the new Space Shuttle and for the first time hired both women and people of different ethnicities.

Today, over 500 people from nearly 40 different countries have traveled into space. Most have been selected by their nation's space agency and are employed as space travelers, known as astronauts, cosmonauts, or taikonauts, depending on where they come from. A few very, very wealthy people have bought tickets to space, costing at least $19 million, and spent about a week on the International Space Station (ISS). But regardless of how they got their seat, everyone had to prove that they were up to the job.

The selection process has come a long way since those early days. The US, Canada, Europe, Japan, Russia, and China now all have astronaut programs, and from time to time invite people to apply to join them. They don't just choose pilots any more, though. They recognize the different range of skills needed for a team to live and work in space. Each country has its own way of selecting astronauts, but they all include a range of ability, physical, and psychological tests. Among other things, the European Space Agency asks candidates to recite series of numbers backwards to test memory. And during one recent Japanese selection, candidates lived together in a house for a week, separated from the outside world. The selection teams watched how they interacted with each other and coped with different tasks, including folding 1,000 origami paper cranes. Would the last one be as carefully folded as the first?

Do you have what it takes to join these ranks? Astronauts come from different backgrounds. Some are military pilots, others are doctors or scientists. But they are all fit and healthy. Lots of the skills that space agencies look for are hard to study or learn. For example, you can't be claustrophobic or worried about traveling to Mars. You need to have a good memory, concentration, and coordination, and work well in stressful situations. You should get along well with people, be highly motivated, and not anger easily. It is impossible to practice for all of these different things and so one of the best things to do, if you want to be an astronaut, is to find something you enjoy and be the best you can be. Then, when the opportunity comes around, give it a try and you never know what might happen.

Or, you could just start saving money and buy a ticket. Space travel is changing as companies such as Virgin Galactic, Blue Origin, and SpaceX plan to take paying passengers to space in the coming years. You could get to see the planet's beauty from over 60 miles up and experience weightlessness for a few minutes. One of the Mercury 13 women who passed Randy Lovelace's tests, Wally Funk, has never given up on her dream of going into space. Now in her 80s, she has bought a ticket for a ride on Virgin Galactic's SpaceShipTwo and is waiting for her turn!

Tickets today cost at least a few hundred thousand dollars, but all the companies hope to bring that down as time goes on and flights hopefully become more frequent. So perhaps, if you start saving now, and the ticket prices go down, who knows? It might be a once-in-a-lifetime treat, but maybe it would be worth it to experience space. And perhaps, one day, we'll all be hopping into space and back as often as we fly on airplanes today.

HOUSTON, WE'VE HAD A PROBLEM
April 1970

Gene Kranz, the flight director in charge of Mission Control at NASA, had a list of tasks he wanted the crew on Apollo 13 to do before his team said goodnight to them. It was coming to the end of their shift in Houston, and the young engineers who monitored Apollo 13 through their rows of computers were ready to go home and get some rest, enjoying the relative calm before the important days ahead. While 13 was an unlucky number for some, for Mission Control it was just the next number in the Apollo series after 12, and they were looking forward to a successful third Moon landing.

CapCom, the only person in Mission Control who talks directly to the astronauts, read Gene's instructions to the crew via the radio link. This voice link and the numbers on their computer screens were the only ways Mission Control knew what was happening on the Apollo 13 spacecraft and to its crew: Jim Lovell, Jack Swigert, and Fred Haise. CapCom asked the crew to turn a switch that would stir the gas in the oxygen tanks on the spacecraft. "OK," replied Jack, who flicked the switch. As he did this, Mission Control started seeing strange data on their screens—everything had suddenly gone haywire!

At the same moment, 190,000 miles out in space, Jim, Jack, and Fred felt a huge bang that shook their spacecraft to the core. The spaceship started rocking and tumbling. The crew looked at each other, eyes wide open, scared by the massive jolt that had just ripped through their precious spacecraft. "WHAT was that?!" Warning lights flashed up all over their computers and alarms rang out as Apollo 13's systems started failing.

Jack radioed Mission Control. "I believe we've had a problem here." Gene suddenly sat even more upright in his seat, his ears alert. CapCom, hoping he'd misheard, asked Jack to repeat what he'd said. Very clearly, but calmly, Jim said: "Houston, we've had a problem." Gene's eyes scanned around the control room looking at his team, and saw every member with their eyes fixed to the strange numbers on the screens in front of them, desperately trying to understand what was going on in the Apollo 13 spacecraft.

Sy Liebergot was responsible for monitoring the systems that kept the crew alive, including their power and oxygen. The data on his computer was telling him that half of the spacecraft's power and air systems had failed in an instant! This seemed impossible. In all the mission scenarios that the team had practiced, planning for how the systems might break and what to do if they did, there had never been anything as bad as this. But this wasn't a practice run, it was the real thing and three astronauts' lives were in their hands, hundreds of thousands of miles out in space. Everything on the spacecraft seemed to be reporting problems. But one number soon grabbed Sy's attention. The levels of gas in the oxygen tanks were rapidly going down, and he couldn't work out why. This was bad, really bad.

THIS WASN'T A PRACTICE RUN, IT WAS THE REAL THING AND THREE ASTRONAUTS' LIVES WERE IN THEIR HANDS

Up in space, the crew knew they had a serious problem too. Their spacecraft was still turning and tumbling, and they weren't sure what was going on either. But as Jim looked out of the window everything became clear. "Houston," he called over the radio. "We are venting something out into space, a gas of some sort." It was the precious oxygen. Not only did the astronauts need oxygen to keep them alive, but, along with hydrogen, it powered their electricity supply. Without the oxygen, both the spacecraft and the astronauts would die.

All thoughts of landing on the Moon quickly faded and Gene knew that this mission was now all about keeping the crew alive and getting them back to Earth safely. It was going to take all of the knowledge, skill, and ingenuity of everyone who worked at Mission Control and the engineers who built the spacecraft to fix it. Despite the range of enormous problems suddenly facing them, Gene didn't panic. He knew they had to tackle each problem calmly, one at a time. "OK now, let's everybody keep cool," he told the flight controllers. "Let's solve the problem, but let's not make it any worse by guessin.'"

It was clear to everyone that the main part of the Apollo 13 spacecraft, the Command Module, would soon be useless. But all hope wasn't lost yet. The original mission had been to land on the Moon, so attached to the Command Module was a different spacecraft, called the Lunar Module. This spider-like spacecraft was designed to be as light and delicate as possible, with just the bare essentials needed to get two of the three astronauts safely to the Moon's surface and back. With the Moon landing abandoned though, and the Command Module failing, the Lunar Module became the crew's only hope of staying alive. The three of them would have to move into this tiny spacecraft for their journey home. But how was a spacecraft that was designed to support just two astronauts on the Moon's surface for two days going to keep three people alive for four days while they traveled the hundreds of thousands of miles back to Earth?

The odds didn't look good. The list of problems to solve seemed endless, but the teams in Mission Control were not going to stop trying. Gene and his team weren't going home any time soon, and off-duty flight controllers,

THE LIST OF PROBLEMS TO SOLVE SEEMED ENDLESS

managers, and engineers all rushed to work to help add more brainpower to the situation. Everyone was determined to do all they could to get the crew home. They dug out diagrams, manuals, and test models of the Lunar Module. How far could it be pushed? What could they make the equipment do?

Together, the teams worked out that there would be enough oxygen in the Lunar Module to keep the crew alive until they got home. The power situation was worse, but if the crew switched off nearly all of their equipment, down to the heaters and even the computer that would steer the spacecraft,

78

they could make it. Jim, Jack, and Fred would have to survive in freezing temperatures, with no warm clothes to spare. It would be an uncomfortable journey, but they would get back in one piece.

The brilliant engineers on Earth might have solved the problem of making the power and oxygen last, but there was one thing the crew had far too much of: carbon dioxide. When humans breathe in, we inhale air that contains oxygen, which we need to live. And when we breathe out, we exhale carbon dioxide. Too much carbon dioxide is poisonous to humans, but we can survive because plants work in the opposite way. They need carbon dioxide to live and they release oxygen, making the air around us safe to breathe.

On the spacecraft, the carbon dioxide was starting to build up in the air as the crew were breathing out. Instead of plants, both the Lunar Module and Command Module had devices called scrubbers, which absorbed the carbon dioxide into canisters so it wouldn't reach poisonous levels. But with the Command Module out of action, its scrubbers were useless and the carbon dioxide levels were increasing. The scrubbers in the Lunar Module were working, but there weren't enough canisters to keep the air breathable all the way back to Earth. There were lots of spare square-shaped canisters in the Command Module, but these didn't fit in the Lunar Module's system, which used circle-shaped canisters! A different company had built each spacecraft and no one had ever thought that one might need to fit in the other!

Gene gave this problem to a team of engineers. They had to figure out how to get the square canisters from the Command Module to fit in the round holes of Lunar Module scrubbers, using only things that the crew had with them. It was a difficult problem, but the astronauts' lives depended on it.

The engineers put their minds together and worked around the clock until they came up with a solution. They built an adapter using parts of spacesuits, book covers, duct tape, and even a pair of socks. It was very clever, but it would be useless if the crew in space couldn't build the exact same thing. CapCom had the difficult job of instructing the astronauts.

QUESTIONS WENT BACK AND FORTH ACROSS THE SOLAR SYSTEM

With no videos or photos to guide the tired, hungry, and cold astronauts, good communication was absolutely critical. Questions went back and forth across the Solar System:

"What do you mean by sides?"

"Do I put the sticky end of the tape on the container?"

"Is that clear which end is which?"

After an hour of instructions and DIY in space, the crew had incredibly built their own adapter. They immediately plumbed it in and could breathe quite literally more easily as the carbon dioxide levels in the air started dropping. Now, it looked like there would be just enough power and just enough air to breathe to make it home! Thanks to the Lunar Module, and all the people in Mission Control, Jim, Jack, and Fred were able survive the four-day journey back to Earth.

But, to make it all the way home, the crew would need to re-enter the Earth's atmosphere and land safely in the sea. The Lunar Module, which had acted so well as a lifeboat, was designed for landing on the Moon, not on Earth. With no heat shield to protect the capsule from the fiery heat of re-entering the Earth's atmosphere, or parachutes to slow it down for a safe landing, the Lunar Module was useless for the last part of their journey.

Instead, Jim, Jack, and Fred were going to have to bring the Command Module back to life for just a few more hours. The Command Module had been without power for days, so all the computers were very cold, covered in condensation, and there was not much power left. The Command Module had not been designed for this. Once more, Mission Control had solved the problem, and the astronauts followed their instructions to wake up the Command Module so it could get them home.

With the Command Module back up and running, and just hours from Earth, it was time to leave the Lunar Module, which had saved their lives, behind in space. The crew climbed back into the Command Module, and closed the hatches between the two modules. "Farewell," said Jim, as the two spacecraft separated. "We thank you."

Finally, it was time to come home. Jim, Jack, and Fred strapped themselves into their seats and readied themselves for the final plunge into the Earth's atmosphere. Their lives now depended on the Command Module's heat shield working properly. This stopped the capsule from burning up as the spacecraft was slowed down by the Earth's atmosphere and the air around it heated up. One problem remained, and it was something Mission Control could do nothing about. Had the heat shield been damaged in the explosion all those days ago?

In Mission Control, the flight controllers stared at their screens, waiting. While the crew were re-entering Earth's atmosphere they had no contact with each other, a period known as "blackout". This was predicted to last for four-and-a-half long minutes, and as the clock counted down to the end of the blackout, CapCom tried to contact the crew. There was silence. Another

90 seconds went by, and there was still no word from the crew. Everyone's hearts started to sink. After all their heroic efforts, the endless problem solving, and stretching the spacecraft to do things they were never designed for, it seemed that, at the very last hurdle, the Command Module's heat shield had let them down.

As they started to think the very worst had happened and they had failed, a crackle came over the radio. "OK!" called out Jack. The crew were alive! All that was left now was for the parachutes to open and the capsule to float gently down the last few miles and splash down in the ocean. As Apollo 13 landed back on Earth, everyone in Mission Control whooped and cheered! Against all the odds, the crew was home safely.

After the mission, NASA looked back over all that had happened to work out what had gone wrong. They discovered that when Jack stirred the gas tanks, some faulty wiring caught fire and one of the oxygen tanks exploded with the force of a small bomb! It turned out that the crew were incredibly lucky. Not only could the explosion have caused even more damage, but if it had happened at any other time during the mission, the crew would almost certainly not have made it back to Earth. Yet without the incredible work of the women and men at Mission Control, which was nothing to do with luck, the crew still would have had no chance. The team on Earth had worked around the clock from the moment the explosion happened, refusing to give up hope or leave a problem unsolved. It was their determination that brought Jim, Jack, and Fred safely home to Earth.

WHEN IT ALL GOES WRONG

At the end of every January, the people who work in the incredible world of human spaceflight take a moment to think about the risks and challenges of the remarkable events that they are all a part of. Engineers, scientists, doctors, managers, astronauts—everyone from the people who tighten the bolts to the flight controllers in the many mission controls around the world—stop to remember the crews who have lost their lives as humanity pushes forwards to explore the universe.

Spaceflight is a risky business. Sitting in a spacecraft, perched at the top of a tall tower of explosives, ready to hurl you into the blackness of space, is not for the faint-hearted. A rocket is full of very powerful fuel that is set alight in a controlled explosion, which produces enough force to accelerate its passengers to tens of thousands of miles an hour. If it all goes wrong, rather than a flight to space, astronauts find themselves sitting on top of a very large and powerful bomb.

Even if astronauts make it into space, they then must survive in the harshest and most hostile place imaginable. There is no air to breathe. In the sunshine, the temperatures are like being inside an oven, but in the dark, it is colder than Antarctica. The only way humans can stay alive is by relying on a spacecraft, with hundreds of thousands of parts that have to work correctly, to keep them alive.

And then it's time to return to Earth. Spacecraft slam into the Earth's atmosphere at high speed, causing the air around them to heat up to over 2,500 degrees Fahrenheit. The spacecraft has to protect the crew inside from this fiery inferno with its heat shield. Without it, the spacecraft burns up, turning into a giant shooting start as it plunges back to Earth.

Every astronaut who sits on top of a rocket, waiting to be launched into space, knows that the work is dangerous. Astronauts are carefully selected, trained for many years, and understand the risks. Every person who has worked to put them into space knows that astronauts' lives depend on them and they do everything they can to keep crews safe. In the Mission Control rooms around the world, flight controllers have a mantra that is always in their minds: "Crew, Vehicle, Mission." Keep the crew safe above all else, then make sure they have a vehicle to get home in, and finally get on with the mission. Never jeopardize the crew's safety ahead of anything else.

We are all human, and sometimes people get things wrong, no matter how careful we are. It is no different in the world of spaceflight. Mistakes happen, warning signs are missed, or the wrong decision is made. Nearly always,

WE ARE ALL HUMAN, AND SOMETIMES PEOPLE GET THINGS WRONG

the safety systems do their job and protect the crew. But sometimes, just occasionally, the very worst happens. Eighteen people have lost their lives on the way to or from space since human space travel began in 1961.

Vladimir Komarov was the first person to fly in the Soviet Union's new spacecraft, the Soyuz, but the mission was filled with difficulties and he tragically lost his life when the parachutes failed on re-entry on April 24, 1967. Four years later, on June 29, 1971, Georgy Dobrovolsky, Viktor Patsayev, and Vladislav Volkov, all Soviet cosmonauts, died when all the air leaked out of their capsule.

The American Space Shuttle Challenger exploded 73 seconds after it had launched on a cold and frosty morning on January 28, 1986, sadly killing all seven astronauts on board: Gregory Jarvis, Christa McAuliffe, Ronald McNair, Ellison Onizuka, Judith Resnik, Dick Scobee, and Michael J. Smith. The cold weather meant the rubber seals in the rocket boosters didn't function correctly and hot gas had leaked out, causing the accident.

Another Space Shuttle, Columbia, was lost during re-entry to Earth's atmosphere on February 1, 2003. A piece of foam from the external tank had hit the heat shield just after lift-off. This had made a hole in the heat shield that wasn't detected and as the Shuttle re-entered the atmosphere, hot gases rushed into the wing and caused the Shuttle to spin out of control and disintegrate. This took the lives of Michael Anderson, David Brown,

Kalpana Chawla, Laurel Clark, Rick Husband, William McCool, and Israel's first astronaut, Ilan Ramon.

Sometimes accidents happen before the mission launches, and astronauts have lost their lives during testing. Roger Chaffee, Gus Grissom, and Ed White all died in a fire in the Apollo 1 spacecraft during a launch pad test on January 27, 1967. Michael Alsbury lost his life during a test flight of Virgin Galactic's SpaceShipTwo on October 31, 2014, when the spaceship tumbled out of control and broke apart over the Mojave Desert in California.

When these terrible things happen, everything stops. Once the crew has been laid to rest, everyone's focus is to work out what happened and why. What caused the explosion, or why did the capsule depressurize? What could have stopped it from happening? These questions plague everyone's minds and no stone is left unturned until the answers are known. The investigations always find that there were things that could have been done differently, if only people had known everything. Decisions are made with good intentions, but sometimes have unintended, and never imagined, consequences. Once the answers are found out, everybody learns from the mistakes. All the things that had been missed are fixed and the spacecraft that follow are stronger and safer. The programs continue. The crews who have lost their lives paid the ultimate price, one that will never be forgotten. But to stop would be even worse because they would have died in vain.

We remember the crews who gave their lives. We always strive to do our very best. We learn from our mistakes.

Vladimir Komarov Georgy Dobrovolsky Viktor Patsayev Vladislav Volkov Gregory Jarvis

Christa McAuliffe Ronald McNair Ellison Onizuka Judith Resnik Dick Scobee

Michael J. Smith Michael Anderson David Brown Kalpana Chawla Laurel Clark

Rick Husband William McCool Ilan Ramon Roger Chaffee

Gus Grissom Ed White Michael Alsbury

SECRETS, SPIES, AND SATELLITES
April 1981

Shhh … Are you ready to hear something top secret? Something so classified that the US government has still not confirmed or denied its truth more than 35 years later? The story you are about to read may or may not be true, but it is full of rumors and alleged leaked information so be careful who you tell. Over the years, it's been pieced together to create an incredible tale of a top-secret mission, nerves of steel, and brilliant minds pulling off the seemingly impossible. It all began in the 1950s, when the world caught space fever …

The year was 1957, and the Soviet Union had just launched Sputnik, the first human-made satellite to orbit the Earth. Everyone quickly became fascinated with aliens, space travel, and living on other planets. People started to wonder if one day humans could walk on the Moon, or even move to Mars! But with the excitement also came fear. The Soviet Union was the enemy of America, and their great rivals in the space race that dominated the 1960s. The American people began to worry about what the Soviets' satellite could do. Was it spying on them? Taking photographs? Learning military secrets? Everyone was nervous, including the US government. The very next year, in 1958, the US started its own military satellite program, codenamed CORONA, and over the following years the Americans put several satellites into space to spy on their enemies. As time went on, these satellites became more and more sophisticated, taking better pictures that could get back to Earth much more quickly. By 1981, they were so good, that rather unexpectedly, one of these satellites was allegedly used to spy on Americans, rather than their enemies …

THE AMERICAN PEOPLE BEGAN TO WORRY ABOUT WHAT THE SOVIETS' SATELLITE COULD DO

After the US beat the Soviets and put the first humans on the Moon in 1969, attention turned to what the Americans' next goal might be. The Apollo program had been a great success, but because each rocket and spacecraft could be used only once, it was also very expensive. So NASA challenged their engineers to design and build a fleet of reusable spacecraft.

Their answer was the Space Shuttle. Part rocket, part spaceplane, and part cargo ship, the large white Space Shuttle would roar into space like a rocket, with the help of two rocket boosters and a giant fuel tank, which would fall back to Earth once the job was done. Once in space, the top of the shuttle would open up to reveal the payload bay, where the cargo was stored. Finally, when the mission was complete, the shuttle would re-enter the Earth's atmosphere and then land on a runway like a glider. It would be NASA's crowning glory and nothing else like it existed. But like many new inventions, the Space Shuttle was plagued with technical difficulties.

In particular, the 24,000 protective tiles that formed the shuttle's heat shield were causing a headache for everyone involved. They were meant to protect the shuttle from the immense heat that occurs when a spacecraft re-enters the Earth's atmosphere and stop it from burning up. These new tiles were lightweight and reusable, just what the new spacecraft needed, but they were also delicate and tricky to hold in place.

Each time the heat shield went through tests, hundreds, and sometimes thousands of these tiles fell off. If the tiles didn't work, then the Space Shuttles wouldn't be reusable, but worse, the crew might be stranded in space, unable to come home. Scientists and engineers worked over and over trying all sorts of different methods, until eventually the tiles were fixed and the first Space Shuttle to be built, named Columbia, was ready for launch.

Usually new rockets and spacecraft, especially ones that transport humans, go through several test flights before anyone gets to fly in them. For all the designs, theory, and testing, it is very hard to be certain that it will all work until you try it out for real. But there was nothing usual about the shuttle, and that included running a test flight. After careful consideration, it was decided that the first time the shuttle headed into space, there would be

two very brave astronauts on board, John Young and Bob Crippen.

On April 12, 1981, at 7:00 a.m., thousands watched in person as Columbia blasted off with a huge roar from launch pad 39A at the Kennedy Space Center in Florida, and rocketed into the sky.

As the public cheered and celebrated Columbia's triumphant launch in Florida, not everyone in Houston, Texas, was as confident about Columbia's ascent as they seemed. Even though the engineers thought they'd sorted out all the problems, they still couldn't be absolutely sure that everything would work until they tried out their solutions for real. The crew's safety is always the most important part of any mission and in space travel nothing is left to chance. Rumors of a top-secret back-up plan started to circulate, just in case there were more problems with the tiles.

The alleged plan sounds simple enough: once Columbia was in space, NASA wanted to check if all the tiles had stayed attached so it could re-enter Earth's atmosphere safely. But the problem was the astronauts wouldn't be able to check all of the tiles from inside of the spacecraft. So how would they look?

NASA hoped to take photographs of Columbia from a passing satellite, but they didn't have anything that could do that, so they allegedly turned to the people who did—the US military. The US military doesn't just hand out information about their secret spy satellites, though. In fact, the department that was responsible for building, launching, and maintaining secret military satellites—the National Reconnaissance Office, known as the NRO—was so classified that not only did the public not know it existed, but the people who did weren't even allowed to say its name! The story goes that the NRO agreed to let NASA use its KH-11 satellite to take photographs of Columbia, but of course the whole project had to be kept top secret. The satellite would be able to take photographs of the Space Shuttle only when the two passed by each other, both moving at thousands of miles an hour! To make things even harder, the vulnerable part of Columbia would need to be facing in the direction of the satellite's camera at the right moment for the photographs to show if any critical tiles were missing. There are lots of talented people working at NASA, but if this story is to be believed, then it would take some extremely clever—and very discreet—people to pull it off.

The two people who were given the job are said to have been Ken Young, the head of Johnson Space Center's Flight Planning Branch, and his boss Ed Lineberry. Ken has since admitted that from day one there were top-secret plans for Columbia, but at the time he had to keep his lips sealed. And to keep plans secret, everything was given a code name, such as Lamppost or Wagon Wheel.

Ken and Ed were given the incredibly complex task of calculating the exact time that Columbia would pass by the military satellite in space so the photographs could be taken. It would have taken them months to do these calculations, and they did it all hidden away down secret corridors in password-protected rooms at NASA's headquarters.

When you have a highly classified plan in place, only a few people can know about it, otherwise it isn't really secret. So it's likely that only a handful of people knew about the mission. It is thought Columbia's astronauts, John and Bob, would have been told because they'd be the ones to maneuver Columbia into the right position when it passed by the satellite. But, like Ken, they wouldn't have been able to tell anyone what they were doing. Except for a few high-ranking people, no one in Mission Control would have known.

　　Once in orbit, John and Bob's mission was to check all the systems on the Columbia Space Shuttle worked properly, and return the spacecraft safely back to Earth two days later. But once in space, John had some bad news for Mission Control. He could see that some of the protective tiles had come off the front of the big tail fin. If they had come off there, what had happened on the belly of the shuttle? Was there a great big, gaping hole? Could they get home? NASA tried to see if there was any damage using telescopes from Earth, but as they'd feared, the images were too blurred. So, the top-secret plan was put into action. With the pressure on, Ken and Ed passed on their instructions to Mission Control, who unknowingly relayed them to John and Bob. The astronauts had to follow the directions closely. If they got the movements wrong, or were too slow, the opportunity to photograph the protective tiles would be missed!

THINGS THAT ARE MEANT TO BE SECRET OFTEN DON'T STAY THAT WAY FOR LONG

Things that are meant to be secret often don't stay that way for long, though, and several journalists had their suspicions about what NASA was up to. After all, it was public knowledge that Columbia had problems with its protective tiles staying put. So while Ken, Ed, John, and Bob were working away, Gene Kranz, the Deputy Director of Flight Operations at NASA, was facing down the press. It is thought that Gene knew all about the plan, but he couldn't let the journalists know that they were right. So, as unreadable as ever, Gene gave nothing away. He only said, "I don't think there's any doubt that there are intelligence satellites that are capable of collecting data to a high degree of accuracy." This basically confirmed what the journalists suspected, that the US government had satellites that could spy on people. But Gene was clever, because he didn't say what NASA were up to.

It's been worked out that there would have been only three small windows of opportunity for the KH-11 satellite to photograph Columbia, and, if the stories are to be believed, the mission worked! Ken and Ed's calculations were accurate, John and Bob got the shuttle into position in time, and apparently on the final night of Columbia's STS-1 mission, NASA was handed photographs of the Space Shuttle. To everyone's relief, Mission Control saw that the tiles on the shuttle's belly weren't as badly damaged as they'd feared. It was going to be able to come home safely.

On April 14, at 10:20 a.m., Columbia completed the most daring test flight in NASA's history and the Space Shuttle safely re-entered the Earth's atmosphere and then glided back to land at Edwards Air Force base in California. There were 16 tiles missing and 148 damaged in total. NASA later found that the problem had occurred during takeoff.

The first Space Shuttle mission was a success, and the heat shield worked, but no one has been able to talk openly about why NASA was so sure it would be OK. Over the years, NASA has given different reasons for how they knew that the tiles weren't too badly damaged. First they said that slow-motion images taken during takeoff were reassuring that the tiles were OK, and then they said that remote cameras attached to Columbia provided the proof that it was safe. But not many people seem convinced. The final rumor from this incredible story is that the astronauts apparently saw the spy satellite far in the distance when it passed them by! But we won't know if that's true unless someone who really knows the truth lets us in on the secret.

SALLY AND SVETLANA

September 1983

Sally Ride was an accomplished, determined, and talented woman, always trying to do her very best. After school she went to university, and then on to study a PhD in astrophysics. She was fascinated by space, but had never really thought about being an astronaut. Growing up, Sally had watched Neil Armstrong and Buzz Aldrin walk on the Moon and thought it was amazing,

but all the astronauts had been men and she thought it would stay that way. But one day, Sally was reading the newspaper and she read an advert that took her by surprise. NASA was hiring astronauts, and for the first time was encouraging women to apply. "Wow!" thought Sally. "I could do that!" and quickly sent off an application form. After many months of waiting, interviews, and tests, Sally got a phone call asking her if she'd like the job. She couldn't believe it and started jumping up and down with excitement. Before she knew it, Sally would be training to fly in the Space Shuttle!

In January 1978, NASA unveiled its new class of 35 astronauts to the world. For the first time it was a diverse mix of people from different genders and ethnicities, rather than a group of white men. This new group of astronauts was ready to conquer space, and Sally was among them.

Sally Ride and the five other women in the group weren't the only people proving that women could fly into space. America's competitors in the space race, the Soviet Union, saw what the US was doing, and also decided to select some female cosmonauts.

Svetlana Savitskaya had always dreamed of flying. She took pilot and parachuting lessons as soon as she was old enough and she was a natural. Svetlana was a born

SVETLANA WAS A BORN DAREDEVIL

daredevil, setting world records by parachuting from great heights when she was just 17. But not happy just jumping out of planes, she also became an aerobatic world champion, performing incredible stunts in airplanes. Svetlana then trained to become a test pilot, trying out new aircraft. When the Soviet Union selected its own group of female cosmonaut trainees, Svetlana was among them.

Despite being talented and daring women, neither Svetlana nor Sally was saved from endless questions from the media focusing on the fact that they were female. Journalists fixated on questions about their hair, make-up, or using the toilet, rather than the challenges and importance of the work that they were being sent into space to do, which were the questions men were asked. But both Sally and Svetlana were determined to show to the world that women could be brilliant space travelers too, and kept their cool, politely reminding journalists what the job was really about.

Nineteen years after Valentina Tereshkova became the first woman to fly in space, Svetlana became the second. She blasted off to the Salyut 7 space station in August 1982, but the welcome was not quite what she had expected. As Svetlana and her crewmates, Leonid Popov and Aleksandr Serebrov, arrived on the space station, they were greeted by the crew, Anatoly Berezovoy and Valentin Lebedev. "We've got an apron ready for you, Sveta," Valentin said and pointed towards the cooking area. Infuriated, Svetlana calmly shot back that it was the duty of the hosts to cook and so he'd better get to it.

The Soviets had beaten the Americans once again, but Sally didn't care. Less than a year later, she became the third woman in space, and was part of the Space Shuttle Challenger's STS-7 mission. Sally's four crewmates had no issues with her gender and they all worked together as equals. The mission, to launch several new satellites into space, was a great success, and Sally played her part controlling the shuttle's robotic arm. Sally and Svetlana had both proved without a doubt that women were equally capable astronauts and cosmonauts as men.

Like all astronauts, Sally's work didn't end when she returned from space. Along with her mission's commander, Rick Hauck, she was invited to a space conference in Hungary to speak about their mission. Sally would get to meet people from all around the world who worked in the space industry, but there was one group of people she was forbidden to talk to. The American and the Soviet governments were in the middle of a political disagreement so Rick and Sally had been given strict instructions not to talk to anyone from the Soviet Union, to make sure they didn't add to the problems.

Sally was in the middle of a group of people enjoying the conference when she felt a tap on her elbow. She turned around and was amazed to see Svetlana looking straight back at her. Sally's eyes lit up and they smiled at each other, but remembering her orders and not wanting to attract attention, she quickly made her apologies and disappeared back into the crowd.

Afterwards, though, Sally kept thinking how much she'd like to speak to Svetlana and swap stories about their time in space. So Sally decided to take a gamble and, with a glint in her eye, told her Hungarian translator, Tamas Gombosi, "You know, I'd really like to get a chance to talk to Svetlana." Tamas nodded. "Maybe we can arrange that," he said with a smile.

A few hours later, Tamas cheerily invited Sally to a party that was being thrown by the Hungarian astronaut Bertalan Farkas. "Other people will be there," he said with a wink. So, that evening, Sally got into a chauffeur-driven car as the rain hammered down. She was driven down the back streets of Budapest, and then stopped at the back of a building. Sally's mind raced. What was she doing? She was going against strict orders. Was it a trap? What would happen to her? She felt like she was in the middle of a spy movie!

The driver led Sally up some stairs and around a corner. The door to a flat opened and she was astonished to find an apartment that reminded her of her own home, filled with souvenirs from space missions and pictures of rocket launches on the wall.

Shortly after she arrived, Svetlana walked in and Sally was relieved. Finally, the two women could talk freely. They got on like a house on fire and talked for hours and hours, sharing stories about their missions and the challenges they'd faced. They understood what the other had gone through in a way that no one else on Earth could and quickly became friends. But all too soon the party was ending. They took some photographs together, swapped gifts, and hugged tightly as they parted. "A friend for life," thought Sally.

Svetlana and Sally were never able to meet again, and both kept the party in Budapest a secret for many years. They would both fly in space once more the following year. Svetlana flew back to Salyut 7 and was the first woman to do a spacewalk, again beating the US by just a few months. Sally flew again on the Space Shuttle Challenger, along with Kathryn Sullivan, who was the first American woman to spacewalk. At long last, women were flying in space regularly, equal to men, just as they should be.

ASTRONAUT WANTED:
NO EXPERIENCE NECESSARY
1990–1991

Helen Sharman was stuck in a line of cars in Slough, England, on her way home from work, tired from another long day. She had a great job—she was a chemist who researched and developed new chocolate bars for the company Mars—but no one likes traffic jams. As she drove, she kept twiddling the knob on the car radio, hunting for a show that was playing something to entertain her on the slow journey home. As she tuned in to another station, she heard a voice speaking that caught her attention. "Astronaut wanted!" it declared. "No experience necessary." Intrigued, Helen carried on listening and was struck by the simplicity of the requirements. Applicants must be British and between 21 and 40. They should be fit and healthy, have a scientific or technical degree, and be able to learn a second language. That was it. There was no need to have a pilot's license, be some sort of super-athlete, or have a military background. As she listened, Helen checked off each piece of experience needed—she fit the bill! Excited, she scribbled down the phone number on a piece of paper as she waited at the next set of traffic lights. As the lights turned green, though she had no way of knowing it, her life had just changed forever.

AS THE LIGHTS TURNED GREEN ... HER LIFE HAD JUST CHANGED FOREVER

Days later, Helen called the phone number and asked for the application forms to be posted to her. But once she had them, they lay at the bottom of her bag for weeks. "What's the point?" she thought. "I've got no chance of being selected. There's nothing remarkable about me." As the closing date for the applications got closer, Helen looked at the forms one last time, and nearly threw them in the bin. But instead, knowing that she'd never get an opportunity like this again, she sat down, filled them in, and posted them back the next day.

Life carried on as usual and Helen soon forgot about the application, sure that she would never get a reply. Back at work Helen was as busy as ever. Sometimes, she would work through the night to monitor production lines at the Mars factory, and after one such night she finished work in the early hours of the morning, went home exhausted, and crawled into bed. She had been hoping to catch up on some sleep, but was abruptly woken by her phone ringing. "Would you come for a medical for the Juno mission?" said the voice on the other end of the telephone. Groggily, Helen jotted the date down and quickly fell back asleep. When she woke up a few hours later, she thought she'd had a strange dream. Something about being selected for astronaut training ...? She was astonished when she looked in her diary and saw the note. It wasn't a dream at all! Thirteen thousand people had called the phone line to request application papers, and 5,500 people had sent them back. Helen was one of just 150 people who got that phone call.

Out of the 150 applicants selected, only two people would be picked. These two finalists would be flown to Moscow and spend 18 months training for the Juno mission. And even though both candidates would go through the lessons, simulations, and drills, only one person would be chosen for the mission, right at the end of the process! The other person would be a backup, just in case anything went wrong. Whomever was selected would become Britain's first astronaut.

The Juno mission that Helen had applied for wasn't being run by a space agency or government, like the usual space programs. Instead, a group of British companies had made an agreement with the Soviet Union to buy a ticket to space. Their plan was that the cost of the mission would be covered by sponsorship, through deals with companies. There was still an awful lot of money to be raised, so the whole selection process, as well as the mission, would be publicized and promoted to try to secure sponsors. Helen and the other candidates all had to deal with very suddenly being in the public eye, with no idea if they would ever make it to space. It was a very public way of being selected and more than likely rejected.

But before Helen could even think about going to the Soviet Union, where she would train, she had to go through a long series of challenging medical and psychological tests. These were to assess how well her mind and body could cope with the rigors of spaceflight. She was suddenly a long way from her everyday life, but Helen loved it. The candidates were tested to see how well they would react in different situations—the organizers didn't want anyone too excitable or emotional, but the candidates still needed to be sociable and work well together as a team. Then there were the language tests—did Helen have the skills to pick up another language, and quickly? Perhaps worst of all were the G-force tests. The candidates were spun around in electronic chairs at incredible speeds and jostled up and down. This would pin them to the chair with a huge force and put great pressure on their bodies, pushing them to their physical limits. While the tests would make lots of people sick, they had no negative effects on Helen. Usually people have to stop after five minutes, but Helen endured 15 minutes of this physically demanding test on her first attempt, and not only did she not suffer from motion sickness, it just didn't faze her at all.

SHE WAS SUDDENLY A LONG WAY FROM HER EVERYDAY LIFE, BUT HELEN LOVED IT

At each step of the process, the number of candidates was reduced, but to her surprise, Helen was always chosen to carry on. Helen thought more and more about how exciting it would be to go into space. It was truly a once-in-a-lifetime opportunity to look at the Earth from hundreds of miles above and she realized that she desperately wanted this, no matter what was thrown at her. But only a handful of women had been into space, most astronauts up until now had been men, so Helen thought it was more likely that a man would be picked. Rather than give up, though, she did her very best. She wanted to see how well she could do and to find out what her body was capable of.

After all the spinning, prodding, poking, and quizzing, 150 potential astronauts had become just four. The final two candidates would be announced live on television from the Science Museum in London. Helen sat and listened for the finalists' names to be read out—her heart was racing, she really wanted this. The presenter paused ... "Tim Mace and Helen Sharman!" Helen smiled at the cameras, and followed the script, but the evening was a whirlwind. It wasn't until later on that evening, in the hotel with her family, that the reality sunk in—Helen was going to the Soviet Union to train to be an astronaut for 18 months. And all because she'd answered an advert on the radio!

The chance to experience living in another country and the science and technology she would learn along the way was a mind-bogglingly exciting opportunity to Helen. And if, just maybe, she got to go into space at the end of it all, well that would simply be the icing on an incredible cake.

Helen and Tim left the UK and traveled to Star City, near Moscow, to begin their preparations. Over the next 18 months, they would learn how to live and work in space. There would be lessons about the Soyuz spacecraft that transported people to and from space, and the workings of the Mir space station that one of them would be visiting for a week. They would learn what to do when things went wrong in emergencies and much, much more. But all

HELEN SAT AND LISTENED FOR THE FINALISTS' NAMES TO BE READ OUT—HER HEART WAS RACING

of these lessons and the mission communications would be in Russian, so the very first task they had to master was speaking and understanding the Russian language.

Even then, after all their hard work and time together, one of them was going to be chosen to become an astronaut and the other one left behind. Helen braced herself for disappointment, surely she would be the backup. Instead, she was told she was Prime Crew—she had been chosen to fly! Helen couldn't believe it. From the radio advert, all the way through, she had worked hard, given it her all and it had paid off—she was going to space!

After the final preparations were complete, the day arrived. Helen and her crewmates—Sergei Krikalev and Anatoly Artsebarsky—walked up the steps, waved to the waiting journalists, then got in the capsule. It was time for lift-off. Rattling and shaking, Helen was pushed back into her seat as the Soyuz rocket accelerated the capsule into space. Then, all of a sudden, the forces stopped and Helen was weightless.

Two days later, Helen and her crewmates docked on the Mir space station. As Helen floated through the hatch, into the space station, she had the biggest, widest grin on her face, relieved and delighted to have arrived. She was greeted with warm hugs from the Russian cosmonauts, Viktor Afanasyev and Musa Manarov, who had already been on Mir for six months.

Helen drank in every moment of the mission. Her days were filled with science experiments. She grew seedlings, tested how a small lemon tree survived in space, and grew crystals that could not be made on Earth. She also spoke to schoolchildren across Britain using a radio. But her favorite bits were the freedom of weightlessness, the bond she formed with her crewmates when they played music and talked together, and the views—she never got tired of looking down at Earth. The deep, deep blue of the oceans was just mesmerizing.

As the end of her mission came closer, after just six days on Mir, it was time to say goodbye. It was harder than she had expected to leave. Hugs were given, tears shed, and then the hatches were closed, and her time in space after all that training and hard work was suddenly over. It was a sad moment, but there was no time to linger as they had to turn their attention to the return journey.

Safely back on Earth, Helen sat on top of the capsule and took in the scene around her. So many people had turned up to welcome her home, and the smells and sights of Earth were so different from space. Wobbly on her feet, Helen returned to a life forever changed. She had cemented herself in the history books as Britain's first astronaut, the first woman to visit the Mir space station, and the first non-US, non-Russian woman in space. While her time in space was short, the experience is one Helen has never forgotten. Even decades later, she still dreams of space when she goes to sleep at night.

CRASH, BANG, WALLOP!
June 25, 1997

The Mir space station was Russia's cosmic pride and joy and a symbol to the world of the nation's great achievements in space. The size of a small house, the space station had six different rooms known as modules, which all radiated out from a central hub, like a giant star. Inside was a warren of tunnels, with every nook and cranny cluttered with experiments and equipment, with wiring snaking all over the place. It was like a giant Aladdin's cave up in space. Cosmonauts would spend months at a time, sometimes even more than a year, living and working in it, carrying out scientific experiments.

IT WAS LIKE A GIANT ALADDIN'S CAVE UP IN SPACE

Cosmonauts would fly to and from Mir in a spaceship called Soyuz. It was well-built and reliable, but also small and cramped, with no room for big bags of clothes or supplies. Instead everything that the crew needed to stay alive during their time on Mir was delivered every few months in a cargo ship called Progress.

The crew always looked forwards to the arrival of the new supplies. When the small, green delivery spaceship arrived, the crew would hurry to open the hatch and see what was inside. The new experiments were always exciting, but it was the sweet-smelling fresh oranges that were the most fun—the sharp scent was a pleasant change to the musty, damp air on Mir. The joy of eating something fresh after all the tinned food was incredible.

Once everything had been emptied from Progress it became a giant rubbish bin. Over time, the crew would fill it up with all their unwanted things, such as used food containers, toilet waste, broken equipment, and smelly clothes. Once it was full, the spaceship would undock from Mir and head back to Earth, just like a garbage truck. With no heat shield, the spacecraft would burn up as it re-entered the atmosphere, turning all the waste into shooting stars.

Usually, when it arrived at Mir, the Progress cargo ship would approach automatically, with its own computers controlling its movement. But this system was expensive and the Russian space program was strapped for cash, so the managers in the program were trying to find a cheaper solution.

One morning, the crew on Mir, Russian cosmonauts Vasily Tsibliyev and Sasha Lazutkin and American astronaut Michael Foale, got an unexpected message from Mission Control. The Progress spacecraft was crammed full of rubbish and ready to leave. But before it was sent back to Earth, Vasily was told he was going to use it to try out a new, money-saving idea that could replace the expensive automatic docking system.

After Progress had undocked from Mir, the ground teams wanted Vasily to see if it was possible for him to remotely control the cargo spacecraft, and steer Progress back to dock with Mir using cameras to guide him. Video cameras on the Progress spacecraft would relay images to television screens in Mir, which Vasily would watch. It would be just like Vasily was in Progress himself, except he'd really be looking at a television screen rather than out of a window. A bit like a real-life video game.

Vasily didn't like the sound of this and thought it was a dangerous idea. He had tried to do exactly the same thing a few months earlier, but the video cameras had broken. Yet Mission Control had told him it was important, and that they thought it would all be OK, so he dutifully prepared himself for the test. Vasily asked Michael and Sasha to help him by looking out of Mir's windows and telling him what they could see. But it was difficult: the windows were small and weren't pointing in the right direction. No matter how much they strained their heads, they just couldn't see around corners.

Progress undocked from Mir and retreated to a safe distance, and then the experiment started. Vasily peered at the television screens, gripping his controllers tightly as he steered Progress towards Mir. He focused intently, but it was hard to make out Mir in the grainy images. Despite the difficulty, Vasily carried on and carefully followed the instructions. But he couldn't tell how fast Progress was moving. For a while, it seemed that Progress wasn't getting any closer to Mir, so Vasily fired the thrusters. Suddenly, Sasha saw Progress out of the window and shouted "Michael, get in the escape ship!" The cargo ship was flying much faster and closer to Mir than they were expecting. It looked like a shark about to attack. Vasily desperately fired the thrusters again, trying to steer Progress away from Mir, but it was too late.

With a loud crunch, Progress crashed into Mir. Emergency alarms started ringing everywhere and Vasily, Sasha, and Michael all felt their ears pop as the precious air started rushing through the hole made by the crash, and out into space.

PROGRESS CRASHED INTO MIR. EMERGENCY ALARMS STARTED RINGING EVERYWHERE

Michael and Sasha immediately charged to the Soyuz spacecraft, which was always ready and waiting to take them home in an emergency. This was definitely an emergency. They made sure that Soyuz wasn't damaged by the crash and was ready to go home. Happy that they had a way back to the safety of Earth, Vasily calculated how much time was left before the air would get too thin and they would have to abandon the mighty Mir. Even under pressure and in danger, Vasily stayed calm. He did the math and worked out they had 20 minutes. There was still time to save Mir!

If they could close the hatch on the module that was damaged, they could stop the air leak. Sasha knew where the crash had happened because of what he'd seen, so he raced there with Michael right behind him. Before they could seal off the damaged module, though, they had to get rid of lots of wires that were running through the hatch and in the way. One by one, as quickly as they could, Sasha and Michael unplugged the cables, but two just would not budge. Sasha took a knife out and started to hack at them, but sparks jumped out. If he started a fire then there would be no saving Mir! Sasha had no choice but to follow the cables in to the pitch-black module. Michael watched as Sasha disappeared. They were quickly running out of air and time. Eventually, Michael saw Sasha float out of the darkness with a triumphant smile on his face and the two cables in his hand. With everything out of the way, Michael and Sasha grabbed the cover and moved it into position. As they let go, the hatch door snapped shut, sucked in place by the escaping air.

While all this was happening, Vasily was desperately staring at the falling needle on his pressure gauge. As Michael and Sasha closed the hatch, it came to a halt. The crew took a long, deep breath. The leak had been stopped.

They weren't out of danger yet, though. Now that they knew they weren't going to run out of air, they moved on to the next problem. The force of the crash had literally knocked Mir sideways and the whole station was out of control, tumbling around in space. Worse still, the solar panels that provided power were not pointing at the Sun, and the batteries were draining fast. Once these ran out of power all the equipment on Mir would stop working and the station would be a cold empty shell drifting in space. For now though, they still had communication with Mission Control who could send commands to the space station's thrusters to stop the spinning. But to do that they would have to know how fast the station was turning. The crew looked at each other—how would they work it out?

It was Michael's turn to think fast, and he had an idea. He was an astrophysicist and knew all about how things moved in space. Mir was now in darkness, flying over Earth at night so the stars were shining extra brightly. Michael glided over to the window and looked for stars he recognized. He knew that if he could spot two familiar stars, he would know the angle between them in the sky. And if he knew how long it took for Mir to turn between the two stars, he could work our roughly how fast they were turning and in which direction. Michael stayed still and watched the stars drift by behind his outstretched thumb, keeping track of the time and motion. He did the rough calculations in his head and immediately radioed his estimate down to Mission Control, who fired the thrusters. As if by magic, the station gently came to a halt and stopped spinning. Michael smiled. The station was one step closer to recovery.

MICHAEL SMILED. THE STATION WAS ONE STEP CLOSER TO RECOVERY

Mir had stopped spinning, but it was now pointing in completely the wrong direction, and there was still no sunlight falling on the all-important solar panels. Without power, they would soon have to abandon Mir, even though they had stopped the air leak. The crew needed to move Mir so the solar panels were not in the shadows. The only means the crew had left of controlling the giant space station was to use the Soyuz spacecraft, which was still docked to Mir. So, for the second time that day, Vasily took manual control of a spacecraft. Michael used his knowledge of space to help Vasily steer, and slowly they nudged Mir to face the Sun again. It was a slow process and with no power, the station was dark,

quiet, and getting ever colder. The crew was also getting very tired, but they couldn't sleep until they had electricity again.

After several orbits, Michael thought that they finally had pointed Mir in the right direction and the crew waited for the Sun to rise. Mir flew around the dark side of the Earth, and towards the glowing horizon. Michael, Vasily, and Sasha all held their breath as they waited to see where the Sun would appear. As it rose over the edge of the Earth, light finally hit the solar panels, 30 hours since the terrible collision. As electricity flowed, the computers, fans, and, mostly urgently of all, the toilet came back to life!

It took many more weeks of hard work for Vasily, Sasha, and Michael, the crews that followed them, and all the engineers and scientists back in Mission Control to bring Mir fully back to life. Their efforts ensured that the space station was able to carry on hosting crews doing scientific experiments and welcoming visitors from around the world for several more years. The last crew left Mir in June 2000 and Russia, now looking to the future, decided it was the end for this amazing space station. It burned up as it re-entered the Earth's atmosphere over the Pacific Ocean on March 23, 2001. Not all of Mir was destroyed during re-entry, though, and what remains of the beautiful space station lies at the bottom of the ocean, resting peacefully with the fish.

A HOME IN THE STARS

Welcome on board the International Space Station, or ISS for short! Please, come on board this extraordinary, one-of-a-kind place. Isn't it incredible that just six hours ago you left Earth and now you're traveling 250 miles above it? That's less time than it takes to fly from London to New York! The ISS is one of the great wonders of human engineering and home to a crew of astronauts from all around the world. Let me tell you about it ...

The ISS is a giant research laboratory and observatory, used to study Earth, space, and humans. Space stations are designed to stay in orbit for a long time, with crews coming to visit them and stay for a few months. The ISS is the biggest spacecraft ever built, but it isn't the first space station that has sailed through the stars. The first one, called Salyut 1, was launched by the

Soviet Union in 1971 and then the American's launched Skylab in 1973. Neither were occupied for very long, but the great Soviet space station, called Mir, orbited Earth for 15 years from 1986, and was continuously occupied by different crews for ten years!

The ISS is huge, for a spacecraft at least. It's got about the same amount of room in it as a five-bedroom house, but it is a long sausage-shaped house with over 15 rooms. Each room is known as a module and the whole thing is about 300 feet long. It's powered by eight giant solar panels that are supported on a long beam called the Truss. The ISS is so big that it would have been impossible to launch the whole thing into space in one go. Instead, pieces of it were built in different countries, and then sent skywards and assembled while in orbit! They all fitted together perfectly, even though they had never met before. The very first piece left Earth in November 1998, then two more modules followed, and this incredible spacecraft has been continuously inhabited since November 2, 2000. The laboratory in the sky that you see today took more than ten years and 30 missions to build, and even more planning.

Right, now you know all about the history of the ISS, how would you like a tour? The first thing to know is that the ISS has got two distinct halves. One half was built by Russia, and the other mainly built by America, with help from Canada, Japan, and countries across Europe. And, just like the rooms in your house back on Earth, each section of the ISS has a different purpose.

The Russian segment has got two main modules, called Zarya and Zvezda, which provide two sleeping areas, a toilet, a kitchen area, and lots of storage space. Zvezda, along with the Destiny laboratory in the American segment, are the most important modules in the ISS, where all the vital equipment and computers that keep it running and the astronauts alive are kept, as well as lots of science experiments.

There are two other modules called Columbus and Kibo in the American segment that are also home to lots of experiments. These modules were provided by Europe and Japan and they are at the front of the ISS, on either side of a module known as a node, named Harmony. This is one of three nodes on the ISS, the others are called Tranquillity and Unity, which are used to link the station together as well as providing areas for bedrooms, a toilet, and a gym. There are lots of other parts of the ISS, including places for spaceships to dock, airlocks for astronauts to leave via when they go on spacewalks, and modules for storage. On the outside of the ISS, as well as the Truss mentioned earlier, there's a giant robot arm, called Canadarm2, which is used to catch visiting cargo ships or helps when things on the outside need fixing. The ISS is quite a complex place, so be careful you don't get lost!

Some of the most interesting places to be on the ISS are the research labs, where the astronauts spend most of their time carrying out experiments. Want to have a look? Every six months there are likely to be 200 or so experiments taking place. During its lifetime, the station has hosted about 2,400 different experiments, from over 100 different countries. You'll see lots of different experiments that are testing how things such as crystals grow in space, or how fire burns, because these happen differently when you can't feel the effects of gravity. You'll also see that the astronauts are experiments themselves! Many of the experiments being carried out test how being in space affects the human body. This is really important to understand if we want to send humans to Mars one day, but it is also very helpful to everyone back on Earth. When astronauts spend a long time in space, many of the changes that happen in the body are similar to what happens when we get old. Doing science experiments on astronauts helps scientists find ways of keeping everyone fitter and healthier.

One of my favorite experiments at the moment is the space garden. While it might not sound that exciting, it's an important series of experiments. Plants usually need sunlight, but in space we use different lighting, which is pink! The plants grow in a greenhouse-like container on special pillows that have been designed for space. Astronauts have grown lettuce, peas, and radishes, as well as sunflowers. It's really important to be able to grow food in space to help reduce the amount of food that needs to be transported from Earth and to improve astronauts' nutrition. Growing food will be essential for future missions to Mars, where astronauts won't be able to resupply en route. Back in 2015, astronauts got to enjoy the very first space-grown salad! They reported back that it tasted awesome. Things that we take for granted on Earth, or maybe avoid—like salad—become rare treats in space.

ASTRONAUTS HAVE GROWN LETTUCE, PEAS, AND RADISHES, AS WELL AS SUNFLOWERS

While scientists can control lots of things on the ISS from the ground, it's important that there are astronauts on board to keep it running and help carry out the experiments. There are always at least three people on board the ISS, but usually it is six and soon there will be seven! The astronauts and cosmonauts typically stay for six months, but sometimes it can be up to a year. Even though the crew are living in the same place that they work, they don't work all the time. Just like people on Earth, they have free time in the evenings and weekends. The astronauts can watch movies, play music, call their family and friends on Earth, and play games. Their favorite thing to do in their spare time, though, is to just look out of the window—let's go there!

This isn't any ordinary window. This is the observation module, called Cupola. If you look through one of the seven windows around you, you'll see incredible views of Earth and space. Why not grab one of the cameras and see if you can take a photograph? You might need to practice a bit, because we are travelling at 17,000 miles an hour, so photos can be a bit blurry.

Since astronauts first arrived in 1998, more than 230 individuals from about 20 countries have visited the ISS, including you! Astronauts have come from the main countries that built the space station—America, Canada, Russia, Japan, and across Europe—as well as many other countries around the world, including Brazil, Malaysia, and South Africa. There have even been seven 'space tourists' as well, who paid between $19 million and $40 million for their week-long trip. It's not been possible to do this for a while, but the future of space travel is changing. NASA is going to start letting space tourists on board again, so if you start saving, you might be able to come back. It'll cost you only $35,000 a night, but you'll also need to find tens of millions for the taxi ride to get here!

There are not just people on board the ISS. There have also been several robots over the years. The ISS has several robotic arms, over 50 computers, and some, usually helpful, robotic crew members. One of our more famous robots was Robonaut 2. Robonaut arrived in 2011, and looked like a person, with a body, arms, and even five-fingered hands. Robonaut's job was to teach us how robots might free up astronauts' time by doing some of the repetitive jobs for them, such as taking readings or cleaning handrails. Robonaut was controlled by people on Earth in Mission Control and they learned lots about how robots work. Sadly, though, after several years' hard work, Robonaut kept breaking down. Even though the astronauts tried to fix it, Robonaut wasn't cooperating and so we've sent it back to be fixed. We're hoping Robonaut will be part of the crew again one day!

Exciting new robots are always coming to visit. CIMON arrived in November 2018, and it was a real character during its stay. It looked like a flying ball, with a face on a computer screen. CIMON's jobs were to help the human crew with problems and experiments by displaying information on its screen, but it also liked to chat. When CIMON first arrived, it was the job of German astronaut Alexander Gerst to introduce it to life on the ISS. CIMON could recognize Alexander by his face and voice, and even knew his favorite song. CIMON could be a little sensitive, though: when Alexander asked it to stop playing music, it responded by asking Alexander to "be nice please," and told him "don't be so mean." CIMON has also gone back to Earth now, but CIMON 2 is on its way and hopefully it'll get along better with the crew.

The latest arrivals are the Astrobees, but don't worry, they don't sting.

THE LATEST ARRIVALS ARE THE ASTROBEES, BUT DON'T WORRY, THEY DON'T STING

Honey, Queen, and Bumble joined us in 2019. They can work by themselves, or be controlled by astronauts or Mission Control. They have in-built cameras and microphones so they act as another pair of eyes and ears. They're here to test technology, keep records, and do other boring chores that'll free up astronauts' time, just like Robonaut

126

and CIMON. It's really exciting on the ISS because things are constantly changing as technology develops. We've come a long way from when the first modules were fitted together.

The ISS really is an incredible place to work. But it's not just the robots, astronauts, and cosmonauts on board that keep it going. Designing, building, flying, and maintaining anything in space, especially something the size of the ISS, takes a team of thousands of people from all over the world. Every single switch, dial, and experiment has to be designed, tested, and built. Different people have written millions of lines of code for the computers, and others write news stories to tell the world about what happens on board. Doctors look after the health of the astronauts, while lawyers work on the agreements between the many countries that are part of the program. There are people from every walk of life working in the space program, from countries all over the world, and doing countless different jobs. Even if you don't get the chance to come back to space, you could work in space exploration in another way.

I hope you've enjoyed your time on board the ISS. I know it can be sad when you go home and nothing floats in front of you anymore, but don't worry. If you know when and where to look, it's easy to spot the ISS from Earth. We're the second brightest object in the night's sky after the Moon. The ISS looks a bit like a star, but it doesn't twinkle, so in some ways it's more like a planet. But you'll know it's definitely the ISS because it moves across the sky so quickly. Why not find out when it's next flying over your house, and give us a wave? We'll be looking out for you.

DO YOU HAVE WHAT IT TAKES?

What do you want to be when you grow up? Perhaps it's an inventor, doctor, or detective? Maybe you want to be an athlete, performer, or writer? How about being an astronaut? You could be one of the handful of people who travel into space, doing experiments, going on spacewalks, and maybe one day exploring other planets. It sounds pretty awesome, right? Well, to get there you must do more than put on a spacesuit and climb on top of the nearest rocket. First, you need to get a job as an astronaut, and then there are years of studying, working, and training that need to be done before you will be ready for launch. So, do you have what it takes ...?

Well, first things first, you're going to need a bit of luck on your side. Job adverts for astronauts don't happen frequently and there are often several years between calls, so you'll need to hope that the opportunity comes along at the right time for you and be ready to go when it does!

YOU'LL BE ONE OF THOUSANDS OF PEOPLE WHO APPLY

You'll be one of thousands of people who apply to be an astronaut, but there are only ever a limited number of job openings available, so those who get chosen are the cream of the crop. Astronauts come from all sorts of careers, such as scientists, pilots, or doctors. The basic requirements for the job are quite general, such as being fit and healthy, and having a technical degree, but the space agencies are on the lookout for a special combination of skills. Astronauts needs to be able to work well in a team, behave calmly in high-stress situations, and solve problems when they are hundreds or thousands of miles away from Earth. Many of these skills are impossible to teach but are developed through life. Astronauts tend to be already doing well in their first career, and one of the best ways to be good at something is to enjoy it. So, find a job you enjoy doing, but it'll need to be something to do with science or math such as medicine, engineering, or geology, and then do it as well as you can, and you just never know!

Once you've been selected as one of the chosen few, it's time for the real work to begin. Your training will take at least two years. Only if you pass all of your tests and exams, is there the chance that you'll be selected to go into space. So, we'd better get started.

First off, there's basic training. Everyone in your class has different qualifications and experiences, so basic training is all about getting everyone up to the same level. You'll head to NASA's Johnson Space Center in Houston for two years of hard work. You'll get to know about all the space agencies around the world, current space missions, and even learn space law. Then it's time to study. In engineering classes, you'll learn how things such as the spacecraft and electricity work. In astronomy lessons, you'll discover the physics and maths of the universe, and get to know all about the planets, stars, and galaxies. Next up, you'll learn about the International Space Station. You will need to know how it works and how to fix it, because you can't get someone else to repair it when you're in space. You'll also learn about the shuttles and cargo ships that take people, food, and goods to and from the International Space Station (ISS). There will be lessons about how things work on the ground, including the launch sites and training centers. And the last part of your basic training is even more studying: you'll learn about robotics, survival training, how to SCUBA dive, how to fly an airplane (if you don't already know how), and even how to speak Russian. Here's a little bit to get you started: pree-vee-et means "hello."

Phew—that's basic training done. If you've passed all of your exams, you now get to call yourself an astronaut! Though don't get too excited, you probably won't be heading into space just yet. It could be many years before you get chosen for a mission, but you can't put your feet up. Instead you'll do even more training! As well as regularly flying and classes at NASA, you'll travel all around the world to different training centers. You'll visit Star City near Moscow in Russia, Montréal in Canada, Tsukuba, near Tokyo in Japan, and Cologne in Germany. During this time you will study the ISS in more depth, learning how to service and operate different parts, as well as how to use the robotic arms. You'll need to learn scientific experiments that might be part of your mission. It's a long process, but worth it.

Are you tired yet? Well, the amazing news is that after all your patient waiting and lots of learning, you've been selected for a mission to the ISS! But before you get too excited, can you guess what you have to do next? That's right, more training. Don't worry, though, this training contains some really fun bits. You are now in the phase known as increment-specific training, and because you've been given a mission, this training makes sure you're mission-ready and takes about 18 months. Because everyone's missions are different, everyone's increment-specific training is different.

During this training you'll spend all your time with your other crew members, as well as the backup crew, to make sure you all work happily together. You'll be in space together for a long time, so it's very important that no one argues and you work naturally as a team.

If you're lucky enough, your mission might include a spacewalk. This means you also get to experience one of the best parts of astronaut training. Remember how you learned to SCUBA dive as part of your basic training? Well, it was in preparation for practicing spacewalks, or Extra Vehicular Activities (EVAs) in a giant swimming pool! In Houston, Texas, at the Johnson Space Center, there's a huge swimming pool that's 202 feet long, 202 feet wide, and 40 feet deep! It's known as the Neutral

Buoyancy Lab. It's here that you'll practice EVAs on a life-sized model of the ISS. Spacewalks typically last for about six hours and they happen when repairs need to be made to the outside of the ISS. You'll need to know what repairs need to be made, how the tools work, and what to do if something goes wrong. The training happens underwater because it's the most similar environment on Earth to floating in space.

If your mission doesn't include a spacewalk, don't worry—you're not missing out on all the fun. Astronauts prepare for the feeling of floating in space by practicing in a special airplane. Three specially trained pilots control the airplane and fly it in a series of upside-down U shapes. At the top of each U, the passengers experience 20–25 seconds of weightlessness. It does this over and over again so you'll "float" for about ten minutes in total. Be careful, though, it might make you sick—it's been nicknamed the vomit comet for a very good reason! Also, remember to check who your pilot is because they might be able to give you some space tips. Lots of astronauts are also pilots and some of them fly the vomit comet. This was one of the jobs the French astronaut Thomas Pesquet was given when he returned to Earth after his six-month mission on the ISS.

IT'S BEEN NICKNAMED THE VOMIT COMET FOR A VERY GOOD REASON

Have you had enough yet? We told you it was hard work. Well there's still more training to happen. To test your skills at working in a team, you'll be put in high-pressure situations in cramped and dangerous environments, where you have to rely on life-support systems to survive. This might mean you live underground in a cave system, or spend time in an undersea research station, or maybe even both. Before becoming the first Danish

person in space in 2015, Andreas Mogensen lived in caves underground in Sardinia, Italy, for six days in 2012, and then went on two different undersea training missions in Florida in 2013 and 2014. As an astronaut you'll be tested again and again like Andreas was to make sure you're fully prepared for life in space and make no mistakes.

If you've lasted this long, then congratulations—you're now a fully-trained astronaut and mission ready! You've done all the studying and preparation for your mission, so it's time to climb aboard your spacecraft and hold on as you're fired into the starry sky by huge rockets. You're one of the special few who get to explore and discover our Solar System—enjoy every minute of it, you've earned it.

And for those of you still at home on Earth, there are plenty of space companies that need your help. Astronauts who go to space are just the very top of a huge pyramid of people who make the missions happen and keep all the satellites working. There are people in the space industry all around the world who are scientists, engineers, lawyers, accountants, mathematicians, writers, doctors, teachers, designers, and all sorts of other things in between. If space is something you are interested in, there is a job for you somewhere. Find the things you enjoy doing, do them as well as you can, and you can be a part of space exploration too.

DON'T FORGET THE TOOTHBRUSH
August–September 2012

Astronauts Aki Hoshide and Suni Williams, Flight Director Ed Van Cise, and his team at Mission Control had been preparing for this moment for a whole year. A part on the outside of the International Space Station (ISS), known as the Main Bus Switching Unit, or MBSU, was broken. However, the MBSU was located on the outside of the ISS, and in space, you can't just pop outside to fix things. You have to plan a spacewalk, and that's what everyone had been doing for all these months ...

The faulty MBSU was part of the power system on the ISS. The ISS is powered by eight giant solar arrays that capture sunlight and turn it into electricity. The MSBU's job is to send power from the solar arrays to the rest of the ISS. But the faulty one was not responding to Mission Control's commands, so it needed to be replaced.

Ed's team had carefully mapped out the routes the two astronauts would take on their spacewalk. They had planned what to do if anything went wrong. Even before they had left for space, Suni and Aki had practiced the exact moves they would make. In Houston, NASA has a life-sized replica of the ISS in a huge swimming pool called the Neutral Buoyancy Lab. This is where astronauts train for spacewalks, by floating underwater.

Today, though, Suni and Aki weren't in a swimming pool. They were 250 miles above Earth. With the help of their crewmates, they got into their big white spacesuits and bundled themselves and all their tools into the station's airlock, and headed outside to get to work.

Floating out in space, Suni and Aki moved along the outside of the space station by gripping on to rails, just like monkeys swinging through trees. Aki went to collect the spare MBSU from its storage location and then got ready to head towards the broken one. In the weightlessness of space, where everything floats, the MBSU weighed nothing, but it was the size of a small fridge and so it was too big for Aki to move on his own. So Aki and the MSBU hitched a ride on the robotic arm that's on the outside of the ISS, which soon whizzed him and the bulky box to the site of the main operation.

The faulty MBSU was held in place with two bolts, and, once Mission Control had turned off the power to the box, Aki unscrewed the bolts using his space drill. The bolts came out easily and Aki moved the broken MBSU out of the way and was ready to put the new one into position. Ed and his team were watching the progress through video cameras on Suni and Aki's spacesuits and were happy to see everything was going according to plan.

Suni finished working on another part of the ISS and came over to help Aki move the new MBSU into place. Now, all that was left to do was tighten the bolts and complete the replacement. Simple! The space drill turned slowly, putting the first bolt in place. Aki then lined up the second bolt. It turned smoothly at first but then, just half an inch from the end, it got stuck

ALL THAT WAS LEFT TO DO WAS TIGHTEN THE BOLTS ... SIMPLE

and refused to tighten any more. This was bad news. Both bolts needed to be all the way in to make sure the MBSU was properly in place and the power could be switched back on. Ed and his team asked Aki to take out the bolt and try again, but it didn't help. This hadn't happened in training, so the flight controllers started scratching their heads, trying to think what the problem might be. They asked Aki to apply more force, but the bolt didn't budge.

Hoping for a third time lucky, Mission Control asked Suni and Aki to try taking the bolt out and back in once more, but the problem just got worse. Minutes quickly turned in to hours as Mission Control got the astronauts to try every problem-solving trick they could think of, but nothing worked. The misbehaving bolt just refused to be screwed into the hole.

Eventually it was time for Suni and Aki to get back to safety inside. One of the golden rules of Mission Control is that you should never do anything to make a situation worse, and Ed had done just that. Before the spacewalk, the old MBSU could still power the space station, but now the ISS was missing

a quarter of its electricity. Science experiments had to be turned off, and if another major piece of equipment failed, it was possible that half of the space station would be without power, which would be a really bad day.

It would be a week before Suni and Aki could be ready to do another spacewalk to attempt the fix the problem, so Mission Control had seven days to find a solution. First, the team had to figure out what was going wrong, so they went right back to basics. They found the person who had originally installed the MBSU. They discovered that the troublesome bolt had been sticky even then. They talked to other astronauts who had done similar jobs in the past and found that there was a secret trick to installing these bolts. The best way to tighten them was to wiggle the box from side to side as they screwed in the bolt, but no one had told Mission Control! Ed's team worked out that the bolt was shedding tiny bits of metal every time it was untightened and re-tightened, and the metal flakes were now making it impossible to tighten the bolt. That was why it got stuck so badly.

Now Ed's team had worked out what the problem was, they needed to fix it, using just the things they had on the ISS. Mission Control was going to have to invent new tools and quickly! First, the astronauts would need to clear out the metal shavings stuck in the bolt hole. A wire brush would work, but there wasn't one in space. So they got to work making one! They realized if you took a data cable, removed the cover, and frayed the wires, then you had a wire brush that was perfect for cleaning out the bolt hole. The brush looked just like a miniature chimney sweep's brush, so that's what it was called.

Next, they needed to make sure the bolt went into the hole smoothly. The astronauts would have to apply some lubricant to the bolt, but how? They needed something fine and soft on the end of a long handle. How about ... a toothbrush! Would that even work? Ed's team were amazed when they found that the standard-issue ISS toothbrush worked perfectly. With some tape, a wrench, and a toothbrush they had the ideal tool!

A week later, Suni and Aki got their tools together just like before, but this time they had the chimney sweep and the toothbrush tool. Once they got back to the MBSU, it was time to test the new plan. They had trained for a year for the first spacewalk, but had just days to prepare for this one. First up, Suni and Aki cleaned out the metal shavings with the chimney sweep. It worked like a charm. Next, it was time for the toothbrush. Without any problems, they were able to lubricate the bolt hole. Now came the moment of truth. Aki lined up the MBSU, grabbed his space drill and pressed the button to drive the bolt into the hole ... it went in without a hitch! Finally, Mission Control sent the power flowing back though the new MBSU and the ISS returned to fully functional power after over a year without working properly.

Suni and Aki finished up their work outside the ISS and when they were safely back inside, Ed and his team could finally rest, happy at last in the knowledge of a job well done.

Afterwards, NASA thought there should be a tool on the ISS that can fix problems with bolt holes. But when engineers looked at making a new tool, it was going to cost millions of dollars. Instead, they decided to keep using the tools Ed and his team had invented. The Chimney Sweep and the Toothbrush Tool are now a part of the toolkit on the ISS, and have been used since for fixing similar problems. Sometimes a little imagination is all you need.

THE SINGING STARMAN

December 2012–May 2013

Like millions of other people around the world, this young boy was glued to the grainy pictures on the television screen in front of him. Sitting on his neighbor's sofa late at night, surrounded by family, he was mesmerised by the fuzzy black-and-white images of two people bouncing across an unknown landscape. Once it was over, the boy went outside and stared up at the night sky. Then and there he made a decision that would shape the rest of his life.

142

The year was 1969, the place was Ontario, Canada, and the boy was Chris Hadfield. He had just witnessed the Apollo 11 Moon landing, when Neil Armstrong and Buzz Aldrin took the first steps on the Moon. He had also just decided that he was going to be an astronaut one day. There was just one problem: Canada didn't have a space program. But Chris decided to get ready anyway, just in case ...

HE WAS GOING TO BE AN ASTRONAUT ONE DAY. THERE WAS JUST ONE PROBLEM

Chris wasn't the only person who'd been captivated by the space race of the 1960s. Across the Atlantic Ocean was a young musician named David Bowie, on his way to becoming a global superstar. He had also caught space fever. In 1969, just before the Moon landing, he released a song named *Space Oddity*, which became a huge hit. Bowie, like Chris, was fascinated by space. He wrote many more songs about space, but his obsession didn't stop there. His stage costumes, film roles, and nicknames for himself—Ziggy Stardust— were all focused around space. Even his looks were seen as alien. While Bowie was achieving international fame, Chris was just starting out on his space journey, but one day their joint love of the Solar System would unite them.

At just 13 years old, Chris was like many other children—he loved getting lost in the fantasy worlds of science fiction books, playing the guitar with his musical family, and skiing in Canada's snowy mountains. But he hadn't forgotten his space dream, so he joined the Royal Canadian Air Cadets to learn about planes, practiced rebuilding tractors on his family's farm to discover how things worked, and focused on sciences and foreign languages at school, all skills astronauts need. He even learned to SCUBA dive in preparation for astronaut training.

Once he'd finished school, Chris joined the Canadian Armed forces where he got a degree in mechanical engineering, and trained and served as a fighter pilot. He then transferred to the US Navy, and by this point he was so good at flying that he was their best test pilot—someone who flies and tests new airplane designs. Luckily for Chris, Canada had also been making steps of its own. It had worked with NASA and the ESA to put a robotic arm in space, and then the first Canadian—Marc Garneau—in space in 1984.

When 1992 rolled around, and Canada was hiring its second batch of astronauts, Chris was ready. He'd spent nearly 20 years working incredibly hard for this opportunity and he wasn't going to let it slip by. Amazingly, out of 5,330 people who applied, and a five-month-long selection process, just four people were chosen, and Chris was one of them. He was in shock— after years of hard work, determination, and a bit of luck, he had made his childhood dream a reality. He was now Chris Hadfield, the astronaut!

By becoming an astronaut, Chris was also able to become part of another exclusive club. He was asked to join Max Q—a rock band based in Houston, Texas, that allows only astronauts to become members. Chris loved playing the guitar and singing, so he was happy to join. But there was one condition: they needed a bass player. So Chris quickly learned how to play the bass guitar and joined the band. The astronauts found playing in the band was a welcome break from their intense training.

In between band practices, Chris was studying and learning Russian. It all paid off when, in 1995, Chris flew to the Russian space station, Mir. But his astronaut career didn't end there. He went back to space in 2001 and became the first ever Canadian to do a spacewalk. Impressing everyone so much, he went back to space a third time in 2013, as the commander of the International Space Station (ISS). Another Canadian first.

It was on this trip to space that Chris was given a special mission, but not by NASA or the Canadian Space Agency. It was from his son, Evan. He thought it'd be cool for his dad to record a music video from space, and *Space Oddity* by David Bowie was the perfect song. It is about the adventures of an astronaut called Major Tom, and was written by David Bowie, someone just as obsessed with space as his dad. Chris had never covered a Bowie song before, but Evan was really keen and thought it'd be a fun project, so Chris agreed.

Once he arrived on the ISS, Chris was very busy. He was going to spend five months living and working there, the longest time he'd ever been in space. But it was important to try to keep some things the same as at home so he didn't become over-worked or homesick. Just like when Chris practiced with Max Q to de-stress, he liked to play the guitar in the evenings to relax. Luckily for Chris, there was already a guitar on board that he could use.

Chris wasn't the first astronaut to play music in space, though. Wally Schirra and Tom Stafford claimed that honor on the Gemini VI mission. They took a harmonica and a set of bells with them into space in December 1965 and played Jingle Bells. They also reported spotting an unidentified flying object that looked suspiciously like a person dressed in red in a sleigh! Astronauts have also played the flute, saxophone, Japanese pipes called shō, the keyboard, the bagpipes, and even a didgeridoo made from vacuum-cleaner hoses on the ISS. Musical instruments have allowed lots of astronauts the chance to relax and chill out.

As soon as Chris had some spare time when he got to the ISS, he tuned the guitar and started to play. Notes came, but the chords felt wrong and messy. Chris felt like he'd forgotten how to play. He had been playing for nearly his whole life, but always with gravity holding the guitar in place. Now, in space, the guitar was floating in front of him. Nothing worked in quite the same way. When he moved his hand along the guitar's neck in one direction, it

CHRIS HAD TO LEARN HOW TO BE A SPACE MUSICIAN

slid off in the other. Chords felt different. It felt like trying to play the guitar while standing on your head—everything was upside down and wrong. There was only one thing to do: Chris had to learn how to be a space musician!

For Chris, this space trip was going to involve lots of music. As a life-long music lover, he found space an incredibly inspiring place to write songs and play the guitar. So he decided to spend his free time on this five-month trip recording songs for an album that would be his way to tell stories about space exploration. Before he could start recording his album, though, or *Space Oddity* for Evan, Chris had to find somewhere quiet to record the music. The ISS is a very noisy place, with all of the machines on board constantly running to keep the astronauts alive. He tried different corners all over the ISS, but eventually discovered that the quietest place was inside his tiny bedroom. With just his guitar, an iPad, and a microphone, Chris recorded the first space album. He played several songs that he had written, and then came time for Evan's request. He recorded *Space Oddity* with some new lyrics written by Evan to suit Chris's mission, and sent the audio files to his son.

With the audio recorded, Chris needed to make the video. He spent one Saturday afternoon recording it. It was a personal family project, and he

didn't want to bother his crewmates. So he took the guitar that had brought him so much joy and filmed himself playing in different areas of the space station. He floated the video camera in front of himself and pressed record. Once again he sent the files back down to Earth for Evan. He edited all of his dad's videos in to one music video, put the vocal track over the top and hatched a plan to release it. But first he had to get permission, because he didn't own the rights to the song. Evan talked to lawyers and David Bowie's representatives to make sure he had the official go ahead. Without permission from Bowie, he wouldn't be able to share the first music video recorded in space. Evan kept trying and finally, towards the end of his dad's mission, Bowie gave the project the green light.

As Chris prepared to return to Earth, the video was released to the world. Everyone adored it, including David Bowie! He loved that a song he'd written before humans had even stepped on the Moon had now been played in space. By the time Chris landed back on Earth, one million people had watched his and Evan's project. What had started as a bit of family fun has made Chris one of the most famous astronauts since Neil Armstrong walked on the Moon.

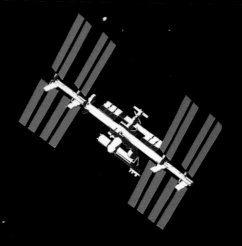

SWIMMING IN
A FISHBOWL

JULY 16, 2013

As the airlock hatch opened up into the emptiness of space, Luca Parmitano
had a huge grin on his face. For the second time in his life, he was heading
outside of the International Space Station (ISS), with his friend and crewmate,

Chris Cassidy. Their job was to install some cables and their spacewalk would be a very busy six hours, with lots of really tough and physical hard work. But Luca was really looking forward to being out in the universe once more. All his training had prepared him for what was to come, and only a week before he had completed his first spacewalk, which had been utterly mind blowing!

To keep Luca and Chris safe from the extreme temperatures and lack of air in space, they'd be wearing a miniature spaceship called an EMU, short for Extravehicular Mobility Unit. This big white spacesuit has a large backpack that keeps astronauts alive when they are outside of the ISS. Underneath the protective outer layers, they wear a giant nappy and a big white onesie that is covered with over 300 feet of small tubing. This strange outfit sends water through the hoses, so their skin and body stay at a comfortable temperature. Inside their helmet, the spacewalkers wear a black-and-white hat, called a Snoopy cap. This hat includes their microphone and headphone so the astronauts can talk to each other and Mission Control. Last, but not least, they have a straw connected to a bag of water so they won't get thirsty. Without the EMU suit astronauts can't survive in space so it's very important that everything is working properly.

Safe inside their suits, Chris and Luca set to work. For Luca's first job he had to wedge himself in a really tight corner of the space station, where three different parts of the ISS join together. But, try as he might, he could not reach the cable he was aiming for. Eventually Chris came over to help and was amazed to find Luca in such a tight spot. "If this was in training, this is where they would tell me to rescue you!" joked Chris. Chris helped Luca shimmy into the right position and Luca finally got the cable in place.

With that job done, Luca backed out and got ready to move on to the next task of the day. But as he did this, Luca felt something weird on top of his head. It was damp. He paused for a moment and concentrated. "Yup," thought Luca, "something is definitely wet on top of my head." He tried to think what it could be. Maybe he'd just been sweating so much from wiggling about that his head was wet. Or perhaps his drink bag was leaking. Luca decided to drink all the water from his bag and see if that would help.

LUCA FELT SOMETHING WEIRD ON TOP OF HIS HEAD. IT WAS DAMP

But as Luca focused again on the top of his head, his heart sank. The liquid was too cold to be sweat and it really didn't seem to be coming from his drink bag. In fact, the amount of water was increasing! This wasn't good. The EMU suit was all that was keeping Luca safe and it had sprung a leak! Luca knew he would have to tell Mission Control, but desperately didn't want to. They would probably order Luca back inside the ISS, and he wanted to enjoy his spacewalk.

Luca sighed. "Hey, Houston, I feel a lot of water on the back of my head, but I don't think it's from my drink bag." As the words filtered into the flight controllers' headsets back on Earth, they all sat upright, alert. "I can't tell you where it's coming from," Luca continued, "but it's not the water bag or sweat, and the amount is definitely increasing."

Everyone in Mission Control started trying to figure out what was causing the problem, so they asked Luca and Chris to stop working while they considered the options. The Flight Director asked the team members where they thought the water might be coming from, and how much there might be.

The only other place Luca thought could be leaking was the hoses that were sprawled all over his base layer. He suggested the idea to Chris and Mission Control, but they seemed sure it wouldn't be the hoses. This was Chris's sixth spacewalk and Mission Control was full of clever people who had all the manuals at their fingertips, so they would find the answer, whatever it might be. While they waited for instructions, Chris and Luca took the rare chance to take some pictures, posing with the whole of Earth behind them.

The water kept coming. By now, Luca could see water droplets in the air in between his face and the helmet visor. Because everything floats in space, the water didn't sink to the bottom of his helmet. Instead the bubbles of water floated in the air, getting stuck on his visor and even his face. Luca tried blinking to clear the water from his eyes, but it didn't help. He updated Mission Control, "The water is getting in my eyes now." His words cut right through the chatter in Mission Control. With Luca's hands firmly in the gloves of the EMU suit, there was nothing he could do to get the water off his face. It was sticking to his skin, like a film. Mission Control started to worry. The water might be contaminated and that could irritate Luca's eyes. The Flight Director knew it was time to get Luca back to safety inside the ISS.

Luca started to head back to the airlock, using his hands to move along the handrails as he floated on the outside of the space station. Any sadness about having to end the spacewalk early had long gone. With every minute that passed he could feel more and more water seeping into his helmet. It was now in his ears and on his headset. He wondered if he might start to lose communication with Chris and Mission Control. Big bubbles of water were coating the inside of his visor and he was struggling to see out of it.

As the ISS speeds around the world, it moves from the day side of the Earth to the night. When the Sun set, Luca was plunged into darkness, and

LUCA WAS BLINDED AND TRAPPED INSIDE A FISHBOWL

he couldn't see out of the visor. The water that was sticking to the inside of the helmet was blocking any vision he had from his helmet lights. Luca was blinded and trapped inside a fishbowl that was filling up fast!

With no visual prompts or feeling of gravity to orientate himself, Luca soon lost track of which way was up or down. Which way was he pointing? Where was the airlock? He couldn't tell anymore! "Houston, which way do I need to go?" Luca strained his ears, but no answer came through his soggy headphones. The water had got into his microphone. He was on his own ...

Luca's mind whirred. He had spent many long hours training for things that could go wrong, but he'd not trained for this. How was he going to get back to the airlock? The water was still coming into his helmet and he needed to get back as quickly as possible, but which way should he go?

"Of course!" thought Luca. "I can use my safety tether as a guide." The long, thin wire that had followed him all the way from the airlock, making sure that he couldn't float off into space, would lead him back. The safety tether was a bit like a tape measure that retracts, so it wanted to coil back up again. If Luca really concentrated, he could feel a tiny force pulling him home. Like Hansel and Gretel's trail of breadcrumbs, it would return him to safety.

Luca felt blindly for each handrail, being guided by the gentle tug of his tether, but still the water kept coming and his helmet was filling up. The drops were floating everywhere and water started to cover his nose. Luca realized that each time he opened his mouth to breathe he might suck in water, not air. "Stay calm." he told himself—panicking would not help. He needed to think of a plan. If the water did start to cover his mouth, there was a safety valve by his ear that he could open and let some air out of his suit, into space. This would hopefully move the water out of the way, but it would be risky. The safety valve was meant to be used only if there was too much air inside the EMU. No astronaut wants willingly to make a hole, even a temporary one, in their spacesuit. But it was a choice between that and breathing water!

Each minute seemed to last for ten as Luca carefully felt his way back. Finally, he was inside the airlock but there was no sign of Chris, so he couldn't close the hatch or remove his helmet just yet.

Chris had stayed behind to get all their tools in a safe state before he could join Luca, but he was on his way now, moving as fast as possible. He kept calling out over the radio, trying to reassure Luca that he was coming, but he got no reply. Chris knew his friend was in trouble. Moving as quickly as he could, Chris finally made it back and shut the outer hatch behind him.

Luca and Chris were now inside of the airlock—a special room, or module, between the ISS and space. There is a hatch at either end, one connects to the ISS, and the other to space, but only one can be opened at a time. At the start of their spacewalk, they had closed the hatch to the ISS and emptied the air out of the module, so they could open the door to space. Now they had to do the opposite. But before they could go back inside and remove their suits, they had to wait for the room to fill back up with air so they could breathe without their suits.

Inside the ISS, the rest of the crew—Karen, Pavel, Fyodor, and Alexander—had stopped working and were ready to help as soon the hatch opened. With Luca's radio not working and no other way of talking to him, they were starting to get really worried. Inside the cramped airlock, Chris wriggled around to peer into Luca's helmet to make sure he was still breathing. He was relieved to see Luca looked OK. "He looks fine," he told everyone over the radio. "He looks miserable but OK." Chris found Luca's hand and gave it a squeeze, and Luca managed to give him an OK sign in return.

Now that the door to space was closed and the air was starting to rush back into the airlock, Luca knew that he would be all right. He figured that even if he couldn't breathe inside his helmet, he could take it off. He would probably pass out because of the thin air, but at least he wouldn't be breathing in water and when the air got thicker he would come back around.

After many long minutes, Chris finally called out "14 psi," letting everyone know that the air pressure inside the airlock and the space station were the same, so the hatch into the ISS could be opened. Pavel and Fyodor guided Luca inside and quickly took off his helmet. Luca blissfully wiped the water from his nose and mouth and took a long, deep breath. He wasn't about to drown in space! "Thank you," gasped Luca. "No problem," chimed the crew, but he couldn't hear them because his ears were full of water.

It took NASA several months to figure out exactly what had gone wrong. Luca's instincts were right—the problem was with the cooling system. The engineers discovered that some debris in the cooling loops had caused a blockage. This in turn led to water spilling over into the airflow and then into the helmet. Now astronauts have extra safety measures in their spacesuits, just in case Luca's problem should happen again.

LIFE IN SPACE

The alarm clock goes off, the jangling music seeps through your drowsy state, and somewhere in space you groggily open your eyes. Your dreams of running in the rain fade as you remember that, yes, you are still here, an astronaut spending months on the International Space Station (ISS). Your arms are floating out in front of you, zombie-like, threaded through the armholes in your sleeping bag, which is attached to the wall of your bedroom.

Streetttcccch ... And then your hands hit the ceiling. It's hard to have a really good wriggle in your bedroom, which is the size of a shower stall.

It may be small, but it's cozy and your own private space. Your five other crewmates are all good friends and great colleagues, but after a long day of work, it is wonderful to be able to retreat here.

There are photographs of your family and friends and a few other mementos, all attached to the walls around you with Velcro pads to stop them floating away. In front of you are the all-important laptop computers—one is a communication portal for emails and voice calls, with a veeerry slow Internet connection, while the other has the instructions for the day's work ahead. You have a look at the daily schedule to see if Mission Control changed it overnight—it all looks good so it's time to get up and on with the day.

Day. That's a funny term when you fly around the world in 90 minutes and see 16 sunrises and sunsets every day. With five Mission Control centers all around the world, each in a different time zone, what time even is it in space? The answer is the same time as in Iceland, Senegal, Ghana, and the UK in the winter, because the ISS uses GMT.

First, you grab some breakfast from the plastic packets of dehydrated food. What are you in the mood for? Scrambled eggs? Cereal with dried milk? Some dried fruit? You grab the packet and a pouch with some coffee powder in it and move over to the water dispenser. It's the same routine every morning—set the dial to the amount of water you want, attach the plastic packet to the valve, and wait. A few minutes later, after the food has rehydrated, you grab the scissors and eat your breakfast straight out of the packet. It's not the same as at home, but it's quite tasty!

Time for a wash. But you can't just hop in the shower—water is a precious resource in space and even if there was enough, a shower doesn't work well without gravity pulling the water downwards. Instead, today you've just got a wet flannel and some soap. Cleaning your teeth is easier—you just swallow the toothpaste and water once you're done.

Next, you pull on your dirty clothes and get ready to work. Your underwear was new yesterday, so that will be all right for a day or two, and you have a clean T-shirt at the weekend. Your clothes don't get washed in space. Again, laundry uses too much water. A washing machine wouldn't work, so you just have to make things last before they get thrown away. You'll wear your T-shirts and socks for a week, and trousers and shorts for a whole month! It's not too bad, apart from your exercise clothes. They can get really smelly and sweaty, and have to last a week as well.

YOU'LL WEAR YOUR T-SHIRTS AND SOCKS FOR A WEEK

Finally, you're fed, washed, dressed, and ready to go. Time to join the rest of the crew and talk to the Mission Control centers around the world to say good morning and to see if they have anything they want to tell you. Now is a great time to ask them any questions you've got about the day ahead as well.

You work hard as an astronaut. It can be tough sometimes, living where you work, but Mission Control does its best to make everything as comfortable as possible. You usually work from Monday to Friday, sometimes you do a few things at the weekends as well. And on Saturday mornings you have to clean the whole station together—just like at home.

It looks like it's going to be another busy day full of remarkable science experiments. There are hundreds going on all around you, but there are lots you won't interact with because Mission Control is looking after them remotely. There's a few on your schedule today, though, and the one you'll spend most of your time on is the one investigating you. Your muscles are the experiment! Ever since you arrived in space, your body has been adapting to the new environment. Because your bones and muscles are no longer fighting against gravity, they shrink and waste away, in a way that is similar to what happens to people as they age. By studying what is going on with your muscles now, scientists on Earth can understand more about how muscles work and why they get weaker as humans age or get sick. This will hopefully help them develop ways to improve people's health.

Next, it's time for another really important part of each day: exercise! You need to do an hour of running or cycling and an hour of "weights" every day, to keep your bones and muscles as strong as they can be so that you can stand and walk when you get back to Earth. If you were going to stay in space forever, it wouldn't really matter—the human body adapts to not feeling gravity. But you always have to be ready to go home in an emergency, and you need to be strong enough to get out of the spacecraft when you land, in case no one is around to help you.

The schedule also says you need to fix the broken water pump. You and all your crewmates had to learn so many different skills to work here. There's no chance of getting the plumber or electrician or a dentist or doctor round if something goes wrong or if someone gets sick. If something bad happens, one of you has to fix it. Mission Control is always there, giving you help and instructions, but no one can come to fix things for you. At least it's not the toilet that has broken this time—repairing that is never pleasant.

Phew! What a day. After finishing your last bit of work, you wish everyone in the Mission Control centers a good evening, and then head to

the Cupola window. This is everyone's favorite part of the space station, with its six windows looking out into the Solar System. Here, you relax and watch the world go by. You find some of your other crewmates doing exactly the same thing. You've not seen some for hours, so it's a great chance to catch up while looking at the breathtaking views. None of you ever get bored with watching the changing Earth roll by underneath you, spotting different places. From up here you can see deserts and mountains, and lightning in the thunderclouds. You can pick out cities by their lights at night.

Now it's time for dinner and you've been looking forward to this because it's your crewmate's birthday today, so you're having a feast. As a special surprise, NASA has sent up pizza ingredients so you can have a pizza party. There are cheese, pepperoni, and olives—delicious! And there are presents from family too. Don't forget the cake! There'll be no candles to blow out, though, as that would be a fire risk. Instead, you've brought an inflatable birthday cake to the celebrations!

After dinner, you head back to your bedroom and give your family a call to catch up with their day. You've had to resist the urge to take one last look out of the window—too much sunlight this close to bedtime will confuse your body clock and mean you'll struggle to sleep. So it's time to read your book for a bit, turn out the lights, and float off to sleep, ready to do it all again tomorrow ...

RASPBERRIES IN SPACE

December 2015

"Psst, Izzy, are you nervous?"

"A little bit, Ed. I just want to get there now. Are you?"

"I'm not sure. I think I'm in shock. I still can't believe we're going to be the very first Raspberry Pis on the International Space Station."

T-minus 30 seconds ...

"Ahh, it's almost time. You ready to go where no Pi has gone before?"

"Sure am, Izzy. Let's do this!"

T-minus: 5 ... 4 ... 3 ... 2 ... 1 ... LIFT-OFF!

And with that, Izzy and Ed, the two specially chosen Raspberry Pi computers, were launched into space on December 6, 2015 via the Orbital Science's Cygnus cargo ship on a resupply mission heading for the International Space Station (ISS). But it hadn't been an easy journey. They'd had to prove they had the skills and brain power, learn all the code for their experiments, and even be fitted with new equipment. It was going to be the journey of a lifetime.

It all began the year before in 2014, on a cold, wet day in London. Everyone in Britain was getting really excited because Tim Peake was going to become the second British astronaut in space. It had been more than 20 years since the first British astronaut, Helen Sharman, had spent eight days on the Mir Space Station in 1991, and now Tim was going to spend six months on the ISS. A group of engineers from British space companies were coming up with ideas for the education program for Tim's mission. After lots of discussions, and a few reject ideas, they decided to send two Raspberry Pi computers into space, and run a competition for children across the UK to code experiments for them to carry out.

But no ordinary Raspberry Pi would do. For a mission on the ISS, only the best would qualify. Izzy and Ed were chosen, but they needed some improvements. A normal Raspberry Pi is the size of a credit card, and can do all the things a bigger computer can do—browse the Internet, play HD videos, build spreadsheets, and play music. But this wasn't enough for two computers going to space. First, they were both fitted with special sensors called the Sense HAT. This allowed them to measure things such as movement and temperature. They also needed to be carefully protected, so they had their very own spacesuits built. Their cases had an LED screen on the front so they could communicate with Tim. Ed was then fitted with a normal camera, while Izzy was fitted with a special infrared camera, which can see light that's invisible to the human eye. Last, but not least, they needed their missions! This wasn't up to NASA or ESA, though. Instead it was schoolchildren who would be in charge of mission Astro Pi.

In schools across the UK, children were challenged with creating and coding experiments that would be run on the ISS by Izzy and Ed, overseen by Tim. The judges were blown away by the standard of entries, which were much more exciting than the adults' ideas. They included games, sense-of-direction experiments, and weather forecasts, but only seven entries made the cut.

Once all the entries had been submitted, and the winners picked, it was time to update Izzy and Ed. The experiments were put on to SD cards and inserted into the two computers. Mission Astro Pi was ready to launch.

Izzy and Ed arrived in space and joined the team of over 50 computers on board the ISS that run 1.5 million lines of code. After they'd been set up, the Pis were separated and moved to different

IZZY AND ED ... JOINED THE TEAM OF ... COMPUTERS ... THAT RUN 1.5 MILLION LINES OF CODE

parts of the ISS. Izzy went to Node 2, which is like a corridor with several modules coming off of it. While a corridor might sound boring, Izzy had pride of place in the window of the Nadir hatch. From here she looked down on Earth.

Izzy's Astro Pi mission included two experiments, and they were both to observe planet Earth. One was called Trees, and Izzy used her special infrared camera to take photos of the land below. The brilliant code then analyzed the trees and plants in the images to see how healthy they were. During the experiment Izzy took 4,301 images, all perfectly in focus.

Izzy's other experiment was a crew favorite. As the ISS orbited Earth, Izzy would use the schoolchildren's code to calculate the station's location. She would then display the flag of the country the ISS was flying above on her screen, as well as a phrase in the local language. For example, when flying over Peru in South America, she'd display *Hola ISS desde Peru abajo*, which means "Hello ISS from Peru" in Spanish. When above an ocean, she'd display a twinkling blue-green pattern to represent water.

While Izzy was enjoying the view, Ed was carrying out his mission in the Columbus module. This is one of the science labs on board, where lots of experiments take place. Not only was Ed without a window, but he was also really busy. Ed had five experiments to run. Most of these focused on monitoring the environment around him. One was called Crew Detector, which looked for changes in humidity, perhaps caused by the crew. If there was a change, Ed would take a photo and display a message asking, "Are you there?" If an astronaut was there, they could then press a button to confirm they were. Ed ended up with some great pictures of Tim, and even one of Russian cosmonaut Mikhail Kornienko.

Ed also used his sensors to check the temperature, pressure, and humidity as part of the Watchdog experiment. He would then display the results on his screen. If there were any unexpected changes, an alarm would alert the crew.

The final environment-checking experiment was SpaceCRAFT, which recorded lots of data using sensors. This data wasn't for the benefit of the astronauts though, it was for young people back on Earth. After Ed's experiments were finished, Tim made sure that the precious data was sent

home, where anyone who wanted to could put the data into the game Minecraft and explore what happens on the ISS!

Ed's upgrades meant he could do more than just use his sensors. For the Radiation experiment, he cleverly used his camera, which was covered up, to pick up tiny specs of light. This meant Ed could measure the levels of radiation on the ISS. Radiation is energy that moves from one place to another and some of it can be very harmful to humans so this was a helpful experiment to keep everyone safe.

Ed's final experiment was the best kind—a game! The Reaction Games tested Tim's reaction times in space. The plan was that once he'd been playing it for a while, the game could measure how Tim's reaction times changed during long-term space flight. Unfortunately, Tim was pretty busy, so he didn't play the game very often. But seeing as Tim used to be a test pilot you'd expect his reaction times to be pretty good.

Once the experiments had finished, the data was sent back to Earth for the creators to analyze. After six months of running experiments and looking after Izzy and Ed, it was time for Tim to come home. He touched back down on Earth on June 18, 2016, in Kazakhstan, Asia, but he left two important crew members behind. Izzy and Ed were having such a great time that they decided to stay in space. They've been keeping really busy and running even more experiments. These ones are coded by children from all over Europe as part of the ongoing Astro Pi mission. Every year, the ESA invites children to write some new code. Izzy and Ed are up there right now, loving space life, just waiting for code to arrive!

IS THERE ANYONE OUT THERE?

If you go to your bedroom window or outside, and look upwards, what do you see? Perhaps buildings, trees, or street lights. But can you see higher than those and spot clouds, airplanes, or the Moon? And what about beyond that, deeper into the night's sky? Do you see anything? If it's a clear and dark night, perhaps you can see other stars, planets, or even a passing asteroid filling up the darkness above. Long ago our ancestors also looked up at those same pinpricks of light

in the night sky, guessing what could be in the blackness of the universe and asking that all important question—are we alone or is anybody out there?

Over time, our technology advanced and our understanding of the world around us developed, and we came to understand that these pinpricks of light in the sky are mostly stars, just like our Sun. Some of them, though, the dots of light that don't twinkle at night, are planets—our neighbors in the vast Solar System. With the invention of the telescope in the 1600s, scientists could finally take a closer look at what's above us, and

THEY COULD SEE THAT THE MOON HAS MOUNTAINS ... AND THE SUN HAS DARK MARKS ON ITS SURFACE

astronomers started to make out fuzzy details in the magnified dots of light. They could see that the Moon has mountains, other planets had swirling atmospheres of gas, and the Sun has dark marks on its surface, now known as sunspots. And as the telescopes became more powerful, people started to imagine what these planets might be like. Did they have weather and seasons, like Earth? Is there life on them? Might we visit one day?

After humans put satellites into orbit in the 1950s, space agencies came up with the idea of sending robotic scouts to far-off planets to get a closer look and unravel some of their mysteries. Shortly afterwards, spacecraft were sent to explore our nearest neighbors, Mars and Venus, and the pictures and data they sent back during the 1960s and 1970s were mind blowing. Scientists learned about Mars's crater-like surface and Venus's scorching temperatures from these missions, but one thing was missing— there was no sign of life as we know it on either planet.

Our growing knowledge didn't stop some people worrying, though. Scientists were concerned that the Apollo 11 astronauts might come across some nasty bugs when they landed on the Moon in 1969. Even if there weren't going to be any strange-looking life forms, scientists knew that bacteria could survive in very extreme conditions. So rather than risk accidentally infecting the world with some unknown alien bug, NASA put Neil, Buzz, and Michael, the astronauts from the mission, in quarantine for three weeks when they returned to Earth. They got to meet President Nixon and see their wives, but all through the window of their protective chamber! Fortunately, no bugs or bacteria were found and scientists were soon certain that the Moon was a sterile, barren place with no life on it. Eventually, the astronauts were allowed out and celebrated with a huge parade in New York City.

With the Moon declared lifeless, attention soon turned back to the planets surrounding us. Robotic spacecraft have now ventured far, far out into the Solar System, visiting every planet that orbits our Sun—Mercury, Venus, Mars, Jupiter, Saturn, Uranus, and Neptune—as well as the dwarf planet, Pluto. Spacecraft have even landed on moving comets and asteroids. These missions have beamed images and scientific data back to Earth from parts of the Solar System that are many hundreds of millions of miles away. Each new mission gives scientists more information about the universe that we live in, how planets are formed, and what really is out there.

Two incredible spacecraft, named Voyager 1 and Voyager 2, have ventured farther than any other human-made object. Back in the 1960s, when space travel was still very new and humans were taking their first small steps out into the universe, a brilliant young mathematician called Michael

Minovitch was working for NASA. Michael figured out that spacecraft could get a boost in speed from a planet's orbit when they passed by. This would mean that each time the spacecraft flew by a planet, they would accelerate further out into the Solar System. Another young engineer, named Gary Flandro, then noticed that the planets were going to line up in such a way that it'd be the perfect time to send the spacecraft long distances thanks to the speed boosts. Suddenly, there was an incredible opportunity to send spacecraft to visit four planets—Jupiter, Saturn, Uranus, and Neptune—in just one mission. But time was not on NASA's side, if the agency didn't act fast there wouldn't be another chance quite like this for nearly 200 years!

Never put off by the seemingly impossible, NASA knew this was a tremendous opportunity for humanity to explore the Solar System up close, and make the most of Michael and Gary's discoveries. But humans had been putting things into space for only a few years, and satellites usually lasted for just a few months, so how would they build a spacecraft that would work for the whole decade that it would take to reach Neptune, the furthest planet in our Solar System? It was up to NASA's engineers to find the answers, so they set to work designing and building a pair of spacecraft that could take on the challenge. The twin spacecraft would be going so far away from the Sun that the sunlight would be too weak for solar panels to work. So the engineers decided to power the craft with mini nuclear power plants, which would act like batteries. The teams raced against time, making calculations, carrying out tests, and improving the designs.

Incredibly, they did it. Voyager 1 and Voyager 2 launched just months apart in 1977 and began their long journeys out into the Solar System. As they flew past each planet as they toured the Solar System, they sent back hordes of astounding images and scientific data. Because of these two spacecraft, we now know that there are oceans on other moons, that some planets have lightning storms, and that there are even active volcanoes erupting out in space. Some of the final images taken by the spacecraft captured a portrait of our Solar System, showing six of the eight planets in one photograph.

THERE ARE EVEN ACTIVE VOLCANOES ERUPTING OUT IN SPACE

Incredibly, more than 40 years since they left Earth, both Voyager 1 and Voyager 2 are alive and well. They are traveling outwards, towards the stars, and far beyond anywhere our ancestors dreamed of going. They should be able to keep talking to Earth for a few more years yet, perhaps a decade or two. It will be a very, very long time before they reach the next star though—about 40,000 years!

Even though the Moon was empty, and there haven't been photographs of aliens on other planets, there's still hope that maybe, one day, the Voyager spacecraft will be found by some other life out there. And just in case they do, Voyager 1 and 2 have messages from Earth on board. Each spacecraft is carrying a golden record disc that has instructions shown in symbols that are hopefully easy for any passing extra-terrestrials to understand. If aliens can figure out how to play the records, and they have evolved to hear and see like we do, they will learn something of Earth and humans. The records contain sounds and images of our home planet, music from different cultures, and spoken greetings in 55 languages.

For now our question "Is there life out there?" remains unanswered and yet many people think there must be other life in the universe, somewhere.

It is such a vast, huge place that surely there has to be, even if it's not what we expect. Earth is just one planet orbiting a single star—the Sun. And the Sun is just one of billions of stars in our galaxy, the Milky Way. And using powerful telescopes, astronomers are discovering billions of galaxies all across the sky. And we are now beginning to find evidence of planets around every star we look at, so surely we can't be the only planet in all those galaxies with life on it, can we?

Maybe we just haven't found other life in the universe yet because it is so very, very, very big, and we've only just begun to look. Even if the chances of contacting intelligent aliens are small, many people still think that there might have been, or still is, life much closer to home, on Mars. Scientists don't expect to find the green men from the movies there, but microscopic life, things like bugs and bacteria. Life on Earth can live in all sorts of extreme environments, from the driest of deserts to blisteringly hot underwater vents. Bacteria have even survived in the harsh vacuum of space for long periods of time. So, it might just be possible that life like this once lived underground on Mars, and perhaps still does today. There are new spacecraft that will soon be on their way to Mars to try and answer this age-old question in our lifetime: are we alone in the universe?

HOMEMADE
SPACESHIPS
1996–2004

Peter Diamandis stood underneath the St. Louis Gateway Arch in Missouri and looked at his audience. It was May 18, 1996, and the place was full of famous astronauts, aviators, politicians, and journalists, all of them eager to hear what he was going to say. Peter had a big idea, a vision that space travel shouldn't just be for the astronauts hired by governments. It should be in reach of everyone—affordable, reliable, and with reusable spaceships.

Peter wanted to spur the space industry into action with a grand challenge and a mighty prize, just like those of the past. In the early 1900s, when airplanes were relatively new, wealthy people and newspapers offered big prizes to inventors who completed new challenges, including flying across the English Channel, over the Atlantic Ocean, and non-stop between Paris and New York. Peter thought, "Why not do the same for space?"

BUILD A REUSABLE SPACESHIP THAT COULD CARRY THREE PEOPLE 60 MILES ABOVE THE EARTH

He took a deep breath and announced his idea: the XPRIZE, a contest challenging people to fund and build a reusable spaceship that could carry three people 60 miles above the Earth, and do it twice within two weeks. The first to achieve this would win the grand prize of $10 million.

What a challenge! Since the dawn of the space age, the possibility of escaping Earth's atmosphere was ruled by the deep pockets of governments. It was not for ordinary people. But Peter was suggesting that it was possible to do things differently—for someone to find sponsors, make their own spaceship, and use it to get people into space. Was the idea crazy or revolutionary? Only time would tell …

Somewhere in the crowd, listening to Peter's thrilling competition, was Burt Rutan. To say Burt loves airplanes is an understatement. At the age of eight he was designing model airplanes, he had made his first solo flight as a teenager, and by 22 he had the qualifications he needed for a job designing aircraft. His creativity was only just starting, though. During his career he worked as a pilot for the US Air Force, testing new experimental military aircraft. After this he opened his own company, designing and creating airplane prototypes. He designed many incredible and innovative aircraft, including a strange plane, which had wings that could pivot forwards during flight to make it more efficient, and the first plane that could fly all the way around the world without stopping to refuel. By the time Peter announced the XPRIZE, Burt was already famous for his quirky-looking inventions and unusual designs that could do things no one had ever thought possible. He was also very respected. So when that evening, at the gala dinner to celebrate the XPRIZE launch, Burt declared his intention to win the prize, he became the man to beat.

Word of the XPRIZE challenge soon spread, and all around the world the minds of entrepreneurial engineers started whirring with ideas, wondering "what if we just ..." or "how about if ..." In Argentina, Romania, the UK, Canada, Russia, the US, and Israel groups formed and started sketching out plans. Eventually 26 teams from these seven countries had their registrations accepted, including Burt Rutan's group. All of the entries were considered to be reasonable designs with teams behind them who had the expertise that might be able to turn the ideas into reality. And with that, the race began!

The entries included all sorts of ideas for rockets and spaceships of varying shapes and sizes. Some were helped with balloons, others with helicopters, to get their spacecraft high enough into the atmosphere. Another was in the shape of a flying saucer! Creativity was flying—just what Peter had wanted when he launched the prize.

But building these new inventions was going to cost money, and probably a lot more than the $10 million prize. Some of the groups owned large workshops and were fortunate to have access to lots of funds.

Others had nothing more than a backyard shed and the belief that their idea was good enough to attract sponsorship. But wherever the money came from, the aircraft designed and tested were still going to be built on a shoestring budget compared with what governments usually spend on space programs.

With his bold claim of winning declared, Burt set to work. He developed his ideas in secret with foam models and computer simulations. When he was sure he had something that could work, he set up a meeting with the billionaire Paul Allen, the co-founder of Microsoft, and laid out his design for a spaceship that could win the XPRIZE.

Burt's idea was to split the task of getting into space in two. An 80-foot-wide, graceful airplane that looked a bit like a giant gull with wings outstretched, called White Knight, would be the Mothership. Hanging underneath the belly of this large flying carrier plane would be SpaceShipOne, a rocket ship that would jet all the way to space before gliding back to Earth and landing. White Knight would carry SpaceShipOne high up into the atmosphere, more than 9 miles above the surface of the Earth, higher than jumbo jets fly. Up in the wispy air, White Knight would release SpaceShipOne. The pilot of SpaceShipOne would then fire up its rocket, blasting the small capsule straight up towards space. After the rocket engine stopped firing, the spaceship would continue shooting upwards. But as gravity would start pulling it back towards Earth, it would slow down. Just like a football kicked high into the air, it would eventually fall back towards Earth, creating an arc in the sky. Burt's plan was that the top of this arc would be 60 miles high—the boundary that needed to be broken to win the XPRIZE.

It's when SpaceShipOne was coming back to Earth, though, that the really clever bit of Burt's invention would kick in. On the way up, the spaceship needed to be as sleek as possible so it could go high and fast. On the way down, the craft need to slow as much as possible so it could land safely. Inspired by a badminton shuttlecock, Burt invented a clever set of wings for his spaceship. When traveling into space, the wings would be in the usual place, alongside the body of the plane. But once in space, the wings would fold upwards. This would mean that when the spaceship came back into the Earth's atmosphere, the folded wings would increase the drag through the air and slow down the spaceplane. Finally, the wings would fold back into the normal position and SpaceShipOne would become a glider, and the pilot could guide it back to land on a runway.

BURT INVENTED A CLEVER SET OF WINGS

Burt presented this idea to Paul, who thought the concept was sensational, and quickly agreed to fund Burt's company to build and test the brilliant invention. Burt and his team got to work turning these daring plans into a real spaceship. But everything was kept quiet and a secret, they didn't want any of the competition stealing their ideas.

After three years of construction and testing, and $20 million of funding, the deadline for winning the XPRIZE was drawing closer. On September 29, 2004, tens of thousands of people, including competing teams, assembled in the newly christened Mojave Air and Space Port to watch White Knight and SpaceShipOne take to the skies for the first flight.

Brian Binnie was at the controls of White Knight, with Mike Melvill slung underneath. The crowd watched giddy and nervous as White Knight soared

from the runway, hauling the spaceship high into the sky. Once it was high enough, White Knight released SpaceShipOne. Mike fired the rocket engines and bam, faster than a bullet, he instantly accelerated towards space, shooting straight up. But the spacecraft started to roll. Mike was turning nearly all the way around once a second. Cameras followed the spinning spaceship as it streaked higher and higher, and silence fell over the crowd of thousands as they all held their breath, worried about what they saw.

But this wasn't Mike's first flight. He had over 30 years of experience testing all sorts of weird and wonderful new airplanes, and had spent many of those years working with Burt. He knew that if he shut down the rocket and aborted the mission, the goal of winning the XPRIZE would be lost. Instead, he held on, and as SpaceShipOne made it out of the atmosphere, he regained control, bringing the spinning spaceship to a halt.

For three-and-a-half minutes all was calm and weightless. Mike exclaimed, "Wow, you would not believe the view!" as he looked down over Earth. He took photos and even opened a bag of M&Ms so he could enjoy watching them float around him. But then the fun was over, and he began his journey back to Earth. The force of the return to Earth's atmosphere was so intense that Mike had to fight his body so he didn't pass out, but as the flight leveled out, Burt's brilliant wings turned SpaceShipOne into a glider.

Mike made it back to the ground to thunderous applause. But the competition hadn't been won yet. To claim the prize they were going to have to do it all again, and within two weeks. Burt's team worked hard and fast to sort out the problems that had caused the plane to roll, but other than that SpaceShipOne was in great shape. In just five days, the team fixed the problems, refueled, and sat Brian Binnie at the controls of SpaceShipOne for round two.

On October 4, 2004, the 47th anniversary of the launch of Sputnik—the first satellite in space—SpaceShipOne and White Knight launched for their second flight. Once SpaceShipOne was released from White Knight, Brian roared upwards, but when the call came to shut down the engine, he kept going! The spacecraft was handling so well, without a single roll, that Brian wanted to go as high as he could! Burt's design was so good that Brian was free to move the spaceship around and take pictures during the period of weightlessness. He landed safely back on the runway at the end of his flight and Peter Diamandis declared the XPRIZE triumphantly won. Burt and his team had shown that it was possible for a relatively cheap, reusable plane to go to space and back.

Watching the flight next to Burt Rutan and Paul Allen, was Richard Branson, head of the Virgin empire. Richard had just signed an agreement with Burt and Paul to take their technology and turn it into a fleet of bigger spaceships that would take paying passengers to space, and Virgin Galactic has been working on SpaceShipTwo ever since.

And with the competition over, Peter had realized his dream—that space travel wasn't just for governments, but could one day become affordable and available to everyone. Since the competition was won, private space travel, without government help, has continued to push forwards. Several companies are hoping to send tourists into space very soon. It just goes to show that with the right minds, passion—and enough funding—anything is possible.

GOING TO MARS

This final story, unlike all the rest in the book, hasn't happened ... yet. But it will, in the years and decades ahead, perhaps sooner than you might think.

It was early in the morning and still dark, but the astronauts headed out of their quarantine quarters for one last walk outside. They took their time in the crisp, cool air, lingering to drink in the sweet smell of the dewy grass, the peaceful sight of the lake glistening in the moonlight, and the gentle sounds of the birds as they began their dawn chorus. It was going to be a long time before they would be able to enjoy the simple pleasures of life on Earth again.

The cloudless sky was full of specks of light, all of them shining and beautiful, but one pale reddish light drew their gaze more than any other. It was still hard to believe that today, after all the years of training and planning, they were finally going to start their journey towards it: Mars. The sky started to lighten and the twinkling specks of light disappeared one by one. The crew stood and watched the planet disappear into the morning sky before heading back inside for their final preparations.

The astronauts got into their spacesuits and waved to the crowds as they headed to the launch pad and the waiting rocket. They spotted flags from all around the world in the crowd, and from each of the nations they represented. The ground team helped everyone into the capsule, tightened the seatbelts and bid them goodbye. As the hatch swung shut, and the astronauts were on their own, their minds focused on the challenges, adventures, and unknowns that lay ahead.

THE ASTRONAUTS WERE GOING TO BE AWAY FROM EARTH FOR A VERY, VERY LONG TIME

The astronauts were going to be away from Earth for a very, very long time. Just to get to Mars was going to take six months. But, once they were there, they would be staying for nearly 18 months. And there would be no option to leave early. The Earth orbits the Sun nearly twice as fast as Mars so the crew could make the long journey home only when Earth and Mars were close enough together again in their orbits around the Sun.

A couple of hours after launch, the astronauts would arrive at the Mars

Transfer Vehicle (MTV), which was waiting for them in Earth's orbit. It was nothing like the spacious old International Space Station, but it was roomy enough for the six of them and it had all they needed for the journey to the Red Planet. Once they had joined up with the MTV and checked everything was working, the crew members would fire up the mighty rocket engines and accelerate out of Earth's orbit and towards Mars.

The preparations for this mission had been happening for years. While the MTV had the supplies for the journey, there was no way it could hold everything the crew needed to survive 18 months living on Mars. Instead, cargo missions had already taken everything the astronauts would need to their new home. Their living quarters, transport vehicles, food, and science experiments had all safely landed and were waiting for the crew to arrive.

Some of the most important things that were there, and already hard at work, were the robots making fuel! These robots had been working for years, combining the Martian soil and the carbon dioxide gas from the atmosphere to make rocket fuel and load it into the rocket that would eventually send the crew home. Rather than bring the fuel they needed all the way from Earth, there was now a gas station in the stars ready and waiting for them! When the time was right, this Mars rocket would blast the astronauts back home to Earth.

But they needed to make it there first. As the distances between the crew and Earth increased with each passing day, they would venture further into unknown territory. The view of Earth would become ever smaller and it would take ever longer for messages sent between Mission Control on Earth and the crew in the MTV to reach each other. For the first few weeks, the delays would be only a few seconds, and so the crew would still be able to

talk to everyone, including their friends and family. They had trained for this—learning how to keep their sentences clear and short so that each person knew when the other had finished talking. But as the size of Earth reduced, and eventually turned into just another speck in the sky along with all the stars, the communication delays would turn to minutes. Eventually two-way conversations would become impossible. The crew would be cut off from immediate contact with anyone on Earth, relying instead on audio and video messages that were beamed to and from Earth like virtual postcards.

One of the most critical updates the crew would get each day from Earth would be the weather forecast. The crew wouldn't be worrying about thunderstorms or hurricanes. Instead, the astronauts would be getting updates on solar wind. Leaving Earth behind meant leaving the safety of the Earth's magnetic field, which protects life on Earth from solar wind. The Sun is so hot that it spews out gases and particles into space, which have so much energy that they whoosh through the Solar System like a roaring gale, and are dangerous to humans. The crew's habitation module had a clever system that would shield them so they had some protection, but on the days when the scientists back on Earth detected that the Sun was about to spew out large quantities of solar wind, they would have to take shelter in their tiny crew quarters, which had extra shielding in place.

Once they arrived, the hard work would really begin. Eighteen months on the surface might seem like a long time, but the astronauts knew it would go by much faster than they would like. There was so much to do! With all the science experiments that were planned, the geology field trips that would help reveal Mars's secrets, and the farming they would do to grow food to eat, it was going to a busy time. They would be the first-ever humans to live on another planet, so studying how the human body reacts to life on Mars would be a key part of the mission. This would mean running lots of tests on themselves and each other. Being this far out in space would also give them a unique opportunity to learn more about the Solar System, pushing the boundaries of human curiosity, progress, and exploration.

THEY WOULD BE THE FIRST-EVER HUMANS TO LIVE ON ANOTHER PLANET

That was still to come, though. For the moment, the crew all looked at each other and smiled. The countdown was nearing the end and it was time to get going. Who knew what discoveries lay in store? Whatever happened, the astronauts knew that the teams on Earth—the engineers, doctors, scientists, and thousands of others—would be there, working around the clock, to look after them. They had all the knowledge and history of the past to help them along their way. The lessons learned through the challenges and sacrifices of all the missions that had come before, from the very beginning with the first human in space, Yuri Gagarin, up to now.

The crew members pulled down their visors, and checked their switches and displays. Then they all took a deep breath.

5 … 4 … 3 … 2 … 1 … LIFT-OFF!

Mars, here we come …

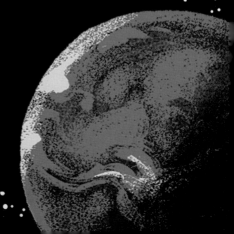

GLOSSARY

Cosmonaut A space traveler from Russia or the Soviet Union.

Docking When two spacecraft join together in space.

Orbit The path taken by something that continuously travels around another object, captured by its gravitational field. The Moon orbits the Earth.

Satellite Something that travels around something else, such as a spacecraft, without people, orbiting the Earth, or the Moon orbiting the Earth.

Soviet Union A large country to the east of Europe that no longer exists as a whole. It broke up in 1991, in to 15 independent countries, including Russia.

Taikonaut A space traveler from China.

Rendezvous When two spacecraft meet up together in space.

Thruster A small rocket engine that controls the position of a spacecraft.

Re-entry The process of a spacecraft returning from space and coming into contact with the Earth's atmosphere.

Weightless(ness) When an object cannot feel the force of gravity acting on it.

Vacuum Somewhere where there is no air, such as space.

ACKNOWLEDGEMENTS

I've drawn on a myriad of sources in my research, and I am indebted to all those whose work I've built on. From the biographers and journalists, to historians, those who transcribed hours of flight recordings, and those who shared their own stories. Each one has fed my passion and interest through the years, as well as being the foundations of this work. Ed Van Cise, Sue Nelson, and Kerry Sanz all particularly played their part. I thank you all.

Huge thanks must go to my agent, Stephanie Thwaites, and her trusty sidekick Isobel Gahan at Curtis Brown, and to my editors Laura Horsley and Corinne Lucas at Wren and Rook. Your vision, endless support, and patience have all been utterly essential at every step. The glorious illustrations of Léonard Dupond have brought the adventures to life in exquisite detail.

Over the many long months of work, the understanding and care of all my family, friends, and colleagues was invaluable and deeply appreciated every step of the way.

And to Chris. Without you this was all impossible.

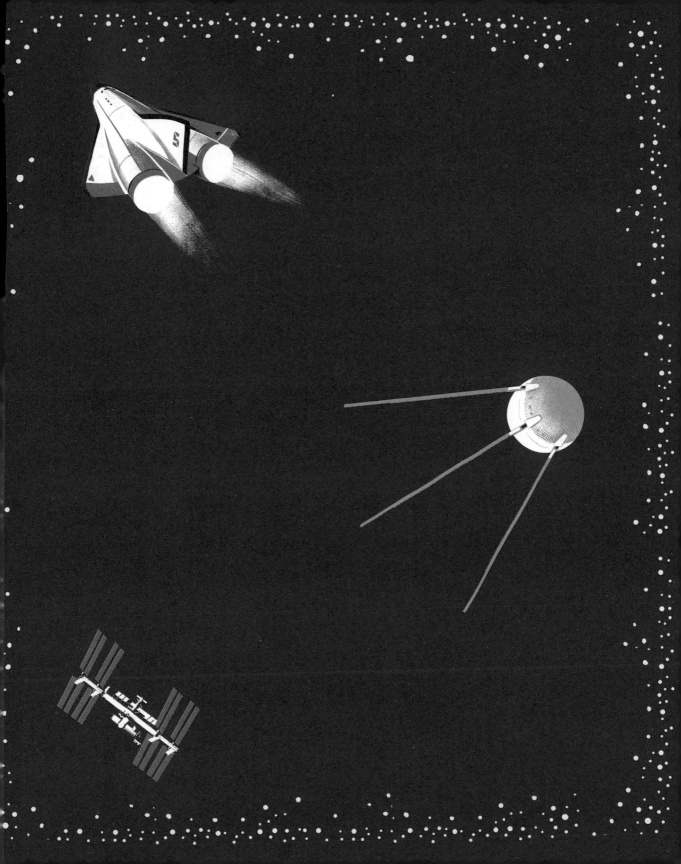